Global Atmospheric Circulations

GLOBAL ATMOSPHERIC CIRCULATIONS
Observations and Theories

RICHARD GROTJAHN

University of California, Davis

New York   Oxford
OXFORD UNIVERSITY PRESS
1993

Oxford University Press

Oxford   New York   Toronto
Delhi   Bombay   Calcutta   Madras   Karachi
Kuala Lumpur   Singapore   Hong Kong   Tokyo
Nairobi   Dar es Salaam   Cape Town
Melbourne   Auckland   Madrid

and associated companies in
Berlin   Ibadan

Published by Oxford University Press, Inc.,
200 Madison Avenue, New York, New York 10016

Oxford is a registered trademark of Oxford University Press

Library of Congress Cataloging-in-Publication Data
Grotjahn, Richard.
Global atmospheric circulations :
observations and theories / Richard Grotjahn
p. cm.   Includes bibliographical references (p.) and index.
Outgrowth of lectures presented to beginning graduate students.
ISBN 0-19-507245-6 (acid-free paper)
1. Atmospheric circulation. I. Title.
QC880.4.A8G77   1993
551.5'17—dc20   91-42360

2 4 6 8 9 7 5 3 1

Printed in the United States of America
on acid-free paper

To Judith Ann
and
my artist grand

PREFACE

I realized a life-long dream a few years ago when I traveled around New Zealand for a month. Lying well away from any other land mass, the islands of New Zealand host many unique and unusual species. I grew particularly fond of several plant species: Pohutukawa, Kotukutuku, and Kowhai, to name a few. Kowhai trees are a beautiful example. Every spring the Kowhai presents a spectacular display of yellow flowers upon otherwise bare branches. January brings long chains of seed pods. Uncounted numbers of Kowhai are planted in private gardens. Despite the local popularity, the Kowhai is little known outside New Zealand. Its seeds are notoriously difficult to germinate. This fact alone would not discourage most nursery growers. However, even after a successful seedling is set out, another 8 years must pass before the first bloom.

The book you are holding evolved over an 8-year period. Thirteen earlier versions of the manuscript were typed and revised. Some reviewers of the early manuscripts encouraged me to persist while others were less complimentary. At times, my dream of providing a comprehensive, cohesive, and notationally consistent description of the general circulation of the atmospere was more like a nightmare. Now that I have reached the end, working through all those drafts was worth the struggle, because the manuscript improved noticeably with each draft. For example, during the sixth year I was able to visit the Southern Hemisphere, to develop some personal contacts, and to gain a far better appreciation of Southern Hemisphere weather than I would have from here in California. A direct result is the emphasis placed here upon balancing the treatment of the two hemispheres, something not always done in atmospheric science books.

The main difficulty in writing this book is not in the gathering of observations from disparate sources, though that took some time. In fact, the main difficulty is in balancing the treatment of both observation and theory in one book. The constraints of nonprofit publishing only exacerbate this chore. Consequently, practicing theoreticians may find the chapters on theories too simple and those who work directly with observations may find the observational chapters too brief. I hope that they and other readers of this text will find all sections accessible.

This book grew out of lectures I present to a class for beginning graduate students. In practice, only about two-thirds of the book is covered in any detail during the ten-week quarter. The students have a variety of backgrounds, but all have completed our twenty-week undergraduate dynamics course sequence or its equivalent. Accordingly, a prime purpose for this book is to be a text. I realized early on in this project that the lack of an existing text in this field means that there is no set fashion in which the subject matter is taught. This book presents all the observational information first, so that a student gets a complete picture of that. Then I discuss theories. Some instructors may not like this ordering, and would prefer to place closely allied observations and theories together. To aid those instructors who prefer a different order than the one I have chosen, I have tried to make many of the sections in the book self-contained. I have also drawn attention to connections between sections in different chapters by extensive cross-referencing.

I believe that this ordering is best for the second audience I have in mind. The book is also intended as a monograph suitable for scientists who want either reasonably accurate observations collected and evaluated in one book, or rather simple but sound theories and tools for understanding the large scale flows, or both. To accommodate this audience it is best to gather observations and theories into separate, coherent sections. A previous monograph on this subject was not organized this way, and I have found it frustrating to consult for that reason.

The first chapter lays out some basic requirements of large-scale atmospheric circulations and introduces some useful jargon while summarizing the earliest scientific papers on the subject. Chapter 2 includes more preparatory information: how and where data are collected, how data are manipulated to form grids of "observed" data, and particularly emphasizes biases that are present in observations described later. Some reviewers found this chapter unnecessary while others praised it. The material is there for the reader to examine as desired.

Chapter 3 presents directly measured zonal average variables: Chapter 4 presents observations of *derived* zonal average variables that arise from energetics and momentum conservation. The diagnostics in Chapter 4 are used for much of the interpretation presented in later chapters, so they are derived and discussed at some length. As elsewhere in this book, an attempt is made to explain explicitly every step of each derivation; you will not find the phrase "It can be shown. . ." in this book! Chapter 5 reviews the observed variables of the previous two chapters with special emphasis upon the *deviations* from the zonal means. The Appendix describes the mean and deviation symbols uniformly applied throughout the book. The trend in recent years has been away from dividing the atmosphere into zonal mean and deviation. The time mean and low and high frequency deviations have been emphasized by some researchers. However, the vast majority of studies available use the zonal mean and deviation partitioning; most variables that had to be discussed have not been published using low and high frequency partitions. Also, the removal of one dimension makes the zonal average state easier to diagram and discuss.

Chapters 6 and 7 discuss various theoretical models and concepts applicable to the zonal average and zonally varying circulation. While some simple concepts are introduced when observations are discussed, these chapters contain the most in-depth interpretation and analysis. The mathematics in these chapters is purposely kept at a simple level and only simple models are derived in detail. Several sections are nonmathematical and emphasize physical interpretation of such circulations as stratospheric motions (§6.2.4), mean meridional cells (§6.3.2), circulation feedbacks and externally forced changes (§6.4), jet streams and their longitudinal variations (§6.5), long wave maintenance (§6.5; §7.2.4), and linear and nonlinear properties of frontal cyclones (§7.3).

The final chapter differs from all the rest in being less formal and more speculative. Some readers may find Chapter 8 as much fun to read as it was to write. That chapter is intended to illustrate some possible uses of the information in this book. Other readers may skip that chapter without concern.

Some readers might expect a book on the general circulation to have a chapter on general circulation models (GCMs). No such chapter is included for several reasons. First, a basic book on GCM designs and applications in general was recently published (Washington and Parkinson, 1986). Second, GCMs are often used to *apply* fundamental concepts developed here to specific situations. Third, relevant citations to GCM studies have been integrated into nearly every chapter and are identified in the index; to collect that material into a separate chapter would disrupt the flow of ideas.

Davis, Calif.   R. G.
December 1992

ACKNOWLEDGMENTS

I should like to thank the many colleagues who made comments on the text, helped provided some of the diagrams, or simply encouraged me to undertake this project. In alphabetical order, I particularly thank B. Barkstrom, M. Blackmon, W. A. Bohan, H. Böttger, C. Brown, M. Chahine, T. C. Chen, R. Daley, E. F. Harrison, I. James, D. Johnson, D. Kann, D. Karoly, V. E. Kousky, N.-C. Lau, L. Li, J. McGuirk, T. R. Nathan, M. Peng, R. Pfeffer, W. Randel, L. Remer, W. Rossow, R. A. Schiffer, A. Semtner, D. Shea, R.W. Tanner, H. van Loon, J. M. Wallace, S. Warren. I am grateful for the comments by the anonymous reviewers, especially the detailed reviews by those who labored for Oxford University Press. I typed the manuscript galleys in plain TEX and appreciate the help I received from J. Alquist, R. Errico and E. W. Lorenz. The page layout was done by Oxford with assistance by E. Sznyter. I drafted most of the figures that needed attention; however, about a fourth of the schematic diagrams were drawn by A. Hipps.

CONTENTS

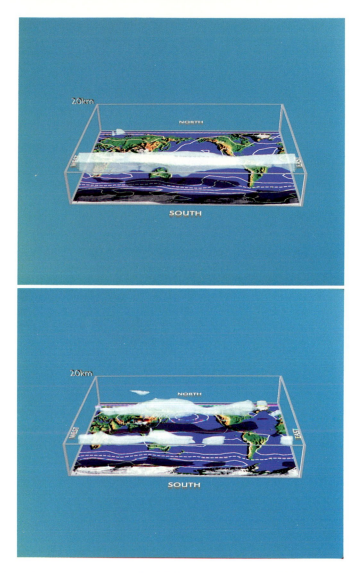

PLATE. 1   Three-dimensional depiction of the global jet streams (January 1979, above; July 1979, below). The cloud-like volumes enclose that portion of the atmosphere where the wind speed exceeds 35 m/s when averaged over the month. The vertical dimension is exaggerated; the box is 20 km deep. Colors on the *flat* box bottom depict elevation with two exceptions: white indicates elevation greater than 3050 m or permanent ice-covered surface; blue indicates water surface regardless of elevation. Contours are of sea level pressure. See text. (See text page 176.)

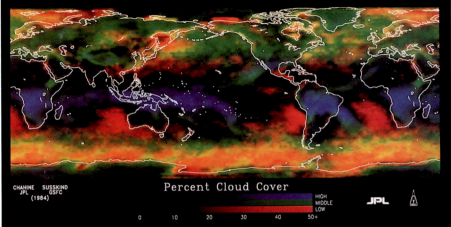

MEAN CLOUD COVER AND ALTITUDE FOR JANUARY 1979
USING HIRS 2 AND MSU DATA

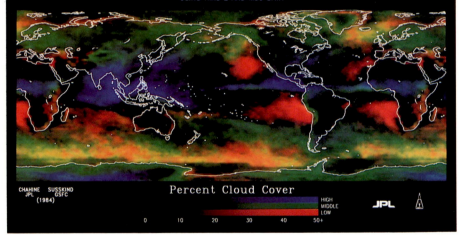

MEAN CLOUD COVER AND ALTITUDE FOR JULY 1979
USING HIRS 2 AND MSU DATA

PLATE. 2   Mean cloud cover during January (above) and July (below) 1979. Percentage of cloud cover indicated by the intensity of the color, as indicated by the scale. The mean height of the cloud top is assigned to one of three categories, each assigned a different color. The data resolution is 2° latitude by 2.5° longitude. Data processed at the Goddard Lab. for Atmospheric Science, GSFC; plate produced by Hussey, Hall, and Haskins at the image processing lab, JPL. Plate courtesy of M. Chahine, personal communication. (See text page 186.)

Global Atmospheric Circulations

1

INTRODUCTION AND EARLY THEORIES

An exact Relation of the constant and periodical winds . . . is a part of Natural History not less desirable and useful, than it is difficult to obtain, and its Phenomena hard to explicate.

Halley, 1686

1.1 A considerable circulation

For more than four centuries, it has been desirable to consider the largest scale motions that compose the "general circulation" of the winds. Today, understanding the general circulation has renewed importance because of our desire to predict the "global change" that may accompany expected global climatic changes in the coming decades. The general circulation is "considerable" in the other sense of the word because it encompasses the largest scales and is comprised of many different processes in the atmosphere.

What defines the general circulation? Monin (1986) defines it as "a statistical ensemble of large-scale components of state of the atmosphere." In the earliest works relating to the subject, general circulation meant the time and zonal average circulation. (The term "zonal" refers to the longitudinal direction.) However, mechanisms are needed to drive the motions, and those mechanisms draw in all the other commonly studied variables: pressure, density, temperature, and moisture. In addition, the mechanisms maintaining the zonal average circulation include zonally varying phenomena (eddies) and time varying (transient) phenomena. So, of necessity, the current definition encompasses much more than the zonal and time average circulation.

The definition of the general circulation is somewhat ambiguous because the atmosphere makes few distinctions between the scales of motion. A scale analysis of the equations provides insight into changing balances in the equations for different scales in time and space. But there is no restriction that phenomena must occur on well-separated length and time scales. Even so, it has been popular practice to identify distinct scales of motion in the atmosphere (e.g., Fujita, 1981, and 14 references in his article). Unfortunately, plenty of phenomena bridge these neat categories. For example, in the tropics the large-scale atmospheric motions are composed of, and driven by, organized small-scale convection.

In this book, the "general circulation" refers to the distribution of atmospheric quantities that are relatively large scale ($>$ 100 km in the horizontal) and tend to exist for more than a few days when averaged over a day. This definition is unsatisfactory because so many phenomena are included in it. Yet such is the nature of the problem. The breadth of the subject is what makes understanding the general circulation intriguing and difficult.

What causes the general circulation is easy to answer: the atmosphere develops motions in an attempt to reach thermal equilibrium with sources and sinks of inhomogeneous radiative heating. Beyond that, matters become very complicated indeed. The *nature* of the motion is a harder story to tell and is the focus of this book. To answer the question properly one must look at a variety of things that are intertwined with the winds, such as radiation, moisture, and landform. The challenge is keeping the discussion manageable. A jigsaw puzzle may be a good metaphor. One does not need all the pieces to be in place in order to see the general picture made by the pieces. Similarly, this book gathers most of the largest pieces and assembles them in a way that makes the giant multidimensional jigsaw puzzle of the general circulation comprehensible. Of necessity though, some phenomena must be neglected or discussed only briefly.

Why study the general circulation? There are a variety of reasons for studying the general circulation; four come immediately to mind. As stated in the opening paragraph, understanding the general circulation has become more imperative now that human activities have begun to have global effects upon our environment. Does global warming mean that all the weather patterns are simply shifted poleward, as suggested in the layperson's literature? From information in this book the answer is, Probably not. A second motivation is provided by weather forecasting. Current forecast models routinely predict weather a week or more in advance. The motion of the weather patterns requires a global model domain to make such a forecast (Smagorinsky, 1967). Tropical, extratropical, and interhemispheric exchanges occur, and all must be considered. Third, a general appreciation of the large scale weather and circulations provides important basic science questions of its own. Parts of the problem are highly nonlinear; some parts are still poorly understood. The large scale circulation defines the environment for other scales of motion and phenomena. A fourth purpose is to illustrate basic dynamical laws and simultaneously establish the degree of consistency between different variables. This book will focus particularly upon the last two items.

1.2 General remarks

There must be a general circulation. The atmosphere is intolerant of disequilibrium, the sort of disequilibrium that is set up by the asymmetric distribution of solar heating and terrestrial cooling. Primarily, there is excess heating near the equator and excess cooling near the poles. Atmospheric and oceanic circulations are created in an attempt to remove the imbalance between the heated and cooled regions.

It is not a simple matter to deduce a workable atmospheric circulation that will provide the required heat transports. The problem is greatly complicated by a number of mitigating factors. In no particular order, these factors include: spherical geometry, land-sea contrasts, rotation of the earth, nonlinear effects of clouds, topographic variations, nonlinear dynamics, friction, dynamical instability, and interaction with ocean transport. Lorenz (1967) emphasizes nonlinear advection in particular. Because motion is not uniform, the velocity field can distort as well as displace patterns. Nonlinearity can cause a large variety of patterns to occur and continuously evolve, with little tendency to repeat a past pattern. This unpredictable behavior is at the heart of the so-called "chaos theory." In fact, Lorenz (1969) later demonstrated that there is little repetition in the atmospheric circulation patterns. The lack of periodicity precludes explicit mathematical representation of the atmosphere by a finite number of symbols or oscillations. So, one is led to studying statistics, the most common statistical application being time averages.

What are the constraints upon the possible atmospheric circulations to accomplish this transport of heat? The obvious constraints are the well-known radiative, hydrostatic, and (outside the tropics) geostrophic balances that predominate for the large-scale motions considered here. Other constraints include momentum and mass conservation. One can anticipate the most basic properties of the general circulation by considering these balances.

- *Radiative balance*. The relation between the radiative imbalance and the circulation is not simple. Radiation alone would set up a strong equator-to-pole temperature gradient that would be much stronger than the observed gradient (§6.1). Such a gradient is incompatible with an atmosphere at rest, so a circulation is set up that transports heat from low to high latitudes. But this heat transport weakens the meridional temperature gradient, destroying both the accompanying height gradient (next item) as well as the source of energy for longitudinally varying motions (§4.5, §4.6). The temperature gradient must be continually reinforced to replenish kinetic energy lost by friction.

- *Pressure*. Obviously, the temperatures in the lower atmosphere are generally colder in the polar regions than in the tropics. From hydrostatic balance and the ideal gas law, one knows that the thickness of a pressure layer is greater in the tropics. The surface pressure has much smaller latitudinal variation than do constant pressure surfaces higher up. The result is an equatorward pressure gradient that increases with height in the troposphere.

- *Zonal wind*. The geostrophic balance can be added to the hydrostatic and ideal gas laws used for the previous item. The resultant relationship is termed "thermal wind" balance. Since the latitudinal pressure gradient increases with height, so must the westerly wind.

- *Surface torque.* There must be an approximate balance between the surface westerlies and easterlies. Otherwise surface friction would apply a net torque upon the earth, slowing or accelerating its rotation. The reader should note that the angular momentum is a function of the distance from the axis of rotation, which is proportional to cos ϕ where ϕ is the latitude. Surface torque balance requires an additional cos ϕ factor to account for the convergence of meridians. Also, the torque is caused by surface wind stress that is not a linear function of wind speed. More details are in Gill (1982; Chapter 2).

- *Mass.* There must be a balance between the mass-weighted northerly and southerly flows (on a zonal average) in order to avoid a net buildup or loss of mass at any location.

In this book, frequent reference will be made to the "tropics," "subtropics," and so forth. While such terms are in common usage, precise definitions are not always made in the literature. The following terms will be used to designate latitudinal bands whose boundaries are somewhat flexible. The tropics, or "low latitudes," will refer to the region from roughly 25 N to 25 S. The subtropics will refer to the latitude belt in each hemisphere from about 20° to 35°. The "midlatitudes" (or middle latitudes) will extend from roughly 30° to 60° in each hemisphere. The "polar region" (or high latitudes) will extend from 60° to 90°.

1.3   Historical background: Part I

This section briefly discusses some early attempts to describe and explain the general circulation. The early history is presented here in order to provide some perspective, and to establish some terminology. In particular, the labels "Hadley cell" and "Ferrel cell" are introduced to identify certain zonal and time mean meridional circulations. These labels will be useful when discussing the observations in later chapters. The historical summary in this book is divided into two parts. The earliest work is quite qualitative and can easily be discussed in this chapter since little background knowledge is required. After 1900, several key works appeared that are more quantitative and are saved until the first chapter on theory (Chapter 6). An excellent and more comprehensive discussion of the earliest works can be found in Lorenz's (1967) book.

The early study of the general circulation was motivated in part by the needs of commerce and, in particular, of sea traders. Indeed, the existence of a general circulation was first recognized by navigators. Many of the colorful names for portions of the general circulation originated with the traders. The "trade winds" of the tropics were so-named because seafarers planned trade routes based upon the remarkable persistence of these winds. The "Horse latitudes" were so named because horses frequently perished on ships that were becalmed in those latitudes. The log books from these voyages were the principal source of observations for the early works on the general circulation.

The weather has always had a strong influence upon human activities. Consequently, many descriptions, comments, speculations, and aphorisms about the atmosphere have been made over recorded history. A summary of these ideas is outside the purpose of this book; interested readers can find numerous examples detailed in Dove (1862). These very early ideas are anecdotal and not based upon sound scientific principles. Therefore our history begins with the report generally recognized as the first quantitative attempt to explain the general circulation.

Descartes (1637) appears to be the first European to apply scientific principles to explain the general circulation. His discourse proposes that solar heating expands the air (greatest at local noon). He assumed that the most dense air was located just before local sunrise, since it has had the longest period of cooling. He thought all motions were density driven. He reasoned that horizontal motions were analogous to thermal convection. Thus, he concluded that air would rush from the area to local noon to local sunrise, i.e. an easterly wind. The science in Descartes discourse can be easily faulted; he incorrectly assumed that density differences were created by water vapor amount. Nonetheless, his work invoked physical principles that were demonstrable in the lab.

A more succesful attempt to explain the large scale circulation is attributed to Edmund Halley. His essay, published in 1686, was concerned specifically with the trade winds and monsoons observed in the tropics. Halley described the winds in various regions of the tropics (Figure 1.1), criticized a popular theory of the time, and presented his own simple theory. Even in his day, it was recognized that the low level northeasterlies would underlie upper level southwesterlies (southeasterlies and northwesterlies respectively, in the Southern Hemisphere). The Indian monsoon was also recognized as a large scale, coherent circulation. Halley's main concern was to explain the easterly component of the tropical winds. One belief at that time was that tropical air would move separately from the earth's surface because it was warmer than air at other latitudes. Since the earth rotates from west to east, it was reasoned, the rotation of the earth beneath the air would be perceived as an east wind by an observer on the ground. Halley ridiculed this idea, proposing instead that the east wind was set up by the diurnal motion of the sun across the sky; heated air would rise and be replaced by cooler air to the north and *preferentially* from the east. The thermal tide aspect of Halley's explanation echoes the mechanism proposed by Descartes half a century earlier. He discounted the eastward motion of cooler air ahead of (west of) the subsolar point, where the heating was presumed to be greatest. A metaphor for his mechanism might be stirring a pot of soup, in which the stirring motion eventually causes the soup to rotate in the direction of the spoon. Later writers rejected Halley's hypothesis. However, a weak form of his mechanism was recreated several centuries later in the laboratory using a pan of water and a hot flame (Fultz et al., 1959).

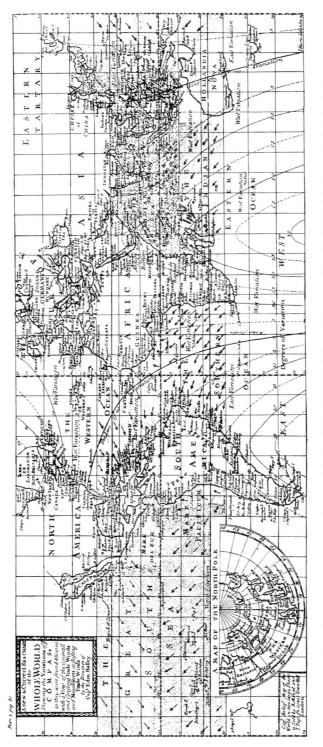

FIG. 1.1 The tropical surface circulation published by Halley (1686) is indicated by arrows. The arrowheads have been enlarged for clarity.

The first published rejection of Halley's hypothesis was written by George Hadley in 1735. Stating that Halley's mechanism was impossible, he proposed that the rotation of the earth caused the easterlies. Hadley's mechanism appears to be much the stronger of the two effects. Hadley thought that the angular *velocity* would be conserved (rather than angular momentum), but his basic idea is sound. He also postulated the existence of a thermally direct cell. Rising motion near the equator would be compensated by sinking motion at higher latitudes (Figure 1.2). As air moves equatorward, its angular velocity would become less than that of a point on the earth, resulting in an easterly wind component. The same reasoning creates westerly acceleration of poleward moving air. While Hadley did not specify where the sinking would occur, it is reasonable to conclude that the sinking would be poleward of 30° for two reasons. First, at that time it was widely believed that the trade winds ended at those latitudes. Second, he was concerned about applying a net torque upon the earth, and knew that the circulation he proposed must develop compensating surface westerlies somewhere else. Presumably, then, the equatorward motion would commence with surface northwesterlies at latitudes poleward of 30°. The circulation just described satisfies other balance requirements demanded by the observations of Hadley's time. There is mass balance between the upper level, poleward moving air and the surface, equatorward flow. This circulation model can even explain some things that Hadley was unaware of: the net transport of moisture to the tropics, the excess in tropical precipitation over evaporation, and the moisture deficit in the subtropics. (One notes that thermodynamics was not well understood at that time.)

In Hadley's scheme the horizontal transports used to satisfy the balances are possibly the simplest—a zonally symmetric meridional circulation. Hadley did not consider longitudinal variations, which were well known before Hadley's paper (e.g. Figure 1.1). The early writers chose to focus on the zonal mean properties. Only after the start of the twentieth century did it become clear that one *needed* to look at the zonally varying phenomena, the eddies, to explain fully the zonal mean.

For several decades after its publication, Hadley's paper was not widely noticed. Meanwhile, there were several "rediscoveries" of his ideas. For example, John Dalton (1793) proposed similar arguments; after submitting his manuscript for publication, he was notified of Hadley's work and revised it to include a comment on the lack of attention Hadley's article had received. Dalton's work goes well beyond the discussion in Hadley's article, with separate chapters on each variable, speculations about latitudes outside the tropics, and an elegant debunking of a popular notion about the formation of high and low pressure. Further, Dalton discusses moisture, which places his work well ahead of his contemporaries. Eventually, Hadley was widely credited with correctly interpreting the main causes of the zonal mean tropical circulation and today we label that circulation the "Hadley cell."

By the 1830s observations consistently indicated that the flow at low levels in middle latitudes was most commonly southwesterly, not northwesterly as predicted by Hadley's theory. Heinrich Dove (1837) reduced the poleward extent of the Hadley

FIG. 1.2   A probable schematic view of Hadley's (1735) description of the general circulation, reproduced from Lorenz (1967). Short arrows show surface wind flow.

cell and introduced flows varying with longitude similar to those illustrated in Figure 1.3. But his notion was based in part upon an erroneous argument: that the south-westerly winds, being warm and moist, originated in the upper branch of the Hadley cell. This idea fails to recognize the drying out of air that has risen to a high altitude before sinking to the surface. Dove did not want his equatorward return current in middle latitudes to occur at higher altitudes, so he suggested a scheme in which currents flowed side by side at the surface. His moist current came from the tropics and the dry one came from the polar regions. His scheme could satisfy a number of observed features, including the wind direction fluctuations in midlatitudes. In a sense, he identified the air flows that we now associate with thermodynamical properties: warm and cold air masses. Dove's scheme could be made to satisfy balance requirements of energy (equatorward current colder than the poleward) and momentum (southwesterly winds carry more angular momentum than northeasterlies). With slight modifications for acquiring water from oceans after the descent in the Hadley cell, Dove's scheme could even explain the observed moisture distribution. But for lack of strong theoretical support, Dove's ideas were not accepted during his lifetime.

Matthew Maury (1855) was also an observer, but he included just the feature Dove had rejected for the high altitude return flow in middle latitudes: an upper level equatorward return current in midlatitudes (Figure 1.4). He based his arguments

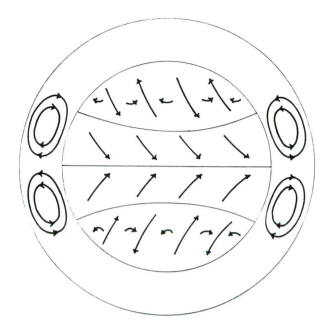

Fig. 1.3   Lorenz's (1967) interpretation of Dove's (1837) description of the general circulation.

partly upon observations and partly on a belief in a "Grand Design." Diffusion or mixing, he felt, were processes of chance and would not really occur in a respectable Grand Design. Maury also believed that atmospheric motions were strongly influenced by magnetism; after all, he reasoned, the magnetic poles were near the regions of coldest surface temperatures. This belief and others were ridiculed by his scientific contemporaries, yet Maury's book remained popular (eight editions in the first eight years). He thought that the crossing currents near the Tropics of Cancer and Capricorn caused mass to accumulate (building surface highs) and opposing air currents to "press against each other" and eject poleward and equatorward surface flow. These ideas seem comical today, but the popularity of his book may have influenced more worthwhile work by Ferrel and Thomson. Of particular importance, Maury publicized the idea of an indirect cell at high latitudes.

In 1856, one year later, William Ferrel published his first treatise. His theory of the circulation incorporated the indirect cell introduced by Maury and has the shape shown in Figure 1.5. Ferrel envisioned three cells in each hemisphere. The Hadley and Polar cells were both thermally direct, with an indirect midlatitude cell in between. The indirect cell is often referred to as the "Ferrel cell." Ferrel thought that his hypothesis explained several observed features: (a) the indirect cell should cause surface southwesterlies to predominate in middle latitudes from angular momentum conservation; (b) dry, calm air near 30° would be produced by the subsidence; (c) the stormy subarctic region would match the location of the rising branch of the Ferrel

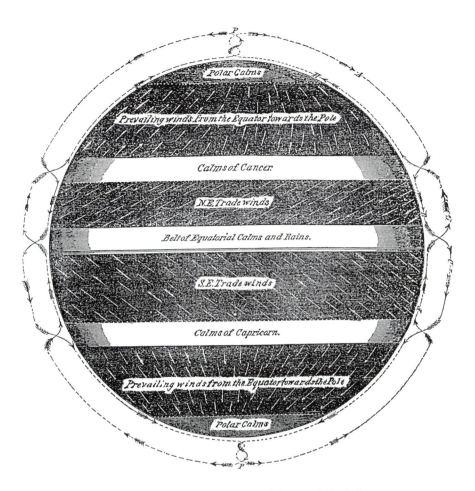

FIG. 1.4   Maury's (1855) diagram of the general circulation.

and Polar cells; and (d) the subtropical highs would arise from deflections of the
tropical easterlies and midlatitude westerlies. His greatest contribution was an accurate introduction of the north-south component of the Coriolis force—which could
explain the turning of the wind direction in the proper way. (He proposed that Coriolis turning causes the deflections referred to above.) His first work was published
in an obscure medical journal. The original three-cell picture was criticized because
it failed to transport heat in the right direction in middle latitudes. To restore the
heat transport, Ferrel shrunk the depth of the indirect cell (Figure 1.6), allowing the
upper branch of the Hadley cell to join with the polar cell. This change was intended
to provide the heat transport while retaining the surface southwesterlies. Even with
modifications, major problems remained with Ferrel's scheme. For example, the indirect cell would tend to transport angular momentum and energy toward the equator,

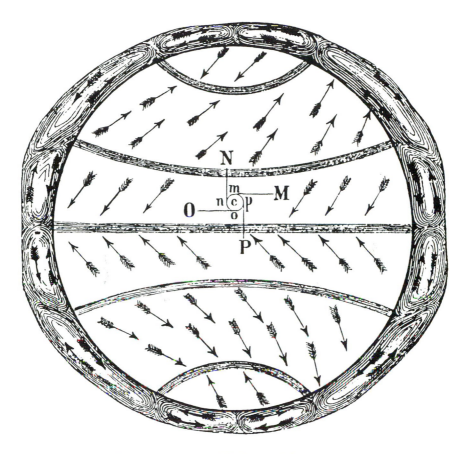

FIG. 1.5 Ferrel's earliest (1856) diagram of the general circulation.

in violation of both balance requirements. Also, Ferrel did not offer a mechanism to drive this indirect cell.

At the same time, James Thomson (1857) independently came out with a scheme (similar to Figure 1.6). He published little else on the subject, whereas Ferrel continued to develop and promote his ideas. But some problems with their scheme eventually proved insurmountable. High altitude (or even middle tropospheric) northeasterlies are not observed in midlatitudes, at least not to the extent predicted by the Ferrel cell. Instead, the observed Ferrel cell is a secondary circulation driven by the eddies through mechanisms that are discussed in connection with the Kuo-Eliassen equation in Chapter 6.

Finally, A. Oberbeck (1888) sought mathematical solutions to the equations that he presumed to govern the zonal average motions. He first investigated a balance between pressure and frictional forces, then included an east-west (only) Coriolis force and (fortuitously?) derived an equator-to-pole Hadley-type circulation. Oberbeck at-

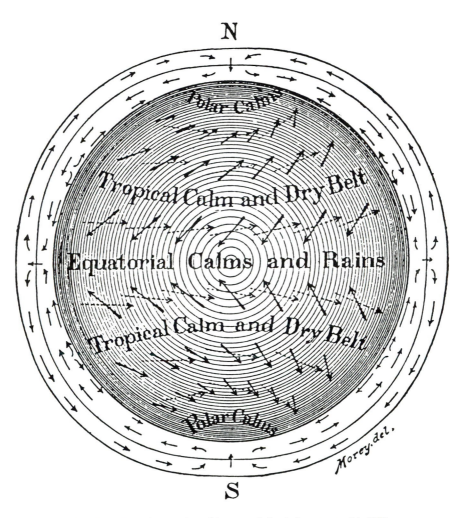

FIG. **1.6**   Ferrel's later view of the general circulation presented in 1893.

tempted to examine more complex expressions, including nonlinearity, but had to make inappropriate approximations. His resultant solutions were not realistic. However, Oberbeck's work is noteworthy because he introduced the notion that analysis of the dynamical equations, rather than qualitative arguments, could yield insight into the general circulation. Unfortunately, his lack of success caused this approach to be ignored for many years. Today, this approach is powerful, often used, and provides backbone for all of the discussions in Chapters 4, 6, and 7.

2

OBSERVING THE ATMOSPHERE

Science and art belong to the whole world, and the barriers of nationality vanish before them.

Goethe, 1813

The relative velocity of the winds may be best ascertained by finding the relative velocity of the clouds.

Dalton, 1793

2.1 An overview

Meteorological observations are the keystone for understanding the atmosphere. A great many observations are routinely taken each day in every country and across every sea. These data are freely shared among the countries of the world through the World Meteorological Organization (WMO). Data sharing is essential to the health of atmospheric science because the atmosphere recognizes no national boundaries.

Sharing data is but the first step toward constructing a comprehensive picture of the atmosphere. Before the data can be analyzed and theories constructed around it, the data must be collected and checked for errors. After data are collected, they are transferred by some means onto a regular grid. Nearby observations for various devices usually disagree. The observations are assigned relative weights based upon the observing devices' reliability and other considerations. An additional processing step may be made to bring the various fields into a specified mathematical balance. In short, "observed" data may have errors and may have undergone much processing between measurement and plotting. The measurement, the processing, and the error distribution of observations are all foci of this chapter. The purpose in discussing these issues is to establish the fidelity of later diagrams showing the "observed" atmospheric structure.

Observing the atmosphere with sufficient detail is a great undertaking. The difficulty and expense of the task have spawned creative solutions to the problem of acquiring accurate data. For example, in Dalton's time, upper air information was obtained largely by watching the clouds: their speed could be determined by tracking the motion of their shadows from a hilltop. Two hundred years later, high technology satellites gather wind data using the same basic idea.

Once the data has been gathered, checked, and somehow interpolated onto a grid, there remains one more problem to consider. The "four-dimensional" space and time-varying atmosphere has exceedingly complex structure and evolution. How can the data be subdivided in useful ways so that important dynamical relationships are seen? On an even more basic level, which averages and slices of the four-dimensional data arrays best illuminate the atmospheric structure? A pie can be sliced many ways, but clearly one way is most common and it is preferred for good reasons. Another focus of this chapter is to examine what means of slicing the atmospheric data make the most sense. In so doing, the organization of the observational data presented in this book is validated.

2.2   The surface-based network and its biases

The surface-based network refers to observations that are made from the same lo-cation at frequent intervals. There are two main components of this network. One component is the network of *upper-air* stations; about 1000 of these radiosonde and rawinsonde stations are operating at present. The other component comprises the 9,500 stations that only make surface observations. The radiosondes measure tem-perature, pressure, humidity, wind speed, and wind direction at *standard pressure levels*. (The standard levels are: 1000, 850, 700, 500, 400, 300, 250, 200, 150, 100, 50, and 10 hPa.) Depending upon the local conditions, additional observa-tions may be reported at *significant* levels. Significant levels are identified where one or more of the measured variables changes rapidly. The principal advantage of station-based data is high accuracy. Table 2.1 summarizes the accuracy and distri-bution of the station-based and other significant observing systems discussed in this chapter.

Station-based observational data have several drawbacks. The main problems are as follows. (1) The spacing between stations is irregular. (2) There is a clear bias for land areas, with poor coverage over the oceans. (3) Errors exist in the measurements and in the data transmission to the central receiving site. (4) There is a bias toward clear skies and light winds, especially for upper atmospheric data (e.g. van Loon, 1972, p.87). (5) The data taken along the track of a balloon may not be representative of the surrounding environment. Nappo et al. (1982) discuss this last point. Bruce et al. (1977) also examined representativeness by comparing simultaneous rawinsonde reports at adjacent stations. The clear sky bias results from visual tracking methods. Once a pilot balloon enters a cloud, it cannot be tracked. When a radiosonde is released near the jet stream, it may be blown over the horizon before it can complete its rise through the atmosphere. This bias toward light winds is one reason why comparatively few radiosondes report information above the lower stratosphere (another reason concerns solar heating of the sonde thermistor). Only about 75% of the radiosondes report above 200 hPa, 50% above 70 hPa, and 10% above 10 hPa (Thomas, 1975). In the past decade, the tracking of some types of radiosondes has been improved by using satellite navigation systems to locate balloon positions.

A good sense for how well sampled the atmosphere is can be obtained from Figure 2.1. Figure 2.1 shows the upper air network as of May 1984. The location of each active radiosonde station is plotted as a star in the figure. Active pilot balloon ("pibal") stations are marked with triangles. Active stations are stations that reported data on at least 10 days during May 1984; 855 (of 1,024 registered) radiosonde stations and 373 pibal stations were active then. Presently, 820 radiosonde stations (600 pibal) provide data from 1,200 (1,000) ascents each day (WMO, 1991). The radiosonde and pibal stations have similar locations with one major exception: most of the upper air stations in equatorial Africa are pibal stations. Accordingly, the winds are sampled better over west Africa than are other variables. The station coverage is excellent over the land areas in middle latitudes. The tropics are sampled poorly, including the land areas. The oceans are sampled very poorly. The lack of station coverage is quite serious in the Southern Hemisphere. In some ways, this map is an optimistic presentation because not all of these stations report twice each day, nor do they all report every day. In fact, only about 75% (Mohr, 1984) of these stations send in data at any one observing time.

Observational studies before the 1960s were based upon many fewer stations. Researchers before then were severely constrained by the paucity of upper air observations. During the past two decades, new stations and observing systems have greatly improved the sampling coverage over much of the atmosphere. Even today, truly excellent data coverage is achieved routinely only for a small part of the globe. However, a major field program was undertaken in the late 1970s to provide greatly-improved global observations, if only for one year. The program was called the Global Weather Experiment (GWE). The program was also known as the First GARP Global Experiment, or "FGGE", where GARP stands for the global atmospheric research program. GWE lasted from December 1978 through November 1979 and took a decade to organize. Because the GWE data have the best global coverage and quality, this book uses many observed fields measured during that experiment.

2.3   Other observing systems and their limitations

2.3.1   *Satellite-based systems*

To obtain reasonable global coverage, other sources of data fill the gaps between the radiosonde stations. Figure 2.2 shows the typical temperature sounding coverage from an operational polar orbiting satellite. Each circle represents a vertical profile of temperatures retrieved by the TIROS-N satellite. About 6,000 temperature soundings

Table 2.1 Properties of various operational observing systems.

Observing system	Fields observed	Error		Time resolution	Coverage	Vertical resolution	Horizontal resolution	Comments and (reports/day)
		Random error	Remarks					
Surface land stations	P_s T_s V_s rh_s	±0.5 mb ±1°C ±1.5 – 3 m/s ±10%	—	≥ 3 hours	WWW-network (fixed locations).	—	≥ 50 km	high resolution over most land areas. 4,095 of 9,500 stations in WWW (20,000/da)
Rawinsonde & radio-sonde soundings	T V rh	±1°C ±3 m/s ±8 m/s ±10%	> ±2°C in stratosphere below 500 mb above 500 mb	6–12 hours	WWW-network (fixed locations)	~50 mb	≥ 300 km (see fig 2.1)	About 885(1200/da) stations. 5 ocean weather ships by 1990. About 600 (1000/da) pibal stations: wind only
Buoys, Ocean platforms (incl. oil rigs, etc.)	P_s T_s V_s	±1 mb ±1°C? ±?	very few measure V_s or air temperature	6–12 hours	mainly used south of 20° S and N. Atlantic		500–1000 km	130–200 were operational during GWE. 300 (2500/da) in 1990. About 170 (600/da) are moored. only they measure V_s
Merchant Ships: (1) Surface Only	P_s T_s V_s rh_s	±0.5 mb ±1°C ±1.5–3 m/s ±10%	assumed same as for surface land station	≥ 3 hours	main shipping routes		varies greatly	About 7,400 ships (3500/da) Few ships in southern oceans
(2) ASAP or other upper air	T V rh	±1°C ±3 m/s ±10%	assumed same as for radiosondes	≥ 12 hours	main shipping routes	~50 mb	varies greatly	About 15 ships (30/da) equipped by 1989
Commercial Aircraft: (1) Post-flight reports	V	±6 m/s	less error for inertial navigation system (INS)	continuous	commercial air routes	200-300 mb level only	500-1000 km	About 100 reports per day, roughly 10 observations per report
(2) ASDAR	T V	±1°C? ±0.5–3 m/s	continuous depends on	commercial air routes	200–300 mb level (see	110 km along flight path		ASDAR on about 40 aircraft (est for

Instrument		Accuracy	Representativeness / notes	Frequency	Coverage	Vertical level	Horizontal resolution	Comments
			INS. ±10 m/s in representativeness		comments)			1986). Also takes "soundings" when plane takes off and lands. Being phased out: Only 3(150/da) by 1990
(3) AIREP	T V	±1°C? ±0.5–3 m/s	location errors. ±10 m/s representativeness	continuous	commercial air routes	200–300 mb level	concentrated along every tenth meridian	About 3,000 aircraft (4,500/da)
Satellite: (1) Vertical Temperature soundings	T	±1.5–3.5° C ±2° C is typical	larger errors in clouds	continuous	global	~200 mb	40 km possible, typically ~500 km	(About 6,000 soundings/da)
(2) LIMS	T	±1.5–3.5°C?		continuous	global	?	?	no longer active
(3) Vertical water vapor sounder	rh	±20 – 30	large errors in clouds	continuous	global	~400 mb	~50 km	
(4) Cloud-track winds (CTW)	V	±4.5–5.5 m/s for low level CTW ±9–10 m/s for high level CTW	In situ measurements of winds suggest much smaller error of 1–3 m/s	continuous	55° N to 55° S	200–500 mb	See Fig. 2.6 30–400 km	(About 6,000/da in 1983). (7,000/da in 1990) Most are over oceans. Speed bias 0.5–1.6 m/s. Newer systems 1–7 m/s better than old.
Constant level balloons	T V	±0.5° C ±1 m/s	drift during 1 satellite orbit, so speed is 100 min. average.	2 hours	see comments	140 mb level only		Used mainly in tropics and Southern Hemisphere. Winds from balloons drift. (About 100 observations/day.)
Dropsondes	T V	±1°C ±2 m/s		12–24 hours	tropics mainly	100–200 mb		Release from carrier balloons or from research aircraft.

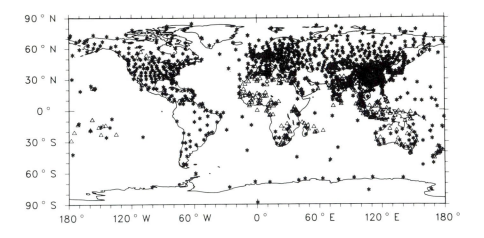

FIG. 2.1   Locations of active upper air stations as of May 1984. Pilot balloon stations are indicated by triangles, radiosonde stations by asterisks. Active stations are those reporting at least ten days during that month.

were made each day by TIROS-N (Halem et al., 1981). Presently, about 10,000 soundings are recorded per day (WMO, 1991).

Temperatures are calculated from the atmospheric long wave emission spectrum. As illustrated in §3.1, the radiation emitted by an object is related to the object temperature. Radiation is absorbed, emitted, scattered, and transmitted through the atmosphere. In general terms, the peak emission from an atmosphere will be near the level where the optical depth (§6.1) equals one. Less emission occurs from other atmospheric levels. The profile of optical depth varies with radiative wavenumber and with the absorbing gas. Consequently, these relationships can be "inverted" to obtain estimates of temperature over broad atmospheric layers. A more detailed explanation of the method is beyond the scope of this text. The schemes used most commonly are based upon an iterative procedure described by Chahine (1970). In addition to the imprecision in the inversion process, the data must be calibrated to accomodate trends in instrument sensitivity, known as "instrument drift."

The emission coming from areas of varying size beneath the satellite is examined in several spectral bands. The TIROS-N spacecraft has three instruments: the High resolution Infrared Radiation Sounder (HIRS), the Microwave Sounding Unit (MSU), and the Stratospheric Sounding Unit (SSU). The HIRS uses 20 infrared channels and scans areas that are 30 km in diameter; it is possible to sample 56 such areas for each 2250 km scan line normal to the flight path of the satellite. The MSU uses 4 channels, scans areas 110 km in diameter, and 11 such areas fit on each scan line. Finally, the SSU has 8 linear segments within the central 1500 km of the scan line.

The calculation of temperature from the radiance is difficult and imprecise. Typical errors are 1.5–3°C. Several authors have examined the errors by comparing satellite-derived soundings with nearby radiosondes (e.g., Bruce et al., 1977; Smith

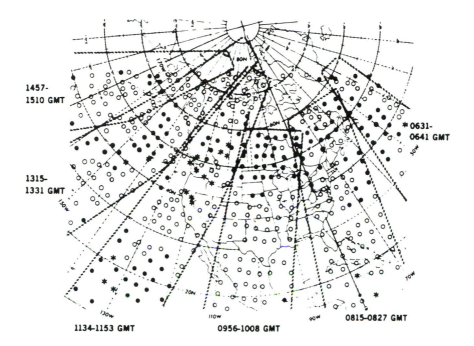

FIG. 2.2 Circles indicate locations of temperature retrievals for the TIROS-N satellite on 6 April 1979. Dashed lines mark the regions scanned by each orbital pass of the satellite. Shaded circles distinguish retrievals through clear sky. From Smith, (1985).

et al., 1979). Figure 2.3 shows root mean square (rms) temperature errors for the HIRS and MSU devices. Figure 2.3 was produced by comparing satellite-derived soundings with those of nearby radiosondes. The HIRS soundings are made in clear or broken cloud areas while the MSU data are used for cloudy regions. The errors in the 700–400 hPa layer are comparable to radiosonde errors. However, the error is much worse in regions where sharp vertical changes occur, such as near the surface and at the tropopause. So, while satellites do provide huge volumes of data—there are many circles in Figure 2.2—the data have rather low quality. Another problem with satellite-derived temperatures is that the errors in cloudy regions may be much greater than in areas of clear sky (Baumhefner and Julian, 1975).

Satellites also provide winds that are commonly labeled cloud-track winds (CTW). Presently, about 7,000 CTW are obtained each day from the five geostationary satellites (WMO , 1991). Ten years ago about 6,000 CTW were logged each day (Halem et al., 1981). The main problem is distinguishing the level of these winds. Recently, some weather forecast centers have reduced or discontinued the use of CTW over the Northern Hemisphere from some satelites (Merrill et al. , 1991). These CTW frequently underestimate wind speeds; Merrill et al. report on improvements to the system for estimating CTW that should correct the problem of assigning a level to a CTW.

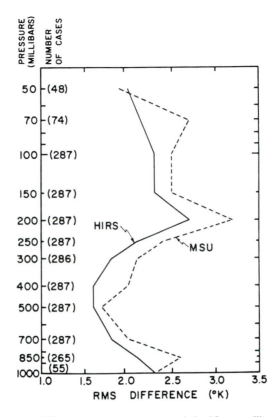

FIG. 2.3   Root mean square difference between temperatures derived from satellite soundings and nearby radiosonde soundings over North America. Soundings are used from the period 22 March through 19 April 1979. From *Bull. Amer. Meteor. Soc.*, Smith et al. (1979) by permission of American Meteorological Society.

A cloud feature is tracked between two or three sequential images (15 minutes to one hour apart) taken by a geostationary satellite. The distance travelled by the feature divided by the time interval between the images gives the velocity. Different operational centers automate the process to varying degrees. Naturally, only one level of cloud is visible at any one point on the globe, as can be seen in Figure 2.4. The vector motion is assigned to an altitude by one of two means. One method is to assign all the winds to fixed altitudes. Though the features tracked undoubtedly varied greatly with height, the winds shown in Figure 2.4 were assigned to two fixed levels. The other method exploits the *brightness temperature* (based upon its maximum infrared emission) of each feature. By comparing this brightness temperature with a nearby temperature profile from a radiosonde, satellite sounding, or grid point analysis, the level of the CTW can be assigned. Operational centers use one or both of these methods. Stereo imaging of clouds is possible in regions where images from two geostationary satellites overlap. Errors from using stereo

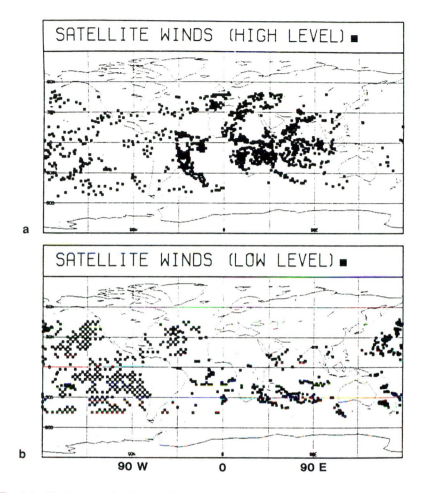

FIG. 2.4   Black squares showing a typical coverage of (a) high altitude and (b) low altitude cloud-track winds (CTW). These diagrams show operational data from 1 February 1979, an intensive observing period during GWE. The upper and lower level reports have very little overlap. From Bjørheim, 1981

altitude estimates are claimed to be about ± 500 m (Minzner et al. , 1978; Hasler, 1981).

Even if the cloud feature's elevation is known, the air speed at that level could be different. The wind may blow around the cloud. The evidence suggests that deep clouds tend to move with the average velocity in the vertical. Several investigators (e.g., Bauer, 1976) have compared CTW with winds derived from radiosonde ascents. The problem with this comparison is that the balloon launch used may not adequately correspond in time or space with the cloud feature studied. Differences as great as three hours and 300 km have been allowed in some tests.

To avoid this problem, Hasler et al. (1979) compared CTW with in situ measurements. Over a five year period, they flew an aircraft through 81 clouds for which

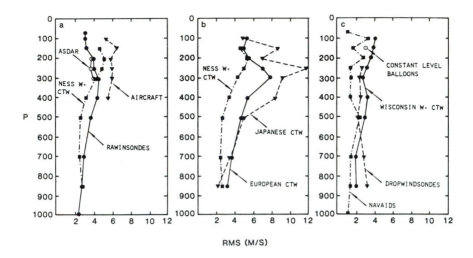

RMS (M/S)

FIG. 2.5   Root mean square differences between Global Weather Experiment (GWE) wind analyses and data from various observing systems. "NAVAIDS" refer to radiosondes released from ships. See Table 2.1 for information about the various observing systems. From *Bull. Amer. Meteor. Soc.*, Halem et al. (1982) by permission of American Meteorological Society.

satellite winds were obtained. Most of the clouds were low-level cumulus. They studied five cirrus clouds. The researchers chose several locations in the tropics and subtropics of the north Atlantic and Caribbean. For low cumulus clouds, the difference between the wind at the cloud base and the CTW was only 0.9 m/s to 1.7 m/s. For the limited sample of cirrus clouds, the difference between the mean wind in the cloud layer and the CTW was 2.3 m/s. These estimates are considerably less than errors found by comparing CTW data with nearby radiosondes. For example, Figure 2.5 compares wind measurements by several observing devices with the final data stored at nearby grid points. This rms error calculation uses the highly-regarded GWE data. The errors range from about 5 m/s for low level winds to 10–14 m/s for high level winds. The large difference between error estimates made using in situ measurements versus *analyses* from radiosondes illustrates the problem of representativeness. Specifically, the CTW averages the wind speed over a larger horizontal and vertical area than the radiosonde balloon. Eventually, it may be possible to supplement CTW in cloud-free areas by using satellite LIDARs. (Light Detection And Ranging instruments use the doppler shift of laser light reflected off atmospheric particulates to measure winds.) Such space-based LIDARs are estimated to have measurement errors of a few meters per second and are currently under development. (See Baker and Curran, 1985, p. 26).

Some weather satellites can estimate the amount of water vapor present in the middle troposphere (600–300 hPa elevation). An electromagnetic spectral band (usually between 5.7 and 7.1 μm) is examined. Water vapor absorbs radiation in this band (between the visible and thermal infrared channels). Estimates of water vapor were first made using instruments on polar orbiting satellites in the early 1970s (e.g., Steranka et al. , 1973). More recently, this band has been available on geostationary satellites (e.g., Morel et al. , 1978). The water vapor distribution also can be scanned for identifiable features that can be tracked between consecutive satellite images. Hence, winds in cloud-free areas can be estimated by tracking these features using the same techniques as for CTW. Stewart and Hayden (1985) estimate the error in these "water vapor winds" to be 4–7 m/s.

Of course, weather satellites see the pattern of clouds and provide some information about cloud heights. Until recently, forecast models did not use this information. In the past few years a new emphasis has been placed upon measuring the clouds. A principal outgrowth was the establishment of the International Satellite Cloud Climatology Project (ISCCP) in 1983. The main goals of ISCCP (Schiffer and Rossow, 1985) include accurate measurement of radiative properties of the earth in addition to establishing an accurate cloud climatology. The cloud amount and type have a crucial effect upon the radiative balance of the earth. The issue has gained more urgency because the cloud details are the highest priority in studies of global climate change (CES, 1989, p. 104).

Finally, the Seasat satellite successfully tested a scatterometer. By measuring the scattering of microwaves emitted by the satellite and reflected from capillary waves on the ocean, the device estimates surface wind speed and direction. Since microwaves are used, the device can make measurements if clouds are present. Operational scatterometers are planned.

2.3.2   Other systems

Other sources of observations include commercial aircraft, drifting buoys, merchant ships, research aircraft with dropsondes, and constant level balloons. Table 2.1 summarizes these measurement systems and the characteristic errors for each.

Commercial aircraft can provide meteorological data by several means. One method is to collect in-flight and post-flight manual reports (labelled "AIREP"). This system typically provides 4500 reports per day from 3000 aircraft. In the 1980s an alternative observing system was tested whereby some wide-body commercial aircraft were outfitted with Aircraft to Satellite Data Relay (ASDAR) units. ASDAR data were quite accurate; recall Figure 2.5. The details about the ASDAR program shown in Table 2.1 are drawn from Sparkman and Smidt (1985). These commercial aircraft systems primarily provide wind data at flight level (300–200 hPa). International flights over the oceans and over the tropics were particularly useful. A "sounding" could be taken during the ascent and descent of takeoffs and landings. Despite its advantages, the ASDAR system dwindled to just three planes as of 1990 (WMO, 1991).

Many merchant ships take surface observations along the major ocean shipping lanes (especially the northern Atlantic and Pacific oceans). Seven thousand of these so-called "ships of opportunity" provide most of the same data as synoptic land stations. Recently, a semi-automated system has been developed whereby radiosondes can be released from merchant ships. Initially tested in the North Pacific, the Automated Shipboard Aerological Program (ASAP) has been expanded to the Atlantic Ocean (15 ships reporting twice per day). The device takes standard radiosonde measurements: wind speed and direction, temperature, humidity, and pressure. As with ASDAR, the data are automatically transmitted to satellites. An operator must inflate the balloon and supervise the automated tracking of the radiosonde. This system is suitable for remote land locations as well. In addition to filling in areas where data are often otherwise not available, these systems have two other benefits. The automation reduces the data transmission time from 12–18 hours to 1–1.5 hours and the error rates from an estimated 30% to 1%.

Buoys are very useful for providing surface pressure in remote ocean areas. Some buoys also provide air temperature and wind speed. The main role of buoys is to fill in major gaps in the surface observational network. Buoys are often placed where merchant ships are infrequent (such as the southern oceans) or where there is a critical need for closely-spaced observations (such as in the Great Lakes). Some buoys are tethered, but most oceanic buoys are set adrift. Ships or aircraft can deploy the balloons. Aircraft deployment allows gaps in the drifting buoy network to be filled more easily than before. The buoys report directly to a polar orbiting satellite. Most buoys are operational for several months. The number of buoys deployed at one time has grown to roughly 300 (WMO, 1991); many of these are in the middle latitudes of the Southern Hemisphere.

Dropsondes are occasionally released from special research aircraft. The aircraft drops a small cylinder containing instruments (similar to radiosonde instruments). Often the instruments descend to earth on a parachute. The data are quite reliable, but it is quite expensive to fly the aircraft repeatedly to remote locations. Generally, these data are available only for special regions during special periods (such as during a scientific experiment or when a hurricane threatens a populated region). Balloons can release dropsondes by radio control.

Constant level balloons (CLBs) consist of a super-pressure balloon and an instrument for relaying temperature data to a polar orbiting satellite. Winds are determined from the drift of the balloon. A super-pressure balloon displaces a fixed volume; therefore it remains at a constant density level in the atmosphere. The level chosen corresponds to the 135–140 hPa pressure layer for the tropics. CLBs fly at a high altitude because precipitation (especially ice forming on the balloon) will rapidly deteriorate the package. Also, dew and ice on the balloon add mass to the device, causing the device to sink to an unknown elevation. Finally, the devices are not allowed to drift below 143 hPa, in order to avoid any commercial aircraft flight path. The lifetimes of the CLBs vary widely, but roughly 60% of the CLBs last longer than 10 days. Most CLBs are released in the tropics (from Canton, Guam, and Ascension Islands); many CLBs drift into the Southern Hemisphere midlatitudes. The CLB lifetimes have been so short because (1) they are launched in the tropics where convection is often vigorous and (2) an on-board cut-down device automatically destroys many. The cut-down device activates if the balloon dips below 143 hPa or crosses northward of an arbitrary geomagnetic latitude.

2.4 Data analysis

After collection, the data are processed in three ways: error checking, interpolation onto a regular grid, and initialization. Data analysis refers to these three steps. The subject is extensive and is reviewed in a book by Daley (1991). The locations and accuracies of the observations are highly irregular. The data are more useful if projected onto a regularly-spaced grid in a systematic fashion. This interpolation step is called *objective analysis*. A common practice is to remove the gross errors during the objective analysis. For a variety of reasons, the objectively analyzed data at the grid points may not be consistent. When other fields are derived from the primary fields, small errors in the primary fields may swamp the signal being sought. The best illustration of this problem is the calculation of vertical velocity. Vertical velocity usually has large errors when calculated by the kinematic method from observed horizontal velocities. When calculated that way, vertical velocity comes from the small difference between two large quantities, a relationship that magnifies the errors in the horizontal wind measurements. While dynamical relationships can be incorporated into the objective analysis, it is more common to balance the data in a separate step known as *initialization*.

A wide variety of objective analysis schemes has been devised. Most procedures begin with a first guess at each grid point. The first guess is either a model forecast (valid at the observing time) or is based on climatology. These schemes differ largely in the way nearby observations are weighted and combined to define the correction to the first guess. Usually several observations are present within a specified distance from the grid point. In areas where there are few observations, the correction may be unreliable because the consistency of a single observation cannot be checked against other observations. Areas with few observations typically rely more heavily upon less accurate observing devices. For example, operational centers place a heavy reliance upon satellite data over the southern oceans because there are few alternatives. One consequence is the rejection of radiosonde data at isolated land stations because those data differ too greatly from the first guess, where the first guess came from a forecast that was largely influenced only by satellite observations upstream. It is quite striking to plot the short-term forecast error at 500 hPa for the forecast model at the European Centre for Medium-range Weather Forecasts (ECMWF). When the zonal mean is removed, contour plots of the ECMWF forecast errors have "bullseye" maxima centered at isolated radiosonde stations on southern ocean islands and down the southern tip of South America (Grotjahn, 1991a). Trenberth and Olson (1988) also suggest that some operational analysis schemes place too much emphasis on the first guess in the Southern Hemisphere. Nonetheless, having satellite observations is far better than nothing; forecast accuracy has greatly improved in the Southern Hemisphere as a result.

After placing the data on a grid and removing obvious errors, additional errors remain for three reasons. The observing and collection system has inherent errors, the observations may not represent the larger region surrounding a station (§2.5), and the atmosphere has a higher order balance (e.g., small scale processes not included in the processing) than the objective analysis scheme. The contemporary solution is to process the data using nonlinear normal mode initialization (NNMI). Machenhauer (1977) and Baer and Tribbia (1977) developed the basic form of NNMI. The idea is to project the observations onto desirable (Rossby wave) solutions and then determine what other solutions (gravity waves) are needed to balance the nonlinear interactions between all the waves. The procedure requires estimation of the wave time tendencies, which is commonly done by making a short run of a forecast or general circulation model (GCM). NNMI acts like a clever filter by removing unwanted (high-frequency) oscillations and keeping all the necessary dynamics to balance the desired waves.

As is the case for objective analysis, NNMI may not improve aspects of the data for which it is not specifically designed. The early designs targeted gravity wave noise in favor of Rossby modes. Consequently, precipitation fields and divergent motions in the tropics would not be expected to improve. Other problems with NNMI are summarized in Errico (1989).

NNMI is worth mentioning here for two reasons. First, the analysis system influences the properties of the data. For example, the analyses made at the National Meteorological Center (NMC) in the 1970s employed Hough function expansions and are thought to be too nondivergent. Kistler and Parrish (1982) reported significant changes when NNMI and the spectral model were introduced at NMC. Changes in model resolution, physical parameterizations, and analysis schemes, as well as missing observations, have all had impacts upon archived gridpoint data (Trenberth and Olson, 1988). Such changes in procedures make it difficult to establish a climatology and to examine interannual variations. Second, the NNMI itself could be a powerful diagnostic tool. This book uses the common, but simplistic, organization of presenting zonal average and zonally-varying data. The projection procedure incorporated by NNMI can be used to interpret the non-zonal data more elegantly (e.g., Tanaka and Kung, 1988). In theory, Rossby wave structures can be isolated that are more physically meaningful than a pure zonal wavenumber.

2.5   Spatial and temporal representativeness

Some error estimates listed in Table 2.1 have large ranges. In situ measurements of winds in the actual clouds used for CTW find quite small errors (1–3 m/s). Yet, when CTW are compared with nearby radiosonde stations, the apparent error is much larger (10–16 m/s). Similarly, commercial aircraft inertial navigation systems (INS) have

Table 2.2 Observational requirements for a global observing network from the WMO (1990) and Mohr (1984).

	Horizontal Resolution	Vertical Resolution	Accuracy (RMS)	Frequency of Observation
Temperature	500 km (1000 km over tropical oceans)	4 tropospheric levels 3 in stratosphere	$\pm 1\,°C$	every 12 hrs
Wind	500 km	4 tropospheric levels 3 in stratosphere	$\pm 3\,m/s$	every 12 hrs
Relative Humidity	500 km	4 layers or 2 parameter vertical function	$\pm 30\%$	every 12 hrs
Sea-surface Temperature	500 km	—	$\pm 1\,°C$	perhaps 3-day averages of instantaneous measurements
Surface Pressure	500 km	—	$\pm 0.3\%$	every 12 hrs
Precipitation	100 km	3km	50%	every 6 hrs

small errors, but AIREP and ASDAR winds have much larger errors when compared to nearby radiosondes and nearby aircraft (Julian, 1983). These discrepancies arise because each system samples different time and space scales. Radiosondes ascend rapidly; aircraft traverse a large region in the horizontal quickly; CLBs have a slow, lengthy drift; CTW follow the ponderous motion of clouds that may be deep; each satellite sounding scans a large area; these are five very different paths to sampling the atmosphere! Therefore, the range in the error estimates given here is due to instrument flaws and to small-scale variability of the atmosphere. The wide range of error estimates gives some indication of the representativeness of the data at a given grid point at a particular time. The representativeness error is likely to be mainly a random error. Hence, time and zonal averages reduce the representativeness error.

The discussion of representativeness begs the question of what time and space scales must be resolved to sample adequately the atmosphere. Not surprisingly, the amount of resolution needed varies with the application. The published World Meteorological Organization *minimum* requirements for a global, large scale observing network are listed in Table 2.2. Mohr (1984) and others propose higher requirements than those listed in the table: perhaps twice as many vertical levels, perhaps half the horizontal spacing and somewhat higher instrument accuracy. Comparing Tables 2.1 and 2.2 reveals important deficiencies in the accuracy of and coverage by the present observing system.

The data coverage varies with the location and the variable being considered. The surface-based observational network provides dense observations over the midlatitude land masses at all levels. *Surface* observations from land stations, merchant ships, and drifting buoys provide good global coverage. *Tropopause level* winds and temperatures over the tropics and southern oceans can be deduced from satellites and constant level balloons. The major remaining gaps in the observational network occur in the middle troposphere over much of the ocean and some of the tropical land area. These remaining gaps are only partly filled by satellite temperature profiles—data that have significant errors.

What effects are caused by the spatial gaps of the observational network? The radiosonde data are given the highest weight by objective analysis schemes. However, Figure 2.1 clearly shows that the radiosonde network has a strong bias toward sampling over land areas, especially in middle latitudes. Rosen et al. (1985) compared zonal averages based upon using only radiosonde data with averages that incorporate all the observing systems available during GWE. They found that the two sets of averages were "generally comparable." The differences that they found were similar in magnitude to the differences between averages calculated using two different objective analysis schemes. Rosen et al. (1985) show a specific example of 200 hPa subtropical jet winds in the vicinity of the east coast of Asia. During January 1979 the maximum winds are 5–10 m/s faster in the ECMWF analysis than in the Geophysical Fluid Dynamics Laboratory (GFDL) analysis. The radiosonde network is very dense upstream (Figure 2.1), yet this basic velocity field is significantly different in the two analyses.

Grotjahn and Kennedy (1986, 1990) try to avoid the influence of a first guess on the estimate of the bias. They calculate novel zonal averages of GCM data first using all model grid points, second using only those grid points near the stations shown in Figure 2.1, and third using only grid points over major land areas. While not realistic, the second average reveals something about the sampling by radiosondes. The first two averages are remarkably similar, while the third set (major land areas only) is very different. The most important conclusion to draw here is that the radiosonde network, especially when supplemented by the other observing systems, is adequate to sample the general circulation features that will be discussed in this book.

Grotjahn and Kennedy (1990) describe the following specific biases in the radiosonde network by studying a GCM. The winter hemisphere Hadley cell is stronger than it should be: the maximum poleward flow is about 30% greater than the correct value and poleward flow extends 10–15 degrees latitude further than it should. The tropical upward vertical velocity is much too big (by a factor of 2) in July. Similarly, the upper tropospheric cloud coverage during July is somewhat too large in the tropics. These errors could be anticipated since strong low-level convergence occurs preferentially over the tropical land areas (§5.3, §5.5, §5.10). Finally, many studies (e.g., Newell et al. , 1972) find zonal average easterlies in the tropical, upper troposphere (§3.3). Both Grotjahn and Kennedy (1990, using a GCM) and Rosen et al. (1985, using GWE data from just one month) show that weak *westerlies* between 400–300 hPa during January may be normal instead.

Daley and Mayer (1986) use a GCM to study observational device errors and objective analysis errors in addition to the spatial bias. Their experiment goes beyond Grotjahn and Kennedy (1986, 1990) in two important ways. First, Daley and Mayer include simulated observations from the *non*-station network in addition to station-based data. The simulated observations also incorporate random and systematic errors for each device. Second, the simulated observations are interpolated using an operational objective analysis scheme. The simulated observations are taken from a 19.5 day (November period) run of a GCM. Figures 2.6 and 2.7 summarize their results.

Figure 2.6 shows errors in the time and zonal mean fields of temperature, geopotential height, zonal wind, and meridional wind. Positive values occur where the objective analysis is greater than the GCM true state. Negative values are shaded. Part (a) of Figure 2.6 shows that the temperature errors over much of the troposphere are less than 2 K. The exceptions are at low levels in the high latitudes. Daley and Mayer feel that these errors partly reflect extrapolation problems in showing data below ground level (i.e., the south pole is at high altitude and has surface pressure much below 1000 hPa.) In the stratosphere the temperature errors are much larger and point to the unreliability of data, especially in the tropics. In part (b) of Figure 2.6 the geopotential errors are shown to be generally less than 10 m except near Antarctica and in the tropical stratosphere. As expected from the thickness equation, since the temperatures are too warm, the heights are too high in the analysis. Figure 2.6c shows that the zonal winds have large errors (stippling in Figure 2.6) in the tropics and south of 70 S. Daley and Mayer do not find tropical, upper tropospheric westerlies in either their analysis or their GCM true state. Apparently, CTW or CLBs are sufficient to resolve the zonal average zonal wind field, but this information cannot be used to determine the tropical stratospheric temperature field because the fields are decoupled there. As shown in Figure 2.6d, the zonal average meridional wind is small in much of the atmosphere, especially outside the tropical troposphere. The errors shown in Figure 2.6d point to problems in the stratosphere and south of 60 S.

Figure 2.7 displays the root mean square of the time and zonal average transient error variance. The format is similar to Figure 2.6. Lightly shaded regions indicate where the ratio of error variance over the transient variance in the GCM is 0.5–1.0; darker shading indicates a ratio greater than unity. Overall, the errors are proportionally greater for the transient fields than for the time mean fields. Again the midlatitude errors are the smallest; transient temperature errors are typically about 20–25% of the true state. The transient temperature fields have quite large errors in the tropics, in part because the temperature field has small variance in the true state (the "native variance"). The velocity field error is plotted using a vector magnitude error. The Northern Hemisphere midlatitudes errors are 15–30% of the true state; the error is about twice that in the Southern Hemisphere midlatitudes. Again the tropics have large errors, in part because the native variance is small. The error in the geopotential height variance monotonically increases with height and quite uniformly with

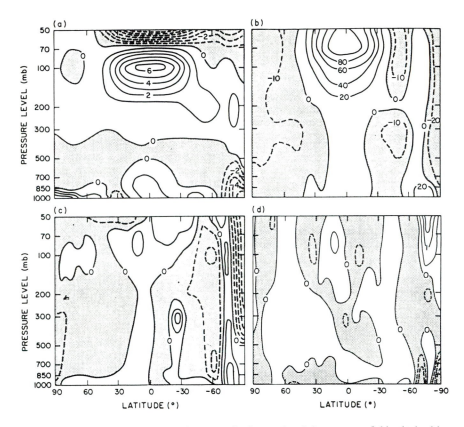

FIG. 2.6   Meridional cross-sections of the error in the zonal and time average fields obtained by an objective analysis scheme. The error shown is the true state in a GCM minus the analysis state derived from simulated observations of the GCM state. (a) Temperature with 1 K interval. (b) Geopotential height with 20 m interval. (c) Zonal wind with 1 m/s interval. (d) Meridional wind component with 0.5 m/s contour interval. From *Mon. Wea. Rev.*, Daley and Mayer (1986) by permission of American Meteorological Society.

latitude. The ratio of error to native geopotential variance is 10–15% in most of the Northern Hemisphere middle latitudes, with slightly higher values in the Southern Hemisphere. In all the variance fields shown by Daley and Mayer, the errors are once again worst above Antarctica and in the stratosphere.

Atmospheric fields vary rapidly in time as well as in space. In some tropical locations the weekly and even the daily variation is comparable in magnitude to the seasonal cycle. How valid is it to calculate climatological, "general circulation" patterns from observations taken at one or two observing times per day? How valid is incorporating satellite soundings that are actually taken over a period lasting several hours before and after the observing time? The diurnal cycle creates an atmospheric tide with amplitude 1 hPa (at the surface) and wavenumber two shape, centered along the equator. Grotjahn and Kennedy (1986) find that the error in sampling the

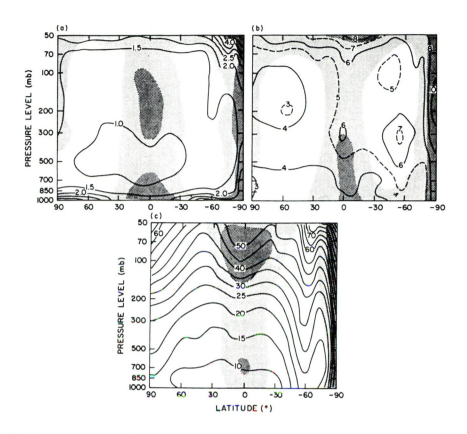

FIG. 2.7 Similar to Figure 2.6 except showing square root of transient error variance (SQEV). Light shading indicates SQEV > 50% while dark shading marks regions where SQEV > 100% of the observed square root of transient variance. (a) Temperature with 0.5 K contour interval; (b) vector wind with 2 m/s interval, and (c) geopotential height with 5 m interval. From *Mon. Wea. Rev.*, Daley and Mayer (1986) by permission of American Meteorological Society.

surface pressure field at just two times (simultaneously) looks like the tide. Assuming geostrophic balance still holds 5 degrees from the equator, then this pressure error pattern corresponds to meridional winds up to 2 m/s. However, the time sampling error was negligible for zonal averages. Oort (1978) found that gaps in the sequence of 12-hourly reports caused a smaller error than that from the unrepresentativeness of a radiosonde ascent and the spatial gaps between stations. Therefore, the diurnal cycle does not seem to alter significantly the results shown in this book. As for observations made outside a standard observing time, modern "four-dimensional" assimilation schemes allow data to be included at any time as part of an endless forecast-analysis cycle.

This subsection has discussed only a few of the many studies of errors in observational analyses. Additional references are cited by Hoskins et al. (1989). Those who archive data and those who study data have devoted much effort to ensure that the data are as accurate and representative as practical. Consequently, the purpose here is to encourage the reader to be aware of the problems, not to reject the observational data. Confidence in the data is gained by two means. First, more than one example of an observed, analyzed field is often shown in the book. Second, dynamical relationships are used to check the consistency of different variables and to reinforce the dynamical notions themselves.

2.6   Compositing observations

The global coverage of observing systems has large gaps. In addition, the atmosphere contains many diverse and variable phenomena. These two facts frequently lead researchers to composite their observations, because compositing improves their ability to draw meaningful conclusions about atmospheric behavior. Zonal averaging and especially time averaging are two common examples of compositing. Table 2.2 shows that the atmosphere can be very well sampled with observations that are rather widely spaced (500 to 1000 km apart). While individual storms move rapidly and vary in intensity and many other properties, their simple statistical properties (like mean and variance) tend to be stable after some period of time. Therefore, time averaging, especially when done separately for different seasons, can usefully depict the general properties discussed in this book.

The first question to ask when compositing is: what length of time (or space) average is needed? For the persistent part of the general circulation, one anticipates that a month is a minimal time interval over which to average. Longer time periods, especially periods longer than several years are even better. Each winter, for example, would be an independent sample. Longer time averages reduce the effect of individual, atypical weather periods; the data and conclusions are not dominated by an unusually severe winter, for example. To separate seasonal variation, Blackmon et al. (1977) average data from 30 Januarys. But time averages throw out some important details. One would not expect time average fields to satisfy all the dynamical constraints. Averaging smears out many small-scale properties that typically meander or change with time. For example, the time and zonal average subtropical jet stream is much less unstable to the formation of frontal cyclones (§7.1) than local regions of the jet on specific days. The reason is that the jet has larger vertical and horizontal shears in the latter case. Hence, the time and zonal mean flow might not predict the formation of the ubiquitous frontal systems. Another problem is that the neglected features may be necessary to maintain the average field. A prime example is the zonal average wind fields; they cannot be explained except when the momentum flux contributions by zonally-varying eddies are included (§4.1, §6.2.1).

Generally speaking, the length of time needed to obtain a representative time average depends upon the area in question. When the average is over the whole hemisphere, then the averaging period can be rather short. James (1983) shows that 15–30 days are adequate for the meridional heat flux, but only *when averaged over the whole hemisphere.* By implication, 15 days is a reasonable minimum time sample for most other variables. When the averaging scale becomes smaller (e.g., a zonal average only), then the averaging period needed usually must be longer. There appears to be no consensus regarding how much time is adequate for time averages and for combined zonal with time averages. Often the averaging period may be dictated by the length of accurate data available. For example, numerous 1–3 month averages have been published from the year-long GWE.

The eddy momentum flux is one major field that is notoriously difficult to measure. The eddy momentum flux is not statistically stable on 30-day averages according to Blackmon et al. (1977). A much longer period average is recommended for this quantity.

The general circulation must conform to some simple dynamics (§1.2). The circulation is in hydrostatic balance except at very high altitudes or on small scales of motion (such as inside a cumulus cloud). With these exceptions and excluding the latitudes close to the equator, the atmosphere is in approximate geostrophic balance. The observational data should reflect these balances, but not too strongly. Ageostrophic effects (that is, non-geostrophic motion of any type) are quite important in maintaining many features. Indeed, the Hadley and Ferrel cells *must* be at least partly ageostrophic since they include nonzero vertical velocity.

Approximate hydrostatic balance has three implications for compositing observations. First, hydrostatic balance greatly restricts the types of atmospheric motion. Hydrostatic balance expresses pressure in terms of the density field *or* in terms of the (virtual) temperature distribution. Therefore, only density *or* temperature need be measured. Indeed, many measurements are taken using pressure as the independent variable for this reason. Second, hydrostatic balance means that the vertical velocity acceleration terms in the vertical momentum equation are small. In other words, vertical motion is much smaller than horizontal motion. Third, since the motion is nearly horizontal, many atmospheric properties can be fruitfully examined in two-dimensional, horizontal slices. Though one should not restrict oneself to this "two-dimensional" view, this viewpoint simplifies the visualization of motion and other properties.

The geostrophic approximation is much less predominant than the hydrostatic, but it is a useful tool for qualitative interpretation. In the past, some of the data could only be estimated by invoking geostrophic balance. For example, upper level motions were sometimes deduced from measurements of the height field. Again, the geostrophic assumption will delete circulations that have vertical motions, and the assumption breaks down near the equator.

This book presents observations by first discussing zonal and time average distributions. Why zonal average? Zonal averaging is useful when a planet's rotation rate is "fast." The rotation rate is fast, according to barotropic turbulence theory (e.g., Williams, 1978), if the meridional gradient of the Coriolis parameter is large compared to the global mean winds, and if interaction with the planetary surface is small.

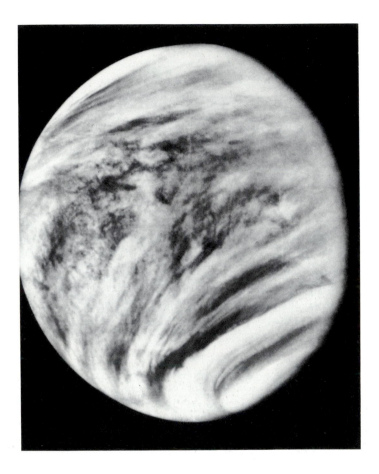

FIG. 2.8   Ultraviolet view of Venus as seen by the Mariner 10. Photo courtesy of NASA.

The planets in the solar system have varying speeds of rotation and atmospheric thicknesses. As one consequence, zonal averaging is more useful on some planets than on others. Of course, other factors are important in determining the circulation of a planetary atmosphere. Here, rotation rate is convenient to make a point that the appropriate averaging to choose depends upon the symmetries characterizing a circulation.

Venus rotates very slowly by the criterion defined in the previous paragraph. Venus is nearly the same size as Earth, but Venus rotates just once every 250 Earth-days. Figure 2.8 shows how the Venusian clouds look in ultraviolet light. The image shows irregular cloud patterns centered near the subsolar point. Figure 2.8 clearly shows that zonal averages would not represent the main character of the Venusian general circulation, except near the poles.

Jupiter rotates very fast for its size; its rotation period is around 11 or 12 hours. The Jovian atmosphere is very deep and one might assume that the circulations seen

FIG. 2.9 Jupiter from a Voyager spacecraft. Photo courtesy of NASA.

from Earth experience little damping due to little interaction with whatever surface may lie below those circulations. On Jupiter the circulation pattern (Figure 2.9) is dominated by features much like zonally symmetric roll vortices. Some important zonal variations exist, such as the Great Red Spot, but on the whole, zonal averages reveal much about the circulation of Jupiter.

Earth rotates swiftly enough so that a significant portion of the atmospheric circulation is zonally oriented. The distribution of solar input is rather evenly spread around a latitude belt before a significant amount of heat can be transported away to the cooler regions. This "even" spreading around a latitude belt is also a function of the radiative cooling rates of the atmosphere. The cooling rates are slow (in the troposphere, around 2 to 4 K/day) compared to the length of a day. Therefore, extreme temperature differences between night and day are not set up. So zonal averages tell us useful things about Earth's circulation. However, the zonally varying features must

FIG. 2.10   The Earth as seen from Apollo 16 during April. Photo courtesy of NASA. Most of the Pacific Ocean is viewed in this image.

also be considered because Earth's atmosphere interacts strongly with the surface. Figure 2.10 shows important features that are not zonally symmetric. For example, two vigorous frontal cyclones are centered over the North Pacific. Important zonal variations in Earth's atmosphere result from longitudinal variations of topography, land and sea area, vegetation type and snow cover, and from the hydrodynamic instability of the currents set up by the latitudinally varying distribution of incoming solar radiation. Chapter 7 will discuss most of these sources of zonal variation.

3

ZONAL AVERAGE OBSERVATIONS

Seas, lakes and great bodies of water, agitated by winds, continually change their surfaces; the cold surface in winter is turned under, by the rolling of the waves, and a warmer turned up; in summer, the warm is turned under, and colder turned up. Hence the more equal temper of sea-water, and of the air over it.

Franklin, 1765

The zonal mean weather patterns are rather symmetric between the Northern and Southern Hemispheres. Most differences that do occur between the two hemispheres can be ascribed to the differing ratios of land to sea area in each hemisphere. The Northern Hemisphere has much more land area, and consequently the seasonal variability is much greater in that hemisphere. The percentage of land area in each five-degree latitude belt is diagrammed in Figure 3.1.

The land-sea contrasts are of two basic kinds: thermodynamic properties (like heat capacity, availability of moisture, and surface reflectivity) and dynamic properties (like mountain blocking and oceanic heat transport). The ocean moderates the seasonal change because the water can mix the heat convectively (as mentioned by Benjamin Franklin) and because the water has higher specific heat than dry soil. The two factors coupled together, heat mixed through greater volume (and thus more mass) of the ocean multiplied by the higher specific heat of water, give the ocean areas much higher heat capacity than the land. Of course, land is at the South Pole and ocean is at the North Pole. But the high reflectivity of year-round snow cover diminishes seasonal change over Antarctica. Arctic sea ice is also highly reflective, but Arctic sea ice is much thinner (about 3 m thick, on average). Heat conduction, through Arctic sea ice and at leads (breaks in the ice pack), reduces the seasonal variability of the Arctic and maintains moderate polar temperatures in contrast to the Antarctic.

This chapter focuses upon the primary variables: radiation, temperature, velocity, pressure, and moisture. Important quantities derived from these primary variables (such as energetic quantities) are treated later, mainly in Chapter 4. Observed longitudinal variations of these primary variables are treated in Chapter 5. Simple dynamical relations (e.g., geostrophic balance) are stressed when discussing the observations; more advanced principles are presented in the chapters covering theories.

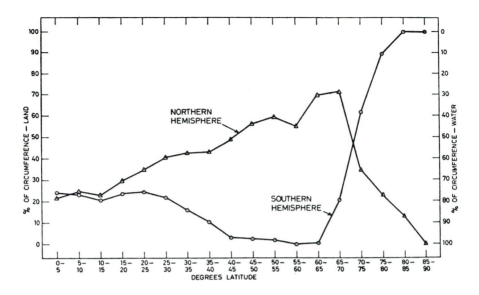

FIG. 3.1   Percentage of land area within each five-degree-wide latitudinal belt. From *Meteor. Mono.*, Taljaard (1972) by permission of American Meteorological Society.

3.1   Zonal average radiation and heating fields

Planck's law of blackbody radiation defines the amount of radiant energy flux passing through a given area (called the *irradiance*, E_i) at any wavelength of electromagnetic radiation (λ) for a given blackbody temperature (T). Planck's law may be written

$$E_i = \frac{C_1}{\lambda^5 [exp(\frac{C_2}{\lambda T}) - 1]} \tag{3.1}$$

where C_1 and C_2 are constants. Planck's law contains two functions of λ. As λ becomes smaller, λ^5 decreases but the term in the square brackets becomes larger more rapidly. Therefore, (3.1) indicates that the irradiance must be a maximum at some wavelength. The location of the maximum is a function of temperature. Planck's law also shows that the higher the temperature, the larger the maximum value of E_i. All these features are evident in Figure 3.2.

   The earth and the surface of the sun have very different temperatures. Accordingly, the spectra of E_i for the earth and the sun have very little overlap. Figure 3.2 plainly shows the lack of overlap. Hence, it is customary to assign one part of the electromagnetic spectrum to solar and one part to terrestrial radiation. The solar radiation is frequently referred to as the "shortwave" radiation while the terrestrial radiation is called the "longwave" radiation. These labels will be used interchangeably.

   Figure 3.2 schematically shows these two spectra. The actual radiation has a complex spectrum due to a variety of atmospheric constituents that absorb radiation

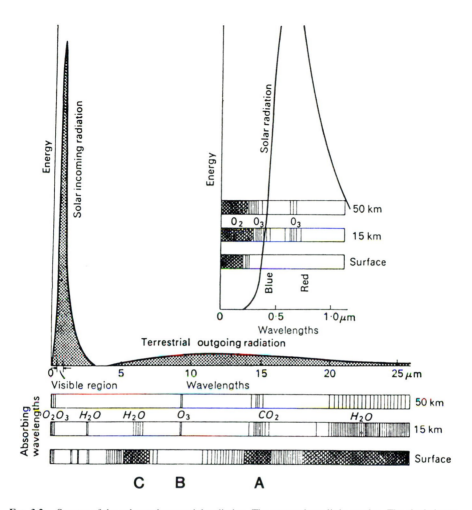

FIG. 3.2   Spectra of the solar and terrestrial radiation. The spectra have little overlap. The shaded area under each spectrum is the same, indicating equal amounts of energy. The inset below the spectra indicates the amount of absorption by various atmospheric gases at three elevations; darker shading implies greater absorption at that level. (For example, ozone absorption is indicated at 50 and 15 km, but not at the surface.) The inset to the upper right of the spectra is an enlargement of the visible portion of the wavelengths. The labels **A**, **B** and **C** at the bottom refer to the same CO_2, O_3, and H_2O absorption indicated in Figure 3.3b. From Iribarne and Cho (1980) reprinted by permission of Kluwer Academic Publishers.

in many wavelengths. The major absorption bands of several atmospheric gases are labelled in Figure 3.2.

    Figure 3.3a presents spectra of the solar radiation reaching the top of the atmosphere and reaching sea level. The difference between those two curves is either absorbed or scattered back to space by the atmosphere. The shaded area in the figure indicates the actual radiation absorbed, while the white area between the two curves shows the amount of radiation scattered back to space. The scattering is comparatively small for the infrared wavelengths, but becomes much stronger in the near-ultraviolet. The human eye cannot see ultraviolet light; the closest wavelengths it can see are perceived as the color blue. Of course, the greater scattering of the short wavelengths explains why the sky is blue.

The longwave terrestrial radiation is presented in Figure 3.3b. Figure 3.3a uses wavelength for the abscissa whereas the abscissa in Figure 3.3b is wavenumber. Figure 3.3b shows the actual longwave emission spectrum of the earth. A polar orbiting satellite measured the spectrum while drifting over Guam. The wavenumber ranges between 800–1000 and 1050–1250 cm^{-1} are called the "atmospheric windows." The observed spectra in these two windows coincide with a blackbody temperature curve for approximately 295 K. Since there is little absorption in the atmospheric windows, 295 K approximately agrees with the surface temperature near Guam at the time this spectrum was taken. Though this is an instantaneous local measurement, most places on the earth have spectra with the same general characteristics. The principal difference in spectra measured elsewhere is revealed as either a slight lowering (for cooler temperatures) or lifting (for warmer temperatures) of the (jagged) emission curve. The general characteristics to note are the carbon dioxide band (600–760 cm^{-1}), the stratospheric ozone band (near 1030 cm^{-1}), and the water absorption (above 1250 cm^{-1}). These bands are labelled **A**, **B**, and **C**, respectively, in Figures 3.2 and 3.3b. The peak wavenumber for emission for a 300 K blackbody is also where CO_2 strongly absorbs radiation. Strong absorption in these wavelengths can be found at most places across the earth.

The atmospheric absorption of the electromagnetic radiation has extremely important implications for the planet. Both the absorption and the scattering of the solar radiation at very short wavelengths ($\leq 0.4\mu$m) greatly favors most lifeforms on the planet. The high energy of ultraviolet light is known to damage many plants and may adversely affect the health of animals. (Life is relevant to our discussion since life is necessary to maintain the atmosphere's present chemical composition. Plants significantly alter the surface budgets of momentum, energy, and moisture, too.) Absorption of longwave radiation also greatly favors the development of life on earth by elevating the temperature of the earth and its atmosphere. Without the longwave atmospheric absorption, the earth's surface would be much cooler. Instead, the average temperature of the earth's surface is kept high enough to permit liquid water to occur. (This longwave absorption is sometimes erroneously referred to as the "greenhouse effect"; in actuality a greenhouse builds elevated temperatures by trapping air.)

The global radiative energy balance is diagrammed schematically in Figure 3.4. Budgets for the earth's surface, the atmosphere, and outer space are given. Of course, the total amount absorbed or emitted must balance, otherwise the earth or atmosphere would experience a net heating up or cooling down. As anticipated above, Figure 3.4 partitions the energy into shortwave and longwave components. To simplify the presentation, the incoming solar radiation has been normalized to 100 arbitrary units.

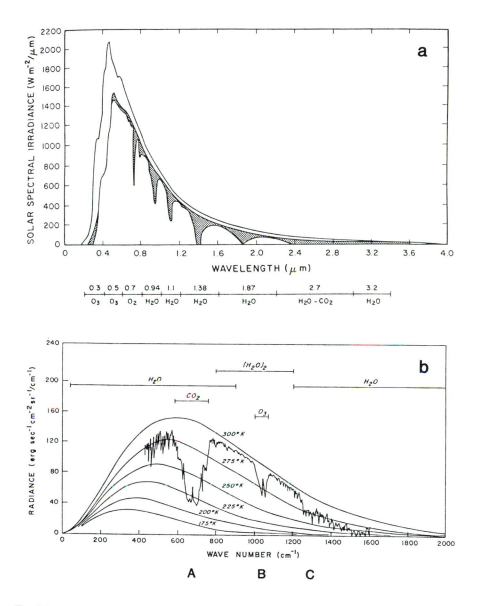

FIG. 3.3 (a) The solar spectral irradiance and amounts of absorption at various wavelengths are drawn. The top curve shows the solar irradiance reaching the top of the atmosphere as observed by Thekaekara (1976). The lower curve shows the irradiance observed at sea level. The area between the two curves is the solar radiance scattered or absorbed by the atmosphere; the amount absorbed is indicated by shading. (b) A terrestrial radiance spectrum is shown as a jagged line. This emission spectrum was measured by a satellite as it passed over Guam. The smooth curves show the blackbody emission spectra for various temperatures. The global average emission spectrum for the earth will be similar to the jagged curve, though the curve may be shifted slightly up or down. The dips along the jagged curve are caused by longwave absorption by the atmospheric gases indicated along the top of the diagram. From Liou (1980).

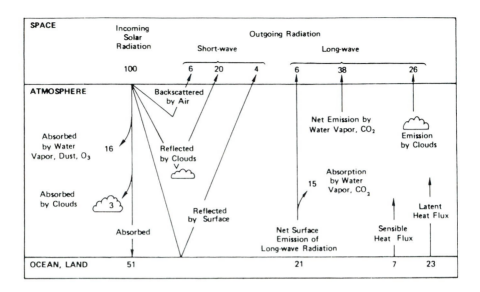

FIG. 3.4   Global annual average radiation and heat balance for the earth, atmosphere, and ocean. The units have been normalized relative to 100 units of incoming solar radiation. From National Research Council (1975).

The amount of reflected shortwave and emitted longwave radiation reaching space totals 100 of these units.

Figure 3.4 indicates how the incoming solar radiation is scattered, absorbed, and reflected by the atmosphere, clouds, and the earth's surface. About half of the solar radiation actually reaches the earth's surface. About a sixth of the solar radiation is absorbed by the atmosphere. The sum total of the reflected radiation amounts to about 30%, a total that is referred to as the *albedo*. The remaining 70 units that are absorbed by the atmosphere and the earth must be emitted back to space. That outgoing radiation is accomplished by longwave emission from the earth's surface (6 units), from the atmosphere (38 units), and from the tops of clouds (26 units). The emission from the tops of clouds has a significant role in the diabatic energy conversions (§4.5.1) and the diabatic forcing of the zonal mean meridional circulations (§6.3). In actuality, the total radiation absorbed by the ground amounts to about 146 units, yet Figure 3.4 only shows 51 units. The missing 95 units are from the longwave emission by the atmosphere. An alternative way of expressing this diagram would be to modify the arrow that shows 15 units of terrestrial radiation being absorbed by the atmosphere. The actual emission from the earth amounts to about 116 units; of that, 6 units go directly to outer space and the atmosphere absorbs the rest. The atmosphere re-emits this radiation, with about 95 units directed back down toward the earth's surface. The 110 units absorbed by the atmosphere, less the 95 units emitted back down to earth, leaves the net of 15 units indicated in Figure 3.4.

Radiative energy is also lost from the earth by sensible and latent heat fluxes. In total, the latent heat fluxes are about three times as large as the sensible heat fluxes. The atmosphere gains the sensible heat at the lowest levels. The atmosphere gains the energy from the latent heat flux where the moisture condenses. The condensation can occur a great distance from the location of the latent heat surface flux. The latent heat fluxes also play a major role in the general circulation; for example, they are the major driving mechanism of the Hadley circulation (§6.3).

As stated above, the actual emission from the earth (116 units) exceeds the amount absorbed from the sun (51 units). The phenomenon occurs because the atmosphere traps part of the energy emitted from the surface. A simple analogy may be useful in visualizing the process. The solar radiation absorbed by the earth is to be represented by a ping-pong ball that is thrown toward the floor once per minute. The floor represents the earth, so that the upward motion of the ball represents the net infrared emission from the earth. The atmosphere might be thought of as a table. When tossed downward, the ping-pong ball bounces off the floor, then off the underside of the table, and again off the floor before it is caught. In this trivial example, the "net radiation" seen by the "earth" is twice the incoming solar value because the ping-pong ball must bounce off the floor *twice* each minute before it is caught.

Of course, to represent the atmosphere with a table is too simple! A slightly more complicated model (using an infinite series) can estimate the observed upward and downward infrared fluxes quite accurately. Figure 3.5 illustrates the first two steps in the series. About 50% of the earth is cloud-covered (Figure 3.23). Clouds are assumed to be black bodies that are slightly cooler than the ground, hence the clouds will emit slightly less radiation downwards than they absorb from below. For this simple model, 90% of the infrared emission absorbed by the clouds is re-emitted downward. The atmosphere absorbs the remainder of the infrared emission (50%); it is assumed that this radiation is scattered and emitted by the air isotropically, so that half of the energy is lost to space and half is emitted back down to earth. The latent and sensible heats are incorporated only to the extent that they create infrared fluxes during the first step. Since the two heat fluxes warm up the air, the infrared emission created by the heating could be equally partitioned between the downward and upward directions. From Figure 3.5, the 51 units of incoming solar radiation result in 29.7 units of downward infrared radiation by the atmosphere after the first step; this is about 58% of the incoming value. These 29.7 units are absorbed and re-emitted by the earth. At step two, 70% (20.8 units) is emitted by the atmosphere back down to the earth, the other 30% is lost to space. At each succeeding step, 70% of the radiation emitted by the earth is re-emitted back down by the atmosphere. The model therefore becomes an infinite series.

$$E_a = 29.7 \sum_{n=1}^{\infty} (0.7)^n \qquad (3.2)$$

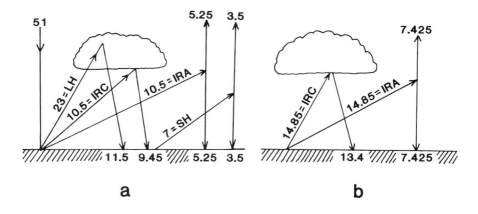

Fig. 3.5   The first two steps in a simple model that explains how the downward flux of infrared radiation at the ground can be much greater than the solar radiation absorbed by the earth's surface. (a) Step one: the latent heat flux (LH) warms up the air when a cloud is formed; the cloud can radiate that energy in all directions, hence only half is assumed to be directed toward the Earth (11.5 units). The infrared terrestrial emission is partly absorbed by clouds (IRC) and partly scattered and absorbed by air (IRA). Assume that 90% of IRC is transmitted back down, but only 50% of IRA. Sensible heating (SH) is treated similarly to LH. (b) In Step 2 all energy is re-emitted in radiative form: IRC and IRA are handled as in Step 1. About 70% of the emitted energy is returned to the earth in Step 2 and at each higher step.

The summation approximately equals 3.28. Consequently, the estimated longwave emission downward from the atmosphere (E_a) equals 97 units. This value agrees well with the observed 95 units. The longwave emission from the earth is just E_a plus the 21 units upward from step one.

So far, the discussion has concentrated upon the global average radiative properties of the atmosphere. Figure 3.6 shows the annual average radiative balance as a function of latitude from Earth Radiation Budget Experiment (ERBE) data taken during February 1985 through January 1986. Figure 3.6 shows the net amount of solar radiation reaching the top of the atmosphere (solid line); when integrated over the entire surface of the earth, this net amount equals the 100 units used in Figure 3.4. The actual amount of radiation absorbed by the earth, atmosphere, and ocean is given by the short-dashed line in Figure 3.6. The absorbed radiation corresponds to the 70 units absorbed by clear air, clouds, and the earth's surface shown in Figure 3.4. The long-dashed curve in Figure 3.6 shows the latitudinal distribution of the longwave emission to space. The integrated amount of the longwave emission must balance the absorbed, so the integrated amount is again 70 of those arbitrary units used previously.

The following general comments can be made about the radiation budget depicted in Figure 3.6:

- *Solar radiation.* The solar radiation reaching the top of the atmosphere varies sharply with latitude. It is much stronger at low latitudes than at polar regions.

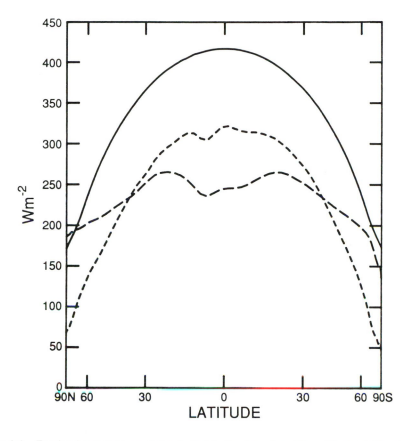

FIG. 3.6 Zonal and annual mean radiation budget for the atmosphere. The solid curve is the net amount of radiation reaching the top of the atmosphere. The short-dashed line indicates the amount of radiation that is absorbed by the earth, atmosphere, and ocean. The long-dashed line is the net longwave emission back to space. These data are from the Earth Radiation Budget Experiment for the period February 1985 through January 1986. See Barkstrom et al. (1990).

On the annual average, this latitudinal difference is greater than a factor of two. The latitudinal variation is even more pronounced for the amount absorbed: absorption varies by a factor of five between the tropics and polar regions. The amount of radiation reflected to space varies strongly with latitude. About 30% is reflected in the tropics, whereas over 50% is reflected in the polar regions.

- *Terrestrial radiation.* The long-dashed line in Figure 3.6 shows the net longwave emission back to space. The terrestrial emission has much less variation with latitude than the absorption. As a consequence, most of the atmosphere is not in local radiative balance. Near latitudes 38 N and 38 S, the atmosphere is in local radiative balance. However, between 38 N and 38 S the amount of radiation absorbed exceeds that being emitted. There must be transport of heat away from these low latitudes; otherwise the atmosphere would experience a net heat gain.

For the high latitudes, emission exceeds absorption, so there is transport of heat into these latitudes. The absorbed radiation minus the emitted radiation gives the net radiation. (The net radiation will be discussed again in connection with Figure 3.8. The locations where the atmosphere crosses over from positive to negative net radiation in Figure 3.8 are slightly different from the locations shown in Figure 3.6. For consistency in this book, latitudes 38 N and 38 S are specified in all the discussions.)

• *Net energy balance.* The areas under the dashed curves must be the same. This follows because the difference between the curves represents a net gain or loss of energy by radiation per unit area. The energy balance can be determined directly from Figure 3.6 because the abscissa has been stretched. The stretching is proportional to the actual amount of surface area within each latitudinal band of the earth.

• *Hemispheric differences.* Absorption and emission for the two hemispheres are roughly symmetric with two exceptions. (1) One noticeable asymmetry is centered near 5 N, where there is a pronounced dip in both the absorption and emission. This dip is caused by deep tropical convection of the intertropical convergence zone (ICZ). Over much of the globe, the ICZ is centered about five degrees north of the equator (e.g., Figure 5.13). The high cloudiness of the ICZ causes the dip in the shortwave absorption. Along the ICZ the longwave emission is mainly from the tops of very deep thunderstorms. The cloud top temperatures are very cold, so from (3.1) the longwave emission is much less than from surrounding areas (where clouds are absent or lower). (2) Another hemispheric asymmetry occurs near the poles. The emission and absorption are both noticeably less over the Antarctic than over the Arctic. The surface temperatures over the Antarctic (especially during winter) are much less than the surface temperatures over the Arctic. The temperature difference is due to the high elevation of the Antarctic plateau as well as the heat conduction through the Arctic sea ice mentioned at the start of this chapter. The actual temperatures will be shown shortly (Figure 3.11).

Figure 3.7, from Oort and Peixóto (1983), shows the seasonal variation of the incoming solar radiation. The seasonal variation is largest in the polar regions. The incoming radiation shown in Figure 3.7a is purely a geometric calculation, mainly including the tilt of the earth's axis relative to the sun during different seasons. A secondary effect is that the earth is closest to the sun in January. The maximum incoming radiation during summer occurs at about 30° latitude when averaged over the three summer months in each hemisphere. A secondary maximum is found at the poles. At the summer solstice the daily radiation *at the top of the atmosphere* is greatest at the summer pole because the sun does not set all day. The polar regions receive zero solar radiation in winter because the sun stays below the horizon. One might expect similar strong seasonal variation in the *absorbed* radiation. Because the earth is closer to the sun, the solar radiation during summer is greater in the Southern Hemisphere than during the same season for the Northern Hemisphere.

As mentioned earlier, about 30% of the incoming solar radiation is immediately reflected back to space. The amount of reflection (the albedo) varies strongly with latitude. The albedo is expressed as a percentage in Figure 3.7c, based upon Campbell and Vonder Haar (1980). Figure 3.7b expresses the amount of reflected solar radiation using the same units as the incoming radiation shown in Figure 3.7a. It is quite clear that the albedo is very much higher in the polar regions; obviously this is mainly due to snow and ice cover. However, with certain surfaces (such as liquid water), a low angle of the sun causes higher albedos than when the sun is directly overhead of . Also, as is shown later in this chapter, the cloudiness during summer is very high in the polar regions. The high reflectance during summer (when the incoming radiation is large), coupled with little or no radiation during winter, gives the curious result that the amount of *annual* average reflected radiation is nearly constant with latitude (Figure 3.7b). Measurements of albedo during ERBE are less smooth than the curves in Figure 3.7c. For example, Barkstrom et al. (1990) show the tropical peak due to the ICZ more clearly.

Subtracting data displayed in Figure 3.7b from data in 3.7a results in the amount of solar radiation absorbed by the atmosphere. The result is shown in Figure 3.8a. The infrared emission by the earth is given in Figure 3.8b. Barkstrom et al. (1990) find higher subtropical maximum values and a lower tropical minimum in one-month averages of ERBE data. If terrestrial longwave emission is subtracted from the absorbed, the net radiation received by the earth-atmosphere-ocean system is obtained. The net radiation calculated by Campbell and Vonder Haar (1980) is shown in Figure 3.8c. In summer, the net radiation absorbed is largest in the subtropics (around 20° to 30° latitude), and there is a net gain of radiative energy from the equator to about latitude 70°. In winter, there is net gain only to about latitude 15°. At latitudes higher than 70°, there is a net loss of radiative energy in all seasons. This loss and gain implies a net transport which, on the annual average, is from the area equatorward of 38° to poleward of 38°. The net radiation in Figure 3.8c changes sign near latitudes 38 N and 38 S. The broad maximum in summer net radiation, between 10° to 40°, is due to the high sun angle and to the general lack of cloudiness at these latitudes during this season. The differences between the Northern and Southern Hemispheres are relatively small compared with the seasonal variation. The variations among estimates by different researchers easily exceed noticeable differences between the hemispheres, as is evident in a figure compiled by Newell et al. (1970). A net radiation excess occurs in the summer hemisphere and a net deficit occurs in the winter hemisphere. Part of the excess is heating up the summer hemisphere while the deficit is cooling down the winter hemisphere. Part of the excess and deficit causes a transport of heat between hemispheres; such cross-equator flow is evidenced by the dominance of the Hadley cell for the winter hemisphere.

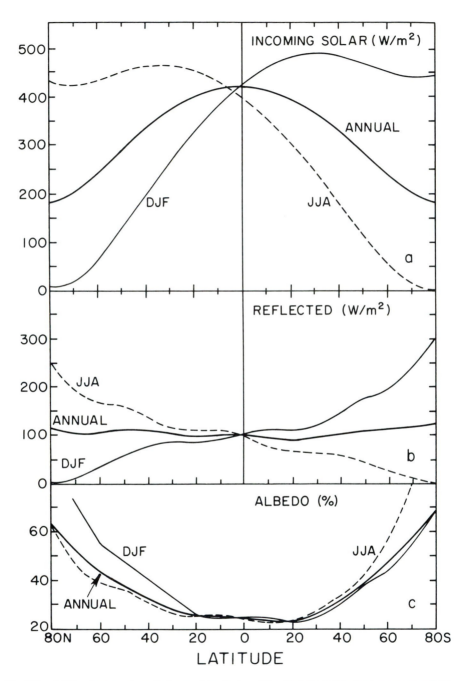

Fig. 3.7 (a) Zonal mean solar radiation reaching the top of the atmosphere showing averages over June–August (JJA), December–February (DJF), and the entire year. (Figure 3.6 has the same annual average curve.) (b) Similar to (a) except showing the amount of solar energy reflected by the earth, atmosphere, and ocean. (c) Similar to (a) but showing the albedo, expressed as a percentage. From Campbell and Vonder Haar (1980) and Oort and Peixóto (1983).

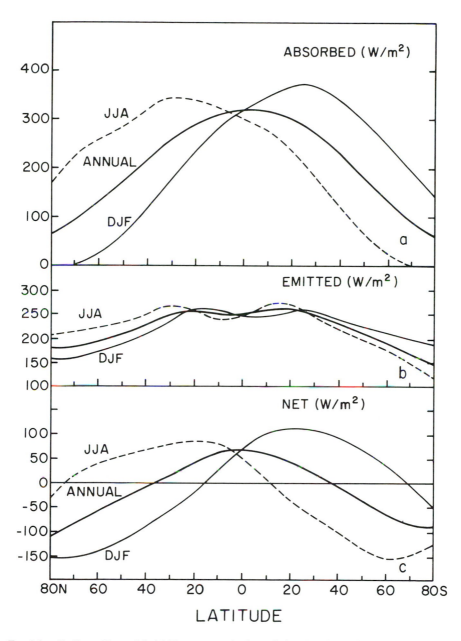

Fig. 3.8 Similar to Figure 3.7. (a) The amount of solar radiation absorbed during three time periods. (b) Similar to (a) except showing the longwave terrestrial emission. (c) The net radiation absorbed is shown. From Campbell and Vonder Haar (1980) and Oort and Peixóto (1983).

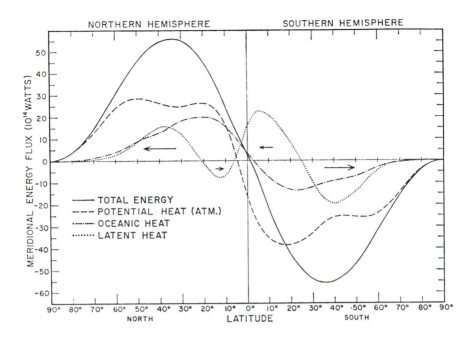

FIG. 3.9   Zonal and annual mean heat flux is shown. The thicker solid line indicates total heat flux. Also shown are contributions to the total by two atmospheric components and the ocean. The atmospheric transport can be in latent (dotted line) or sensible (dashed line) forms. The oceanic transport (dot-dashed line) is calculated as a residual from the other three curves. From *Meteor. Mono.*, Newton (1972) by permission of American Meteorological Society.

Starting at the South Pole, the heat transport is obtained by taking the running sum of the net radiation (normalizing by the surface area around each latitude band). Negative values indicate southward transport. The heat transport required for radiative balance is presented in Figure 3.9.

The oceans and atmosphere both transport heat. Atmospheric heat transport is traditionally divided into two forms. Sensible heat transport includes advection of warm tropical air to higher latitudes as well as equatorward flow of cold polar air. Much heat is transported in latent form. The latent heat of water vapor is released when the water vapor condenses or freezes. Thus, the atmospheric energy transport can be expressed as

$$\text{heat transport} = \frac{2\pi a \cos \phi}{g} \int \overline{[v(C_p T + \Phi + Lq)]} dP \qquad (3.3)$$

where the square brackets indicate zonal averaging and the overbar indicates time (annual) averaging. (Averaging notation is presented in the Appendix.) The acceleration of gravity is g; a is the earth's radius; v is meridional velocity; T is temperature; Φ is the geopotential (gravitational potential energy); L is the latent heat of condensation; q is the mixing ratio, P is pressure and ϕ is the latitude. The following general comments can be made about the three types of heat transports shown in Figure 3.9.

- *Net heat transport*. The solid line in Figure 3.9 is the annual mean meridional flux of total energy. Its maximum is between 35 to 40°latitude. The maximum occurs there since that is where the net radiation curves (Figure 3.8c) cross zero. For latitudes lower than about 38°there is an excess of net radiation, implying that heat must be removed from each of those tropical latitudes. As one integrates the heat flux proceeding poleward from the equator, more and more heat must be transported until one reaches a latitude where there is a deficit of net radiation. At those higher latitudes more and more of the transported heat is used to counter the deficit in net radiation. Hence, the maximum heat transport occurs at the latitudes where the net radiation crosses from positive to negative values. Figure 3.9 is therefore consistent with the zero crossings found in Figure 3.8c.

- *Sensible heat transport*. The largest of the three contributions is the sensible heat transport by the atmosphere (labelled "potential heat" in the figure). Sensible heat transport has a broad maximum from about 15°to about 60°latitude in both hemispheres.

- *Latent heat transport*. The latent heat transport in the atmosphere shows some intriguing complexity. While the latent heat transport in midlatitudes is poleward, in the equatorial zones the transport is clearly *equatorward*. Equatorward transport occurs because the air gains moisture as the trade winds blow equatorward over the ocean. The latent heat transports infer a specific direction to the motion of the low-level air because specific humidity, q, is not negative and q is large only in the lower troposphere. In general, a layer of moist, well-mixed air exists between the surface and the trade wind inversion (at 1 to 2 km elevation); the trade wind inversion deepens as one heads equatorward. (The mechanism for deepening the moist layer is successive development of cumulus clouds. The clouds pump moisture from below to above the inversion.) The air flowing towards the intertropical convergence zone becomes more and more moist; eventually it rises near the ICZ and the water precipitates out, releasing the latent heat.

- *Oceanic heat transport*. The oceanic heat transport has a deceptively simple distribution in Figure 3.9. The oceans transport heat poleward at nearly all latitudes in this figure. The zero crossing for oceanic transport is close to the equator. Much of the oceanic transport is accomplished by great gyres centered in the subtropics of each ocean basin. The transport is quite complex because the ocean is confined to basins, unlike the free atmosphere. The poleward flow tends to be confined into narrow western boundary currents such as the Kuro Shio and the Gulf Stream; some of the equatorward flow is also confined, while some occurs over a broad longitudinal band.

The oceanic heat transports shown in Figure 3.9 are calculated as a residual. That procedure assumes that the atmospheric and total heat transports are accurate. However, there is some controversy regarding the actual oceanic heat transport. Anderson (1983) points out that oceanic heat transports calculated as a residual have been much larger than transports oceanographers calculate from oceanographic data. Figure 3.10a shows several direct calculations of the oceanic heat flux in each major ocean basin. In the north Atlantic all the estimates show poleward heat transport though the magnitude varies by a factor of two. In the south Pacific the three estimates show poleward transport, but the magnitude varies widely. In the north Pacific and Indian Oceans, the estimates disagree in the direction of the transport. Finally, in the south Atlantic the estimates all seem to agree that the oceanic heat transport must be equatorward. In Figure 3.9, Newton estimates the oceanic heat transport at 30 N to be about 2×10^{15} W; in contrast, Figure 3.10a shows that only about 1×10^{15} W is transported by the Atlantic, while the Pacific transport is indeterminate at this latitude. Some more recent studies find even worse disagreement between residual and direct calculations. Other estimates have less disagreement; Hsuing et al. (1989) estimate oceanic heat transport that is close to Newton's estimates between 20 S to 10 N, but about 5×10^{14} W less from 20 N to 50 N. Vonder Haar (Ohring, 1990) reports the global mean residual to exceed direct measurement by 10^{15} W, an amount equal to 20 to 25% of the total ocean transport. The remaining inconsistency may be resolved by recent ideas about the configurations of ocean circulations.

The oceanic circulations include small scale eddies and large scale overturning motions. Density differences drive the large scale overturning circulations (crudely analogous to atmospheric Hadley cells), and they are referred to as the *thermohaline* circulations. The eddies can transport heat toward polar regions, but the global thermohaline circulations have large heat fluxes that can explain the peculiar and conflicting estimates compiled by Anderson (1983). Using observations, Gordon (1986) proposes thermally direct overturning motions south of the Antarctic circumpolar current (ACC) and a tortuously meandering global circulation to the north of the ACC. The principal path of the latter circulation is as follows (Figure 3.10b). Cold surface water forms in the far north Atlantic (and Arctic) sinks, and flows southward along the bottom edge of the continental shelf of North and South America. Upon encountering the ACC both currents flow eastward until the northern circulation branches off, heading northward to the east of New Zealand. Some of the water moves northward along the coast of Asia, while much of the water is upwelling along the Pacific equator. (The surface easterly winds cause surface divergence of the ocean waters, a process that drives the upward motion.) The circulation has now reached the upper ocean; solar radiation heats the water and winds push the water westward. The flow may split and rejoin: some water may go around southern Australia, some passes through Indonesia; both cross the equatorial Indian ocean. The surface flow hugs the south African coast, crosses the subtropical south Atlantic, and flows northward along the eastern ocean becoming the Gulf Stream. This

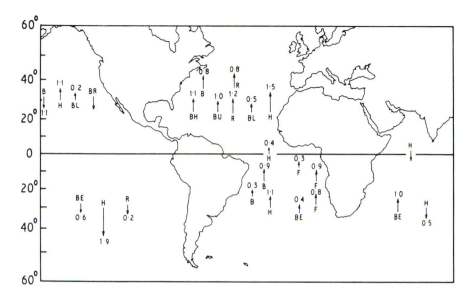

FIG. 3.10a Calculations of the heat transport within each major ocean basin are compiled in this figure from Anderson (1983). The units are in 10^{15} W. The letters near each arrow refer to a specific article as follows: **B**, Bryan (1962); **BE**, Bennet (1978); **BH**, Bryden and Hall (1980); **BL**, Bryan and Lewis (1979); **BR**, Burridge (1982); **BU**, Bunker (see Bryden and Hall, 1980); **F**, Fu (1981); **H**, Hastenrath (1980); and **R**, Roemmich (1980). Some researchers provide more than one estimate; the same observations have been used by some researchers. These estimates disagree with the ocean transport presented in the previous figure.

thermohaline circulation can explain the northward heat flux in the north and south Atlantic, the southward flux in the south Pacific, and the indeterminate flux in the Indian ocean shown in Figure 3.10a. Recent simulations by a coupled ocean and atmosphere global model have reproduced this tortuous circulation (Semtner and Chervin, 1992).

The atmospheric heat fluxes in Figure 3.9 have some interesting differences between the hemispheres.

- *Sensible heat flux.* The sensible heat flux across the *tropics* is much greater in the Southern Hemisphere than in the Northern. Newton states that this asymmetry is due to a more vigorous Hadley circulation in the Southern Hemisphere. A better description may be that the vigorous meridional circulation is enhanced by the Asian summer monsoon. The sensible and latent heat transports for the monsoon are oppositely directed. Warmer, dryer air exists over the Indian subcontinent while cooler, moister air is found over the Indian Ocean. The Asian summer monsoon surface winds are southwesterly, hence cooler air moves northward, across the equator toward India. Therefore, the northward sensible heat flux for the summer monsoon is negative.

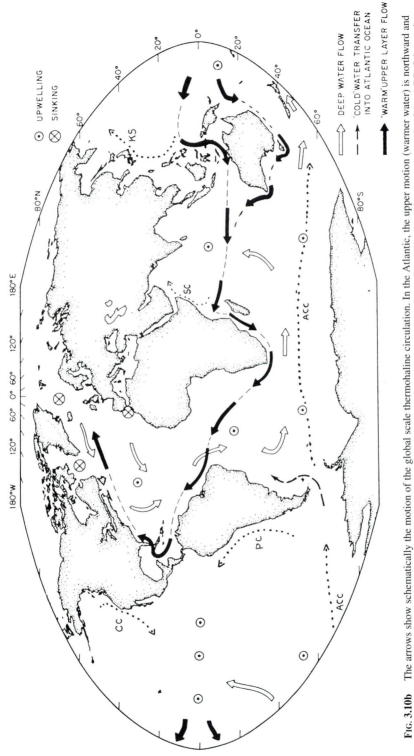

FIG. 3.10b The arrows show schematically the motion of the global scale thermohaline circulation. In the Atlantic, the upper motion (warmer water) is northward and the lower motion (colder water) is southward in both hemispheres. Consequently, the heat transport is northward in the Atlantic to the north of the ACC. Other major ocean currents are shown for reference: **ACC** Antarctic circumpolar current; **PC** Peru current; **SC** Somali current; **KS** Kuro Shio; **CC** California current. Circles indicate upwelling and sinking associated with the thermohaline circulation shown. The figure is based upon one in Gordon (1986), with modifications suggested from model simulations by Semtner and Chervin (1992).

- *Latent heat flux.* The latent heat fluxes, both in the midlatitudes and in the tropics, are larger in the Southern Hemisphere. This distinction is especially true for the tropical regions and can be explained from the differing amounts of land areas in each hemisphere (Figure 3.1). First, the Southern Hemisphere has proportionally more ocean area than the Northern Hemisphere, so more water is available to be evaporated into the air. The Indian summer monsoon is also a significant factor. The monsoon includes strong cross-equator flow (from the Southern Hemisphere into the Northern) which reinforces the asymmetry in Figure 3.9. The monsoon flow towards India is counter to the trade wind flow over the rest of the Northern Hemisphere. In contrast, the monsoon transport is additive to the trade wind flow of the Southern Hemisphere.

3.2   Zonal average temperature fields

The radiation and heat transports are closely linked to the patterns of temperature. The first crude zonal average large scale temperature fields were published by von Humboldt in the first half of the 19th century (reproduced in Hildebrandsson and Teisserenc de Bort, 1907). Upper air observations were made from balloons or approximated from observations on mountains. According to Hartmann (1985), Teisserenc de Bort discovered and named the tropopause and stratosphere, publishing his balloon measurements in 1902. Figure 3.11 shows the zonal mean temperature distributions for December 1978 through February 1979 and June through August 1979. The data in these figures are the final GWE datasets compiled at the Geophysical Fluid Dynamics Laboratory (GFDL) in Princeton. One may compare these diagrams with more readily available figures, such as those published in Newell et al. (1970). Newell et al. use a longer averaging period (several years), but restrict themselves to radiosonde network data. The GWE data have better global coverage but the averaging period is short, and interannual variability will influence the generality of the depiction in Figure 3.11. Despite these concerns, the GFDL temperature diagrams (Lau, 1984) agree with the figures found in Newell et al. (1970, 1972), at least in the general aspects treated here. Differences between the temperature diagrams in Lau (1984) and Newell et al. (1970) are pointed out.

Five general features of the temperature field are seen in Figure 3.11. (1) The troposphere and stratosphere are easily recognized. The temperature generally decreases with height, with less change with height in the stratosphere. Over much of the stratosphere, the isotherms are oriented nearly vertically, indicating nearly isothermal lapse rates. The vertical temperature gradient is often greater in the troposphere. The major exception to this rule occurs in the polar regions, especially during winter. This exception is most pronounced over Antarctica. Newell et al. (1970) also find an inversion in the lower troposphere during winter in the Arctic, but they fail to show any enhanced static stability over Antarctica. This discrepancy may be an unusual condition during 1979, but more likely it reflects the lack of radiosonde sta-

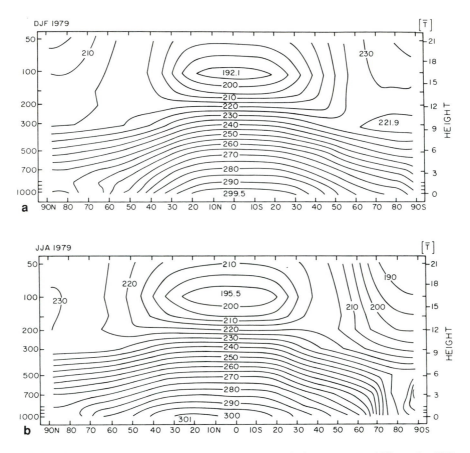

FIG. 3.11   Zonal mean meridional cross-sections of temperature during two seasons. (a) December 1978–February 1979. (b) June–August 1979. The figure is redrawn from diagrams in Lau (1984); the height scale is approximate. The contour interval is 5 K.

tions over Antarctica. (2) The tropopause elevation varies with latitude. The tropical tropopause, at 14 to 17 km, is much higher than elsewhere. In polar latitudes the tropopause is at 8 to 10 km. (3) The coldest temperatures are found at the equatorial tropopause and at high altitude near the pole in winter. The latter is obviously due to the lack of sunlight at the winter pole. The former is harder to explain and provides a platform from which to discuss wave-mean flow interaction in §6.2.4. (4) Figure 3.11 shows various meridional gradients of temperature. In the troposphere, most of the meridional gradient of temperature is found in the middle latitudes. There are local concentrations, particularly noteworthy is the strong gradient near the coast of Antarctica. Since Newell et al. (1970) do not show an inversion over Antarctica, they do not find a strong meridional gradient near the Antarctic coast. Because the tropopause has a meridional slope, the temperature gradient is reversed over much of the lower stratosphere. By reversed it is meant that warmer temperatures are found at higher latitudes so that the gradient is directed poleward. The poleward temperature gradient is found in the lower stratosphere outside of the polar latitudes of the winter hemisphere. Higher up in the stratosphere, the gradient is directed from the winter hemisphere toward the summer hemisphere. The warmest temperatures at 20 km elevation are located at the summer pole; the coldest are at the winter pole. The upper stratosphere temperature pattern is most consistent with the incoming solar radiation (Figure 3.7a). (5) Figure 3.11 shows a noticeable seasonal shift of the pattern. The maximum surface temperatures occur in the summer hemisphere, directly beneath the average position of the sun during each season. As mentioned above, the warmer stratospheric temperatures occur over the summer pole. The seasonal change from winter to summer patterns is often quite abrupt, occurring in a couple of days. The rapid change is called a sudden warming and is the subject of §7.4.6.

Figure 3.11 reveals a complex pattern. One reason for the complexity is that the annual mean temperature at a point is influenced by at least four things: (1) radiational processes like incoming solar absorption, emission from the atmosphere and earth, and how these are affected by chemical composition and cloudiness, (2) advection by the prevailing wind, (3) adiabatic heating by ascent or descent, and (4) latent heat release and any other diabatic process. These four processes are not independent, so it is difficult to assign areas on this figure where an individual process is the prime determinant of the temperature field. The coupling of dynamics, thermodynamics, and radiation make the explanation of the temperature pattern a challenging undertaking. The initial theoretical explanation here (§6.1) will only include simple radiation physics and local convective motions.

Averaged over the globe, the annual mean surface temperature in the Southern Hemisphere is about 1 to 2 K cooler than that for the Northern Hemisphere. It is not clear whether this is a consequence of the greater areal coverage of ocean in the Southern compared with the Northern Hemisphere. The cooler mean temperature may be a function of the lack of ocean near the South Pole, leading to much colder temperatures over Antarctica compared with the Arctic. This interhemispheric temperature difference is difficult to see in Figure 3.11, but it is evident in Figure 3.12 from van Loon (1972). The analysis by van Loon over Antarctica agrees better with Lau's data (Figure 3.11) than with Newell et al. (1970). The surface air is very much colder (greater than 20 K) over the Antarctic region than over the Arctic on an annual average. In the tropics, the Southern Hemisphere is again cooler (by 1 to 2 K). In midlatitudes the opposite is true. Figure 3.12 also compares interhemispheric temperature differences at 500 hPa and 100 hPa. At 500 hPa, the Southern Hemisphere is cooler at all latitudes, whereas the pattern at 100 hPa is broadly similar to the surface temperature difference pattern.

In Figure 3.13, van Loon (1972) details some of the main differences between the zonal average temperature in each hemisphere. He includes variation with season at three levels. His differences are between comparable seasons. For example, he subtracts the Northern Hemisphere July data from the Southern Hemisphere Jan-

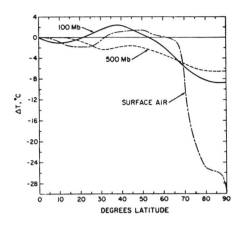

FIG. **3.12**   Annual mean temperature difference at three levels and as a function of latitude. For a given level and latitude, the temperature in the Southern Hemisphere minus the temperature at the corresponding location in the Northern Hemisphere equals the difference plotted. This difference will be labelled the "interhemispheric temperature difference" in Figure 3.13. From *Meteor. Mono.*, van Loon (1972) by permission of American Meteorological Society.

uary values to get the "summer" differences as a function of latitude. Shaded areas correspond to regions and times when the Southern Hemisphere is warmer than the Northern Hemisphere. Important items worth noting are as follows.

- *Surface seasonal temperature change.* Between the equator and 65° latitude, the sizes of the shaded and unshaded areas are about the same. However, for higher latitudes the Southern Hemisphere is much colder, and is colder all year. That is not surprising since much of the Antarctic surface is a high-elevation ice field. The story is different in the middle latitudes. There, compared with the Northern Hemisphere, the winter is much warmer and the summer is cooler in the Southern Hemisphere. Averaged over the course of a year, the Southern Hemisphere midlatitudes are warmer, as noted in the previous figure. The greater oceanic coverage of the Southern Hemisphere easily explains this interhemispheric difference. The explanation is partly given in the statement by Benjamin Franklin that commences this chapter. The main missing part, the greater specific heat of water, was known to Franklin's contemporaries; Dalton's 1793 meteorology book refers to smaller heat capacity as the cause of greater seasonal surface temperature changes over land. The seasonal change is less over the oceans. The temperature difference pattern is not symmetric because the land areas are colder in winter by a greater amount than the land is warmer in summer. The asymmetry arises from atmospheric stability. In summer, the surface heating destabilizes the atmosphere and consequently, convection can mix the heat in the vertical, which moderates the surface temperatures. In winter, the stronger surface cooling stabilizes the atmosphere near the ground, and

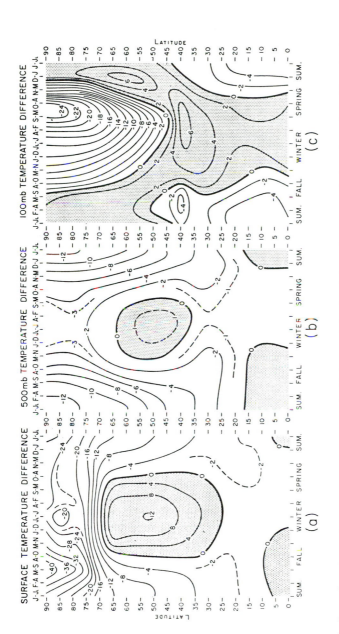

FIG. 3.13 Similar to Figure 3.12. The zonal mean interhemispheric temperature difference is plotted at three levels and as a function of season. The levels are (a) surface, (b) 500 hPa, and (c) 100 hPa. Comparable seasons are used in each hemisphere; for example, values for "winter" use July data for the Northern Hemisphere, values for "winter" use January data for the Northern Hemisphere. Shaded regions indicate instances where the Southern Hemisphere is warmer. From *Meteor. Mono.*, van Loon (1972) by permission of American Meteorological Society.

large drops in surface temperature can result from the restricted vertical mixing. Between 10 and 30 degrees latitude, the Northern Hemisphere is warmer
as noted on the previous figure. The subtropical belt is warmer all year. Deserts
occur at these latitudes in the Northern Hemisphere, while there are primarily
oceans in the opposite hemisphere. The latitudes are low enough so that the
seasonal changes are rather small, hence these regions are warmer throughout
the year.

- *Middle troposphere (500 hPa level).* The differences between the hemispheres
  are not as great in the middle troposphere as they are at the surface, but the general
  pattern is similar. Summer is still a bit colder in the Southern Hemisphere, but
  the winter is no longer much warmer. Perhaps the best explanation for this
  asymmetric seasonal response again pivots on atmospheric static stability. In
  winter, the lapse rate is more stable over the continents than over the oceans. As
  mentioned above, wintertime radiational cooling at the base of the atmosphere
  increases the static stability of the surface air, and the increased stability restricts
  the vertical mixing of the cooling air. The result is that the greater cooling over
  the land is mainly limited to the lower troposphere. The effect of surface heating
  is quite the opposite. Convective activity in summer may mix the stronger surface
  heating over land area through a larger depth of the atmosphere.

- *Lower stratosphere (100 hPa level).* Very low temperatures occur in the Antarctic
  stratosphere. Several explanations are proposed by van Loon. Ultra-low temperatures are possibly a complicated result of the surface temperatures being so low
  that ozone in the stratosphere cannot absorb much terrestrial radiation. Hence,
  the stratospheric heating by this mechanism is reduced. On the other hand, chemical and dynamical processes are now known to reduce the stratospheric ozone
  concentrations over Antarctica. The ozone reduction occurs over the course of
  a winter, with minimum concentrations being reached in spring.

The temperature field undergoes cyclic variation. If the temperature were only
influenced by the change of seasons, then the time variation would be accounted for
by the annual cycle. (That is, variation with a period of one year.) In addition, it seems
logical to expect the amplitude of the annual variation to be rather small in the tropics
and progressively larger at higher latitudes. At the equator the sun is overhead twice
each year, so a semi-annual cycle would be expected there. To a first approximation
this is what we find in Figure 3.14 from van Loon (1972). On closer examination
the expectation is more valid in the Northern Hemisphere. Also, it is more valid
near the surface than in the middle levels of the atmosphere. The smaller change in
the Southern Hemisphere temperatures is easily seen and has already been identified
at several places in this chapter. The reduction of the annual change in the middle
troposphere may be explained by the static stability arguments just made. A related
factor is that the equator to pole temperature difference is simply less at 500 hPa than

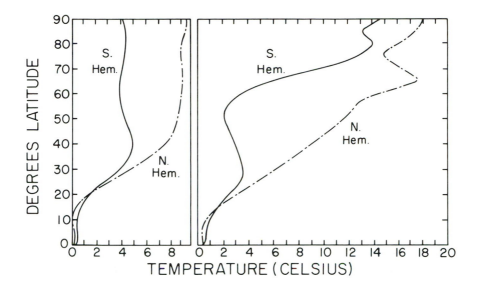

Fig. 3.14 Zonal mean amplitude of the annual cycle of temperature in °C. Left diagram is for 500 hPa, right diagram for the surface. From *Meteor. Mono.*, van Loon (1972) by permission of American Meteorological Society.

at the surface. Finally, the latitudinal variation in Figure 3.14 would be less at high latitudes if a significant fraction of the solar radiation in summer is melting snow and ice. Overall, the annual cycle accounts for most of the temporal temperature change outside the tropics. In the tropics the semiannual cycle (two cycles per year) is strong. In the middle and upper troposphere, between 10 N and 20 S, the semiannual cycle may account for more than 50% of the total temperature variance (van Loon, 1972).

3.3 Zonal average velocity fields

Figure 3.15, from Hoskins et al. (1989), gives the observed zonal mean zonal wind. Most of the features shown in Figure 3.15 agree with similar figures published by Lau (1984) using three-month averages of the GWE datasets and Newell et al. (1972) using only radiosonde data. Acceptably small differences were expected based upon the author's theoretical calculations (§2.5). Such differences as exist between Figure 3.15 and those by Lau and Newell et al. are mentioned in the following discussion.

- *General pattern.* For the troposphere, Figure 3.15 reveals easterlies (shaded) in low latitudes and westerlies elsewhere. In the stratosphere the tendency is for westerlies in the winter hemisphere and easterlies in the summer hemisphere. The maximum low-level easterlies are twice as strong in Figure 3.15 as in Newell et al.

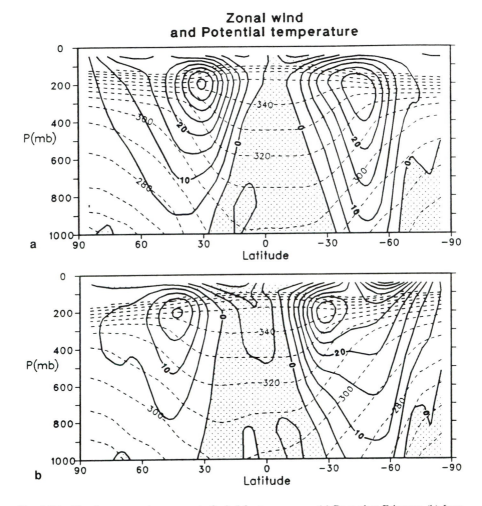

FIG. 3.15   Zonal mean zonal component of wind for two seasons: (a) December–February, (b) June–August. Shaded areas indicate easterly winds. The contour interval is 5 m/s. Dashed lines indicate potential temperature with a 10 K interval. From Hoskins et al. (1989).

• *Jet streams.* Large velocity maxima occur at midlatitudes near the tropopause level. These midlatitude westerly jet streams are called the subtropical jets. Lau (1984) finds similar locations and magnitudes for the subtropical jets. Newell et al. (1972) found the Southern Hemisphere subtropical jet wider than Figure 3.15a and 5 m/s slower than in Figure 3.15b. Newell et al. found the Northern Hemisphere subtropical jet 5 m/s slower in both seasons. In the stratosphere at high latitudes is another jet, called the polar night jet, whose center is located at too low a pressure to be adequately contoured in this diagram. The location of the axis of the polar night jet varies strongly and rapidly with the seasons; its elevation typically ranges from 35 to 60 km (5 to 0.3 hPa) during the winter months (e.g., Hartmann, 1985). The polar night jet is not well identified in Figure 3.15 due to the choice of vertical axis. During June through August 1979, Lau finds the maximum wind speed at 50 hPa to be about 52 m/s instead of the 37 m/s shown by Newell et al. . The polar night jet is much stronger in the Southern Hemisphere, reflecting the much colder polar temperatures. A weak easterly wind maximum occurs in the subtropics of the lower summer stratosphere. All these velocity maxima occur above regions where there exist strong meridional temperature gradients, as one expects from thermal wind balance. The processes that produce these local enhancements in the temperature gradients are different between the stratosphere and the troposphere. Photochemical processes predominate in setting up the middle stratospheric temperature gradient. The tropospheric jets arise from the mean meridional cells and the midlatitude eddies, as will be detailed in Chapters 6 and 7.

• *Seasonal variation of the jets.* The seasonal variation of the subtropical jet stream is quite noticeable, the winter jet being stronger than the summer jet. As with other variables, the seasonal variation of the Northern Hemisphere jet stream seems to be much greater. The jet stream core changes latitude over the seasons, being centered at about 30° latitude in winter and around 40 to 45° in summer. The strong seasonal variation in the stratosphere was already noted. This stratospheric variation could be easily anticipated from the zonal average temperature fields shown earlier.

For the zonal velocity above the boundary layer, the local, instantaneous zonal velocity is comparable to the time and zonal average zonal velocity. Such is not the case for the meridional velocity. The meridional velocity has a magnitude that is much smaller (by a factor of 10) than typical velocities measured at individual stations. Therefore, the zonal mean meridional velocity is difficult to estimate by direct means because the observational errors and biases are exaggerated in the calculation. An early attempt by Starr and White (1951) illustrates the problem; the uncertainties in their estimates exceed the magnitude of the mean meridional velocity at many locations. The emphasis upon land stations (see critique by Zipser, 1974) precluded an accurate direct calculation of the mean meridional wind by Newell et al. (1972). However, Newell et al. were able to make an accurate indirect calculation, as will be shown shortly. Given the intense effort to provide global data coverage during GWE, it is worth examining the zonal mean meridional velocity fields calculated directly by Lau (1984) and presented as Figure 3.16.

In Figure 3.16 the largest mean meridional velocities occur in the tropics. Southerly velocities are indicated by negative numbers as well as by shading in this figure. In the upper tropical troposphere, the wind is blowing away from the equator; below, near the ground, the wind is blowing toward the equator. This pattern, of course, is the circulation of the Hadley cells, one for each hemisphere. It

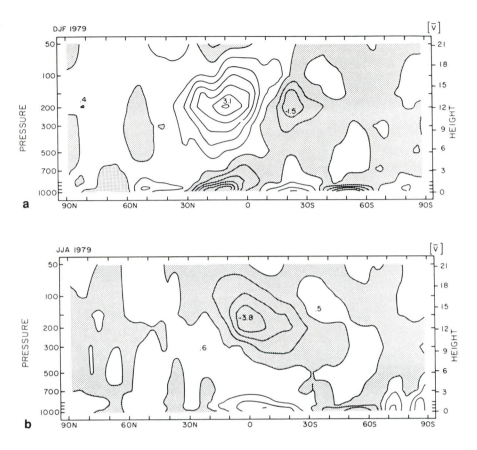

FIG. 3.16   Zonal mean of the meridional component of wind for two seasons: (a) December 1978–February 1979 and (b) June–August 1979. Shaded areas denote southward velocities. The contour interval is 0.5 m/s. Redrawn from diagrams in Lau (1984).

should also be clear that the Hadley cell for the wintertime hemisphere is much the stronger. In midlatitudes the zonal mean meridional velocity is generally weaker. At the surface in midlatitudes, the meridional wind does reach a maximum in the poleward direction. One may identify that midlatitude surface maximum as the base of the Ferrel cell in each hemisphere; but no compensating upper-level return flow is evident in this figure.

As hinted at above, the inaccuracies of the direct calculation of the mean meridional wind can be avoided by an indirect technique. At the same time, the indirect method shows the mean vertical velocity; both velocity components together clearly reveal the mean meridional circulations. The indirect technique used by Newell et al. (1972) deduces the mean meridional wind from the angular momentum budget. The procedure employs a vertical stream function from which vertical and meridional velocities are calculated.

The indirect procedure is as follows. u and v indicate the zonal and meridional wind components, respectively. (1) The vertical and longitudinal average of v is at most a few mm/s, even though there is some net seasonal shifting of mass across some latitudes. For the purpose of calculating the zonal mean of v (designated $[v]$) this net shift is assumed to be zero. (2) The net shift of mass can be expressed as a net transport of angular momentum (see equation 4.5). Setting the net (vertically and zonally averaged) mass transport equal to zero gives an expression involving $[u]$, Ω, $[u'v']$ and $[v]$, where the primes denote deviations from the zonal average (see Appendix). The angular velocity of the earth is Ω. The first three quantities are easy to measure and thus can be used to estimate $[v]$. (3) The relationship used in the second step, equation (4.5), breaks down in the tropics. But the Hadley cells are strong enough so that direct measurement of $[v]$ is adequate there. Newell et al. (1970) use direct measurements of $[v]$ for latitudes 20S to 20N. (4) Once $[v]$ is found, the indefinite integral

$$\psi = \frac{2\pi R}{g} \int_P^{P_0} [v]dP \tag{3.4}$$

is solved for a stream function ψ. g is the acceleration of gravity, R is the earth's radius (r) times the cosine of latitude, and P_0 is the surface pressure. (5) The vertical velocity is deduced from the stream function and the continuity equation. Applying the zonal average to the continuity equation eliminates the term involving u, leaving two-dimensional flow in the meridional and vertical plane. (Therefore, the zonal mean continuity equation only involves $[v]$ and $[\omega]$.) The stream function represents a two-dimensional flow:

$$[v] = \frac{g}{2\pi R}\frac{\partial \psi}{\partial p} \quad \text{and} \quad [\omega] = \frac{-g}{2\pi r^2 \cos\phi}\frac{\partial \psi}{\partial \phi} \tag{3.5}$$

Hence, the vertical velocity can be deduced from the stream function. This procedure gives different people somewhat different results and the velocity magnitudes calculated are only approximate, but it does give a useful picture of the circulation patterns.

The stream functions calculated in this way are depicted in Figure 3.17. A constant amount of mass flows between each pair of contours, hence more closely spaced contours indicate faster speed. (The contour interval is not constant in this figure.) The direction of the motion is indicated by the arrowheads drawn on each contour.

The most obvious pattern is a three-cell circulation similar to that envisioned by Ferrel. A similar three-cell pattern is found in GFDL GWE data (Lau, 1984). Six further comments about the detailed structure of the mean meridional circulations follow. (1) In the December-February plot the meridional circulation is dominated by a thermally-direct Hadley cell in the Northern (winter) Hemisphere. In the other extreme season (the June-August plot) the circulation is dominated by a Southern Hemisphere winter Hadley cell. In both cases the Hadley cell in the summer hemi-

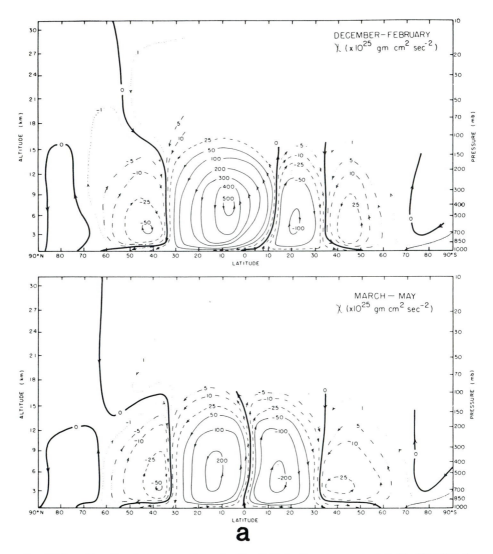

FIG. 3.17a and b   Zonal mean mass stream function fields for four seasons, as plotted by Newell et al. (1972). Arrows on the contours indicate the direction of motion. The contour interval varies within each

sphere is almost nonexistent. During the equinoxes the two Hadley cells have roughly equivalent magnitude. (2) Aside from the seasonal dominance of the winter hemisphere Hadley cell, the latitudinal position varies seasonally as well. The winter Hadley cell has a poleward extent (on the winter side) to around 30° and crosses the equator to rise at about 10 to 20° latitude in the summer hemisphere. (3) A thermally indirect Ferrel cell is seen during all seasons in both hemispheres. The Ferrel cell has a significantly weaker circulation than the Hadley; a similar result is apparent from the direct measurements such as Figure 3.16. (Though the precise magnitude of

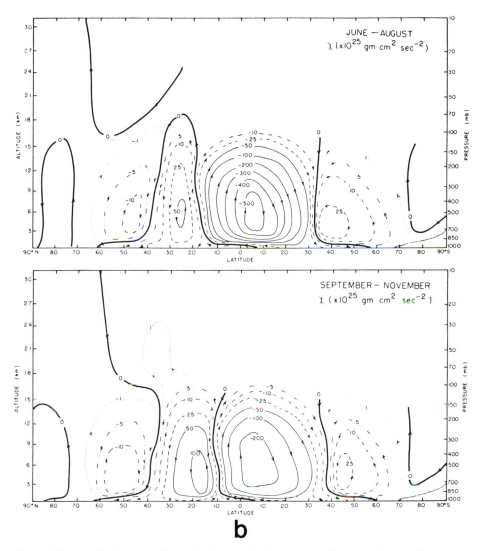

b

diagram. The mass flux between a given pair of contours is the same everywhere along that pair of contours, hence velocity is proportional to the gradient of the stream function.

the velocity is hard to determine from Figure 3.17, it is obvious that the vertical and horizontal gradients are much weaker for the Ferrel cell than for the Hadley.) (4) The Ferrel cell does not have much variation from season to season, though the Ferrel cell seems strongest in winter in the Northern Hemisphere, while there is no significant seasonal variation in the Southern Hemisphere. (5) Figure 3.17 has slight evidence of thermally direct polar cells. The polar cells are very weak and this apparent verification of Ferrel's three-cell hypothesis may only be fortuitous coincidence. Some calculations of the mean meridional circulation have shown polar cells, others have not. (6) As a final comment, the stream function pattern leads one to conclude that the lower horizontal flow in the cells takes place in a thin layer near the earth's surface while the upper-level return flow is somewhat concentrated near the tropopause. The concentration of the flow near the ground and near the tropopause agrees with the direct measurements shown in Figure 3.16.

In two respects these figures may create misleading impressions of the atmospheric general circulation. The first problem was very familiar to William Ferrel. Namely, how does one account for the thermally indirect cell in midlatitudes? The "Kuo-Eliassen" equation (§6.3) demonstrates that indirect meridional circulations are driven by the eddies and the diabatic heating fields. For now, the picture can be clarified by looking at the mean meridional circulation using potential temperature as the vertical coordinate. A second feature that may be misleading concerns the vertical motion, especially the upward motion of the Hadley cells. That upward motion cannot occur as broad, gentle ascent. Instead, the motion must be concentrated into narrow updrafts. These two aspects of the mean meridional circulation are discussed next.

The mean meridional circulation looks different when potential temperature (θ) is used as the vertical coordinate. θ has the great advantage that adiabatic flow is confined to surfaces of constant θ. In contrast, adiabatic flow is not limited to surfaces of constant pressure or height. Diabatic processes, such as radiative heating and cooling, will cause air parcels to migrate from one θ surface to another. Constant θ surfaces have a corrugated appearance in middle latitudes near frontal cyclones: their elevation is low where warm air is drawn ahead of surface low pressure; their elevation is higher in the cold air behind the low pressure center. Since air tends to follow these corrugated surfaces, then the mean meridional circulation will be more clearly depicted by zonal averaging along constant θ surfaces.

Figure 3.18 schematically shows the motion of air relative to a θ surface near a developing low. This figure reveals how a Ferrel cell in pressure coordinates is actually thermally direct in θ coordinates. Such a Ferrel cell can be created by diabatic processes and by eddy heat fluxes. Only the diabatic effects and horizontal heat fluxes are discussed at this point. Eddy heat fluxes and the Ferrel cell are discussed in §6.3.2. How eddy heat fluxes drive Stokes drift motions that oppose the observed Ferrel cell is explained in §7.4.1.

In the warm air sector (east of the surface low pressure center) the air moves poleward until reaching the warm front. Encountering the front, the warm air parcel is forced to rise, releasing latent heat in the process (hence θ of the parcel increases). Air in the cold air sector to the west of the surface low center follows a complementary path. A cold air parcel moves equatorward while sinking. A simple potential vorticity conservation argument (as in §7.2.1) can explain why a parcel will sink as it moves equatorward. In broad terms, trajectories follow paths much like the schematic illustration in Figure 3.18 (Palmén and Newton, 1969; their Figure 10.20). If the paths traced out by the warm and cold air parcels are projected onto the meridional plane, then the paths look like a Ferrel cell when height is the vertical coordinate. If θ is the

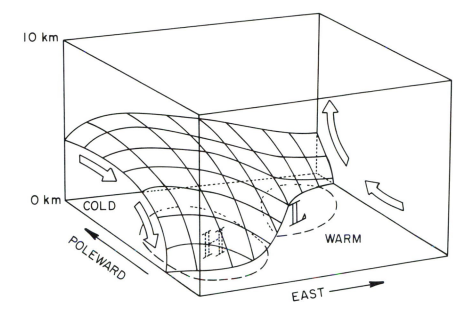

FIG. 3.18   Schematic diagram showing a sloping, three-dimensional isentropic surface and its relationship to developing high (H) and low (L) pressure centers. Schematic contours on the box bottom illustrate the surface pressure pattern. The double-shafted arrows show the meridional motion of air parcels in the warm and cold air sectors.

vertical coordinate, the parcels trace out a thermally direct circulation with three key parts. (1) As cold air (low θ) sinks behind a cold front, it can be warmed by absorbing solar radiation, by exchanging heat with the earth's surface (if it is near the surface), or from penetrative convection (if it is not near the surface). Hence, the flow for low θ values will be toward the equator with θ increasing along the path. (2) As warm sector air (middle θ values) reaches a front, precipitation forms and consequently latent heat is released. The latent heating increases θ. (3) At high altitudes and latitudes, radiational cooling is the dominant diabatic process, causing θ to decrease. These three parts are identifiable in the next figure from observations.

The observed meridional circulations in θ coordinates are shown in Figure 3.19. The most striking feature about Figure 3.19 is that the indirect, Ferrel cell is missing. Instead of a Ferrel cell, the Hadley cell extends further poleward. This single, thermally direct cell seems more intuitive since it transports heat poleward. As anticipated, the rising motion in midlatitudes is from the latent heat release. The upward motion at 40 N in Figure 3.19a corresponds very well with the location of maximum winter precipitation (Figure 3.29b).

The schematic description given by Figure 3.18 is applicable to the growing stage of a frontal cyclone. During the decaying stage, the motion in θ coordinates could look like a Ferrel cell. Based on a later discussion (§4.5.2; Figure 4.22), the meridional motion in the decay stage may be largely *ageostrophic* and it may have smaller

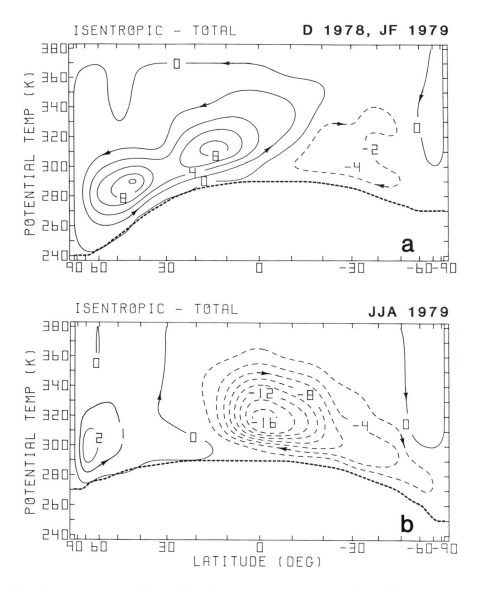

Fig. 3.19 Zonal mean meridional cell circulation as seen in isentropic coordinates. Contours are mass stream function; negative values are dashed and indicate clockwise circulation as shown by the arrows. (a) December–February 1979. (b) June–August 1979. From *J. Atmos. Sci.*, Townsend and Johnson (1985) by permission of American Meteorological Society.

velocities than the largely geostrophic motions in the growing stage. In addition, the decay stage ageostrophic velocities (in pressure coordinates, at least) are nearly cancelled by upstream meridional motions when a zonal average is taken (Figure 6.20). Consequently, the most prominent contributor to the zonal average motion in θ coordinates is probably the motion during the frontal cyclone's developing stage.

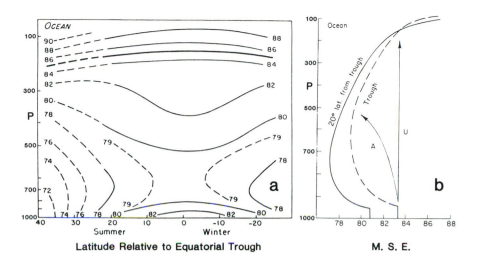

FIG. 3.20 (a) Zonal average meridional cross-section of the moist static energy in the tropics. The contour interval is 2 cal/gm. (b) Schematic profiles of the moist static energy at 20 degrees latitude from the center of the equatorial pressure trough, at the center of the trough, and within an insulated updraft (U). Original figures are in Riehl and Malkus (1958).

As mentioned above, the upward motion in the rising branch of the Hadley circulation cannot occur as large scale gentle ascent. Figure 3.20 illustrates the problem. These diagrams show climatological profiles of moist static energy (MSE). MSE is another quasi-conserved quantity. When moisture is condensed, the amount of latent heat released is latent heat of vaporization (L) times the change in specific humidity ($\triangle q$). From the first law of thermodynamics, that heat is converted into heat energy (C_pT) for a pseudo-adiabatic process. A third form of energy is gravitational potential energy (or geopotential, Φ). The moist static energy combines these three forms:

$$\text{MSE} = \Phi + C_pT + Lq \qquad (3.6)$$

For adiabatic motions, dry static energy is conserved: C_pT balances Φ. Moist static energy extends the conservation to include latent heat of vaporization for water.

Near the equator, a rising air parcel that starts at the surface would first proceed down the gradient of MSE until reaching an elevation near 700 hPa. To rise above that point, the parcel must proceed up the gradient of MSE, which requires work. Yet, the previous figures (e.g., 3.16) clearly show that the poleward transport of the Hadley cells occurs mainly at high elevations (near 200 hPa). In order to reach high altitude, Riehl and Malkus conclude that the upward motion must be confined to strong updrafts embedded within cumulonimbus clouds. Such updrafts are insulated from entraining low MSE air from outside the cloud. In addition, such thunderstorms need to form preferentially along the equatorial trough, where the surface air (that feeds these updrafts) has the highest MSE. Figure 3.20b compares the vertical profiles of MSE along the equatorial trough. Profile **A** is located in clear air, while profile **U** is representative of the air within an insulated updraft. Riehl and Malkus calculate that only one-tenth of one percent of the area between 10 N and 10 S need be covered by these vigorous updrafts. Later work (e.g. Houze, 1982) has demonstrated important roles in tropical convection for latent heat release caused by precipitation from cumulonimbus anvil clouds and for radiation (§6.2.4). However, the basic idea remains that upward motion is concentrated into small scale updrafts.

Broad-scale subsidence associated with the Hadley cells does not have this problem. Subsidence occurs in the subtropics and is thus down the gradient of MSE. In fact, this subsidence is occurring over the subtropical surface high-pressure cells.

3.4   Zonal average pressure fields

Figure 3.21 summarizes the zonal mean heights of three pressure surfaces. Five general comments about Figure 3.21 are as follows. (1) As anticipated in §1.2, the meridional gradient of pressure in the troposphere increases with height. From geostrophic and hydrostatic balance, the geopotential height gradient reflects the meridional gradient of temperature and the westerlies that increase with height in the middle latitudes. (2) In the tropics the pressure is low at the surface compared to the same elevation in the subtropics; the opposite is true in the upper troposphere. The reversal of the horizontal gradient of pressure is due to the great amount of latent heat released in the middle and high troposphere by the deep tropical convection (e.g., Figure 3.27). (The released heat makes the *high altitude* air relatively warmer than air in the subtropics. The thickness of a pressure layer is proportional to temperature and the upper level pressure pattern has much larger amplitude than the surface pressure pattern. Hence, the 100 hPa height field is generally quite similar to the 1000 to 100 hPa thickness field.) (3) The meridional pressure gradient at high altitudes is less in summer due to strong solar heating that is rather latitudinally uniform (Figure 3.7a), especially higher up in the stratosphere. The horizontal temperature gradient (Figure 3.11) reverses sign above the tropopause in the summer hemisphere. The winds in the lower stratosphere are generally light during that time of year, and the vertical shear changes sign, as well. (4) The hemispheric average surface pressure is less in the summer and greater in the winter in both hemispheres. This indicates a net shift of mass between the two hemispheres. (5) The low surface pressure associated with the belt of tropical convection moves seasonally into the summer hemisphere. The subtropics have higher surface pressure. Low pressure lies at the midlatitude storm belts.

More specific topics about Figure 3.21 concern the differences between hemispheres. The most obvious differences between the two hemispheres are found at the middle and high latitudes. There, the surface pressure is very much lower in the Southern Hemisphere, much lower than could be accounted for by gaps in the observational network. Again, the areal coverage of land plays a role in the interhemispheric differ-

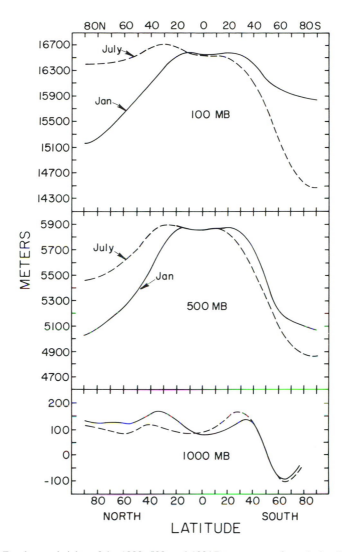

FIG. 3.21 Zonal mean heights of the 1000, 500, and 100 hPa pressure surfaces during January (solid line) and July (dashed). Figure based upon data diagrammed in van Loon (1972), with 1000 hPa height used in place of sea level pressure. (The vertical scale is different for each pressure surface.)

ence. The continental regions tend to have higher sea level pressure in winter and the oceans lower pressure. Therefore, when averaging around a latitude circle there are regions of high and low pressure that partly compensate in the Northern Hemisphere (Figure 5.10). Between 40 S and 70 S one finds a chain of low pressure centers and few high pressure cells (e.g., van Loon and Shea, 1988), thus the zonal mean pressure is very low. The difference between the hemispheres is largely removed as one goes to higher altitudes; it is almost gone by 100 hPa. The thermally-caused summer surface low and winter surface high over land are shallow; the opposite pressure pattern prevails in the mid and upper troposphere.

The distribution of surface pressure in the tropics and subtropics can be explained easily by the following arguments. First, one can imagine a wall, perhaps at 15° latitude, as shown in Figure 3.22a by the dashed line. The surface pressure is initially the same in both regions. In the middle troposphere of the tropics the deep convection releases much latent heat, causing the air to be warmer there than at the same altitude in the subtropics. The warmer temperature causes the pressure at some high altitude, for example 18 km, to be greater in the tropics, that is: $P_1 > P_2$. Of course there is no such wall, so the high altitude pressure gradient causes some mass to be shifted from the tropics as indicated by the arrows in Figure 3.22b. The subtropics have gained mass, so the surface pressure $P_{02} > P_0$; conversely $P_{01} < P_0$. The upper pressures will be changed proportionally. At the surface, higher pressure in the subtropics provides an ageostrophic acceleration of the air toward the equatorial surface trough. These ageostrophic motions are of course a Hadley cell circulation. The surface air circulation will likely compensate for the mass transport occurring at higher altitudes, but the surface flow is not likely to be so large as to eliminate the surface pressure gradient since the driving force for the motion would then be lost. In short, the pressure patterns in the troposphere agree well with the Hadley cell pattern shown earlier. Nonetheless, the pattern arises from a delicate balance between the latent heat release and the pressure fields.

The relative minimum in surface pressure in middle latitudes identifies the frontal cyclone tracks. As a cyclone passes by a station, the pressure will fall and rise. Therefore, cyclone activity could be measured by examining time variance data. Figure 3.23 shows the variances obtained by van Loon (1972). Four general features can be discussed. (1) The variability is greater in midlatitudes than in the tropics. Persistence is characteristic of weather in the tropics. The increase of sea level pressure variance with latitude has been well known for a long time; van Loon cites Lockyer (1910) and Köppen (1882). Another source of the low variance in the tropics is the fact that pressure gradients are generally weaker there. (2) The activity is highest during winter, but the seasonal change is again much smaller in the Southern Hemisphere. The frontal cyclones derive much of their energy from baroclinic energy conversions (§4.5.2 and §7.1); such conversions are proportional to the horizontal temperature gradient. The horizontal temperature gradient is simply much less during summer than during winter in the Northern Hemisphere. In contrast, the horizontal temperature gradient for the opposite hemisphere remains strong throughout the year, due to the very cold temperatures over the continent of Antarctica (Figures 3.11 or 5.4). The persistent, strong latitudinal temperature gradient encourages comparable activity during winter and summer in the Southern Hemisphere. (3) The maximum values of variance are centered around 50 to 60° latitude. This location fits the characteristic life-cycles of midlatitude cyclonic storms. Generally, frontal cyclones progress eastward and poleward as they develop. The largest variability will tend to be where travelling

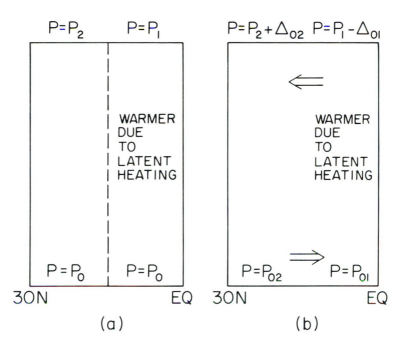

FIG. 3.22   Schematic diagram illustrating how the Hadley cell circulation, equatorial trough, and latent heat release at the intertropical convergence zone (ICZ) are linked. (a) A wall (dashed line) divides the region into equal masses, both with surface pressure P_0. Latent heat release in the equatorial troposphere causes the air to have greater pressure at the top of the domain near the equator. An upper-level ageostrophic motion is set up toward the subtropics. The upper-level motion causes a net loss of mass from the tropics, with a net gain in the subtropics. Hence, the surface pressures change such that P_{02} becomes greater than P_{01}, resulting in ageostrophic motion toward the equator at the surface.

storms have reached their maximum amplitude. So, in the Northern Hemisphere the deep, "cut-off" lows that frequent the environs of the Aleutian and Iceland islands lead to the maximum variance between 50 to 60°latitude. In the Southern Hemisphere there are maxima near 50 to 60°and near 75°S. The latter maximum reflects very deep lows that migrate along the shore of Antarctica. (d) The maximum variance is spread out more broadly in latitude in the Northern Hemisphere. The greater spread reflects the greater meandering of the storms as they follow the long wave pattern. In contrast, the Southern Hemisphere storms tend to track along a latitude circle more closely.

3.5   Zonal average moisture and cloudiness

The final group of zonal average fields considered in this chapter are all related to moisture. Moisture has been implicated in several key processes already, including (a) moisture affecting the transmission and emission of radiation in clear sky (Figures 3.2; 3.3), (b) clouds reflecting and emitting radiation (Figures 3.5; 3.6), (c) transport of energy in the form of latent heat (Figure 3.9), (d) clouds required for the

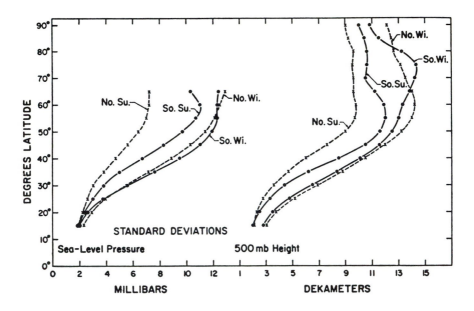

FIG. 3.23   Zonal average profiles of the standard deviations about the monthly means of sea level pressure (left curves) and 500 hPa height (right curves). The figure is reproduced from van Loon (1972, *Meteor. Mono.*) by permission of the AMS and is based upon several references cited in that book. Profiles for the Northern and Southern Hemispheres in winter (Wi) and summer (Su) are presented.

upward branches of the Ferrel (Figure 3.19) and Hadley (Figure 3.20) circulations and (e) causing the equatorial trough and subtropical higher surface pressure (Figure 3.22).

Moisture is one of the least well observed atmospheric quantities. One reason is that humidity is much more difficult to measure reliably than temperature, pressure, or horizontal wind. Another reason is that the vertical extent of clouds is often difficult to measure. Another problem is that clouds closer to an observer obscure other clouds beyond. Current cloud climatologies are generally regarded as inadequate (Hughes, 1984; Schiffer and Rossow, 1983). To improve our knowledge, the International Satellite Cloud Climatology Project (ISCCP) was initiated in 1983. ISCCP is a multiyear project to gather accurate, global-coverage data on cloud cover and radiance (Schiffer and Rossow, 1983).

Measurements of zonal and annual mean cloud cover made by several researchers are shown in Figure 3.24. Hughes (1984) complied most of the curves in this figure. Beginning with the annual average data, several features can be seen in all the curves. The general pattern of cloudiness includes a relative maximum near 5 N. The equatorial maximum is the ICZ, the rising branch of the Hadley cells. The cloudiness is a minimum in the subtropics. The sea level pressure is highest in the subtropics, and one expects the sinking motion above the surface highs to inhibit the formation of clouds. At higher latitudes the cloudiness again increases with maximum values near 60 to

ANNUAL

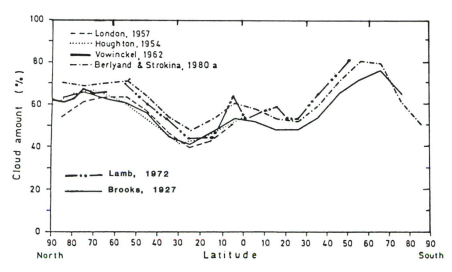

FIG. 3.24 Zonal and annual average distributions of cloud amount found by several authors and compiled by Hughes (1984) are shown. Surface observations were used by: Brooks (1927), London (1957), and Houghton (1954). Berlyand and Strokina (1980) used both satellite and surface data. The curve from Lamb (1972) was added by the author; it is based on satellite data from March 1962–February 1963. Cloud amount is expressed as a percentage. From *J. Climate Appl. Meteor.*, by permission of American Meteorological Society.

80° latitude. The ISCCP data have two obvious differences from the other curves: the ICZ is more sharply defined, and the cloudiness in the Southern Hemisphere subtropical and middle latitudes is higher in the ISCCP data.

From the data compiled by Hughes, 1984, one estimates the global average cloud cover to be about 50%. Preliminary ISCCP results enlarge the estimate to about 60% (Rossow and Schiffer, 1991). The discrepancy may be due to incomplete sampling over the oceans in the earlier studies; the southern oceans are heavily cloud covered in the ISCCP data. However, even the ISCCP data for the three years reported by Rossow and Schiffer typically cover just 70% of the globe. ISCCP data show little seasonal variation for global averages; the standard deviation (SD) is 1.2%. Rossow and Schiffer estimate other global average properties for the July 1983 through January 1986 period as follows. The monthly mean cloud optical thickness value is about 9 with SD of 1.0; albedo is about 50%; surface visible light reflectance is 14% (SD 1%) but albedo (all light wavelengths) is about half that of clouds. Surface and cloud top temperatures are 270 and 290 K, respectively, both with SD of 1.0 K. Rossow and Schiffer confirm that clouds have a net cooling effect on climate (observed clouds reflect more solar radiative energy than they trap terrestrial energy). The conclusion confirms results from radiative convective theoretical models (§6.1) such as Manabe and Strickler (1964).

Rossow and Schiffer (1991) describe interesting regional and temporal variations in ISCCP data. Over oceans, the optical cloud thickness is less (9 versus 13), the cloud tops are lower (3 versus 4.5 km), and there are more overcast and fewer cloud-free occurances. The diurnal variation of cloud cover is small over oceans except in the tropics. Tropical oceans have the greatest cloud cover near dawn and the least in mid-afternoon. Tropical land areas have maximum cloudiness after local midnight with minimum 12 hours later. In contrast, midlatitude land areas have "very large" diurnal cloud changes, especially in summer, with greatest cloudiness in mid-afternoon. The mid-afternoon maximum in cloudiness seems to match the time of maximum surface solar (shortwave) heating. Such heating would destabilize the lower atmosphere leading to convection, including moist convection. The oceanic tropical areas have maximum cloudiness at night and early morning because terrestrial radiation (longwave heating and cooling) plays a major role in destabilizing all levels of the troposphere. A net absorption of radiation at cloud base and net emission at cloud top can have a positive feedback with moist convection (see review in Remer, 1991).

Significant interhemispheric differences are apparent in Figure 3.24. The principal hemispheric asymmetry is higher cloudiness in the subtropics and midlatitudes of the Southern Hemisphere. Higher cloudiness in both regions is easily explained. The Saharan, Arabian, and Sonoran deserts occupy much of the subtropics of the Northern Hemisphere. The subtropical deserts in the Southern Hemisphere are the Australian Outback and the Kalahari; their areal extent is much less than the deserts to the north. Clear sky prevails over the deserts, while small cumulus or low stratus clouds prevail over the subtropical oceans (e.g., Figure 5.13). Some of the higher cloudiness in the midlatitudes is due to the persistent storminess in the Southern Hemisphere (Figure 3.23). The other reason is that the cloudiness is just much more prevalent over oceanic areas as mentioned above (Rossow and Schiffer, 1991). The effect is seen by examining Plate 2.

The curves agree in general shape, but the annual mean values disagree by as much as 10%. The disagreement is a good indicator of the uncertainty in current cloud climatologies. Many factors cause this uncertainty. Hughes (1984) states that surface observations of cloud cover and type are notoriously poor at night and suffer from a strong land bias (§2.5). Until recently, it had been thought that the surface observations may also overestimate cloudiness; but the ISCCP data shown (Rossow, 1989) contradicts this conclusion. The cloud observations from satellites use visible brightness and infra-red measurements. Visible measurements have difficulty distinguishing clouds from bright surfaces such as snow cover or deserts; recent advanced algorithms that couple radiative temperature information with visible brightness partly alleviate this problem.

The seasonal variations in cloudiness are illustrated in Figure 3.25. The figure shows climatologies found by a greater number of researchers than are shown in the previous figure. Consequently, the spread in the estimates of cloud amount is greater than in the previous diagram. The maximum disagreement is nearly 40%! Most

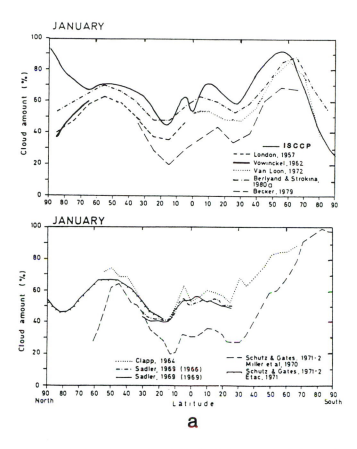

FIG. 3.25a   (a) Zonal mean cloud amount for January as found by several authors and compiled by Hughes (1984); recent ISCCP data have been added. London (1957) used surface observations. van Loon (1972), Berlyand and Strokina (1980) and Schutz and Gates (1971, 1972; labelled "ETAC" (1971)) all used a mixture of surface and satellite data. Becker (1979), Clapp (1964), Sadler (1969) and Schutz and Gates (1971, 1972; labelled "Miller et al. 1970") used satellite data. The solid line in the upper chart is ISCCP data from Rossow (1989). From *J. Climate Appl. Meteor.*, by permission of American Meteorological Society.

researchers who estimate cloud cover by season find the middle latitudes to be 5 to 10% more cloudy in winter. The researchers also find that the equatorial maximum cloudiness shifts into the summer hemisphere. However, an equatorial maximum occurs north of the equator during both seasons; this follows because the ICZ lies north of the equator across much of the Pacific all year round (e.g., Figure 5.13).

The vertical distribution of the clouds is even less well known than the horizontal extent of cloud cover. For example, low clouds may obscure the view of higher cloud layers. On average, the middle levels of the troposphere are visible only 75% of the time, while high levels are visible only 65% of the time (Hahn et al. , 1982). High clouds can obscure the view of lower clouds from a satellite. Hahn et al. (1982) report the intriguing result that high clouds rarely occur alone over the ocean. The frequency

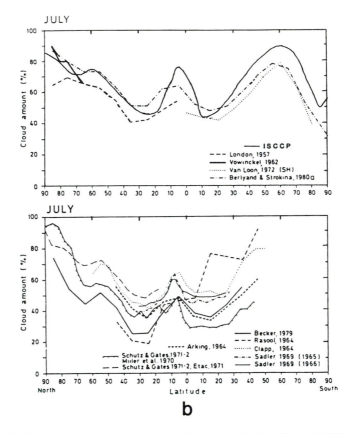

FIG. 3.25b Similar to (a), except showing average cloud amount for July. Brooks (1927) and London (1957) used surface observations. van Loon (1972), Berlyand and Strokina (1980) and Schutz and Gates (1971, 1972; "ETAC" (1971)) used a mixture of surface and satellite observations. Becker (1979), Rasool (1964), Clapp (1964), Sadler (1969) and Schutz and Gates (1971, 1972; "Miller et al." 1970) used satellite data only. Reproduced from Hughes (1984, *J. Climate Appl. Meteor.*) by permission of the American Meteorological Society with modifications to include ISCCP data from Rossow (1989).

of cirroform or altoform clouds alone is 1 to 2% near the ICZ, increasing to 20% in the middle latitudes (examples are the cirrus cloud shield ahead of a front and cirrus along the subtropical jet). The cloud base can be estimated with comparative ease (by triangulation as done by Dalton and others beginning in the late 1700s, or by timing the disappearance of a rising balloon). Satellites can now estimate the cloud tops accurately. An ISCCP goal is to prepare climatologies of different cloud types. Rossow and Schiffer (1991) discuss frequencies of cloud top pressures and depths. Plate 2 shows monthly-average cloud tops. Figures 3.26a and 3.26b show the zonal average percent cloud amount for various cloud types reported by observers. Warren et al. (1986, 1988) use a statistical correction to remove the possibility of low clouds obscuring higher clouds. Altitude information may be inferred from the diagrams: low clouds (St, Cu), middle clouds (As), high clouds (Ci), and deep clouds (Ns and Cb).

The figure shows several cloud properties of note. (1) Convective clouds are more common in tropical and subtropical regions, while stratiform clouds are more common at higher latitudes. This relation holds for land and ocean areas. (2) Over all of the oceans there is some tendency for cloud frequency to decrease with height. This rule does not hold for most land areas. (3) The subtropical desert areas have fewer clouds of nearly all types, but the tropical land areas are more cloudy than the surrounding seas (as is seen again in Figure 5.13).

Several estimates of zonal mean precipitation are given in Figure 3.27. The largest disagreement among the estimates occurs at the ICZ. Shea (1986) feels that the data used by Jaeger (1976) and by Sellars (1966) underestimate oceanic rainfall, particularly in the eastern Pacific. Elsewhere, the data in Figure 3.27 agree better. The largest precipitation occurs in the tropical convergence regions with the annual average maximum located just north of the equator. This location lines up well with the equatorial maximum in cloud cover noted previously. Comparatively little precipitation falls in the subtropics as would be expected from the cloud cover, sea level pressure, and mean meridional circulations discussed above. Rain is unlikely where there are few clouds and large scale subsidence. Precipitation has a relative maximum in the midlatitudes that follows from the cyclonic storm activity noted before. Little precipitation falls near the poles (despite high cloudiness) because the air in polar regions is too cold to hold much moisture.

As for interhemispheric differences, the midlatitudes receive significantly more precipitation in the Southern Hemisphere. This difference reflects the greater cyclonic storm activity that was noted in Figure 3.23. The result is reasonable if pressure variance identifies frontal cyclones and the storm activity correlates with precipitation amounts. As for the subtropics, they are cloudier in the Southern Hemisphere, yet the precipitation is the same in both hemispheres. The greater subtropical cloudiness is primarily a low stratus type cloud like that found frequently off the west coasts of the continents e.g., (Figure 5.13). These low subtropical stratus clouds are thin and do not produce much precipitation.

The final two figures include evaporation and the meridional transport of water vapor. One can deduce reasonable arguments to explain the features found in these figures, but the details are slightly suspect because of inaccuracies and unrepresentativeness of the data. To a great extent, long period time averaging can minimize these sources of error except when the latter source is due to topographic effects. Accordingly, the remarks are confined to cautious generalizations. Figures 3.28 and 3.29 show curves for precipitation, evaporation, and water vapor flux. Figure 3.28 shows annual average values; Figure 3.29 shows seasonal averages.

As one might anticipate, evaporation is greatest in the subtropics of both hemispheres. The sky is clearer (relative to other latitudes) and air near the surface is warmest there. The relative minimum near the equator is due to the shading by the greater cloudiness associated with the tropical deep convection. At high latitudes the cloudiness and lower temperatures reduce the amount of the evaporation there. The

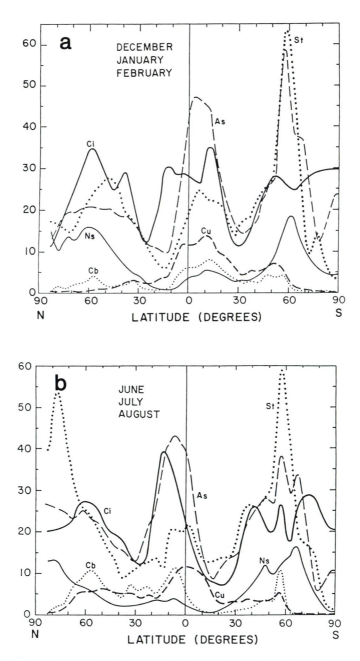

Fig. 3.26a and b Percentage of zonal average cloud amount by various cloud types. (a) and (b) Over land areas during the 11 years of 1971–1981. Diagram redrawn from Warren et al. (1986, 1988).

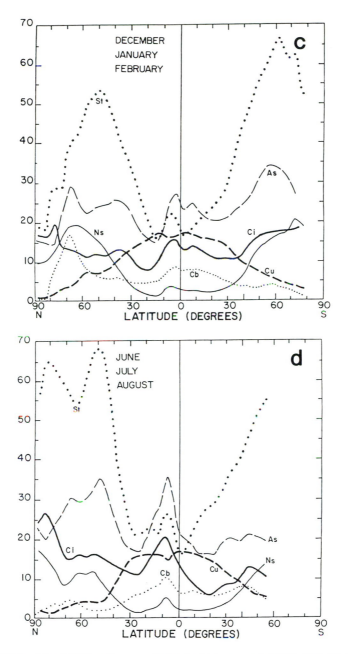

FIG. 3.26c and d (c) and (d) Over the ocean during the thirty-year period 1952–1981. High clouds are all forms of cirrus (Ci). Middle clouds include altostratus and altocumulus (As). Low clouds are stratus and stratocumulus (St), cumulus (Cu), nimbostratus (Ns), and cumulonimbus (Cb). The latter two types, Ns and Cb, extend through a large depth. Obscuration of high clouds by lower clouds has been removed by an approximate, statistical technique. Diagram redrawn from Warren et al. (1986, 1988).

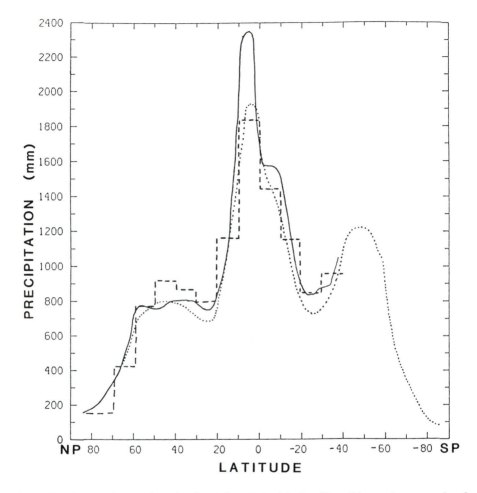

FIG. 3.27   Three estimates of zonal and annual mean precipitation. The solid curve is a composite of data in Shea (1986). The dotted curve is an estimate by Jaeger (1976). The dashed curve is data in Sellars (1966).

greatest evaporation occurs in the Southern Hemisphere subtropics. Even though the Northern Hemisphere is warmer (and the skies clearer), less water is available to be evaporated because the desert area is larger.

The net transport of moisture required for mass balance can be deduced from the precipitation (P) and evaporation (E) data. The net excess or deficit of water is defined as P-E, where positive values indicate an excess and negative values are a deficit. Figure 3.28 also shows the net excess of water. The net water vapor transport is found by summing the deficit or excess of precipitation as one proceeds from one pole to the other. (A similar method was used to find the heat transport, Figure 3.9). The water transport calculated in this way is likely to magnify those errors that occur in the profiles of P and E. However, the general pattern shown in Figure 3.28 is consistent

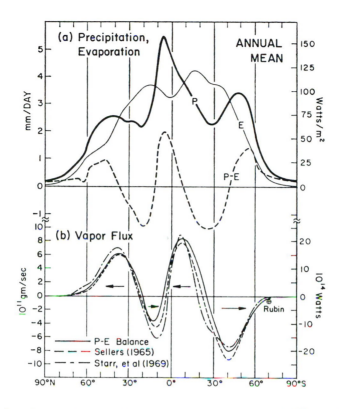

FIG. 3.28   Annual and zonal mean profiles of (a) evaporation, precipitation, and (b) meridional moisture transport. Evaporation (E) values are from Kessler (1968) and Budyko (1963). Precipitation data are from Meinardus (1934) and Möller (1951). Three estimates of meridional transport are shown including one deduced from chart (a), others from Sellars (1966) and Starr et al. (1969).

with our earlier discussions of other variables. The vapor transport is toward the ICZ in the tropics and subtropics. In middle latitudes the transport is poleward.

The seasonal variations of these three quantities, precipitation, evaporation, and water vapor transport, show some interesting features. The seasonal precipitation pattern near the equator maintains a similar peak value in all seasons, but shifts position slightly. The ICZ is quite persistent over the oceans and lies just north of the equator (Figure 5.13); consequently, the peak precipitation persists just north of the equator. This peak is further enhanced by additional convection over Asia during June through August. A secondary peak occurs near 12 S during Southern Hemisphere summer and is due to enhanced convection over the Amazon and Congo river basins. In the Northern Hemisphere midlatitudes, the winter and summer precipitation are about the same. In the Southern Hemisphere the precipitation is significantly larger in winter than in summer. If this is true, then possibly the winter cyclones are stronger and precipitate more readily than those in summer.

In the Northern Hemisphere, the seasonal variation of evaporation is more pro-

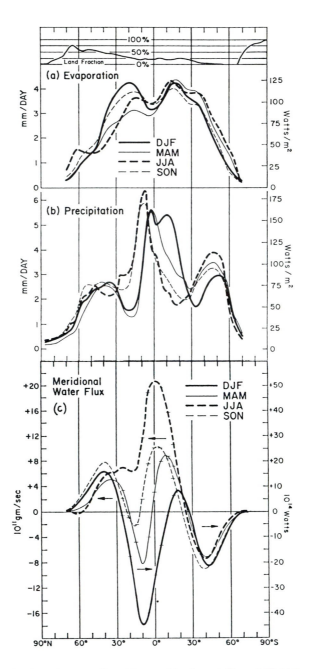

FIG. 3.29 Seasonal and zonal mean profiles of the quantities shown in Figure 3.28. (a) Evaporation values found by Kessler (1968). (b) Precipitation data from Meinardus (1934) and Möller (1951). (c) Meridional transport deduced. From *Meteor. Mono.*, Newton (1972) by permission of American Meteorological Society.

nounced than in the Southern Hemisphere. The Northern Hemisphere subtropics have *less* evaporation in summer than in winter. The drying out of the desert regions plus the formation of stable stratus clouds over the oceans that persist during that time of year partly account for this fact. An additional factor is the stronger surface winds over the oceans in winter.

The required moisture transports show little change in the midlatitudes between seasons, but large changes in the tropics. The transport at the equator is northward during July and southward during January. This result agrees with the low level flow of the Hadley cell in the winter hemisphere.

4

MOMENTUM AND ENERGY RELATIONSHIPS

Our *northerly* winds in the beginning of the winter may arise from the weight of the cold northern air overcoming the warmer southern air, which as the heat lessens, is less loaded with vapours, and therefore more easily gives way to the cold northern and denser air.

Mills, 1770

The inequality of heat in the different climates and places, and the earth's rotation on its axis, appear to me the grand and chief causes of all winds, both regular and irregular; in comparison with which all the rest are triffling and insignificant.

Dalton, 1834

This chapter is a lengthy discussion of some energetic properties of the atmosphere. Mathematical tools are developed to provide one with the background needed to interpret properly the zonal and time average observed energy-related quantities. Such energy-related quantities include energy conversions, as well as the angular momentum balance discussed first. These topics are quite important for at least four reasons. First, as mentioned in §1.2, any reasonable general circulation must satisfy the balance requirements of mass, moisture, momentum, and energy. Second, momentum fluxes and the energy conversions that help drive the fluxes are necessary in order to maintain the circulation against frictional decay. Third, the principal phenomenon that maintains the westerlies turns out to be the midlatitude frontal cyclones, a fact that requires the subject of the general circulation to include the cyclones. Fourth, the cyclones' structure and their existence are determined and understood best by looking at the energetics of those storms.

Equations describing large scale energetics are derived in this chapter and zonal mean observations are displayed. Zonally-varying observations are given in §5.9. Because of their importance for interpretation, energetics terms enter discussions in many places in Chapters 6 and 7. Readers desiring even more detailed information on energetics are directed to Chen and Wiin-Neilson (1993).

4.1 Zonal average momentum equation

In looking at the angular velocity of a parcel of air, one must know what torques and forces act on that parcel to change its angular momentum. The two important ones for the atmosphere are pressure forces and friction.

The symbol M designates the *absolute* angular velocity of a parcel. M has two parts because the coordinate system is fixed with respect to the earth and thus rotates.

$$M = R^2\Omega + RU \tag{4.1}$$

where

$$R = r \cos \phi$$

The first part is planetary angular velocity while the second part is the velocity relative to the rotating coordinate system. r is the earth's radius; ϕ is latitude; Ω is the angular speed of the earth's rotation; and U is zonal velocity (positive in direction of earth's motion).

Angular momentum equals ρM. The rate of change of angular momentum depends on pressure forces along the zonal direction and friction acting on motion along that same direction.

$$\rho\frac{dM}{dt} + R\frac{\partial P}{\partial x} + RF_x = 0 \tag{4.2}$$

ρ is density, $\partial P/\partial x$ is pressure gradient, F_x is friction in the longitudinal direction x. Equation (4.2) is easily derived from the zonal velocity equation; the Coriolis term is within the $v\partial M/\partial y$ term.

Expanding the total derivative in (4.2) yields:

$$\frac{\partial(\rho M)}{\partial t} + \nabla_3 \cdot \rho M\mathbf{V}_3 + R\frac{\partial P}{\partial x} + RF_x = 0 \tag{4.3}$$

Integrating (4.3) over the volume of the atmosphere between latitude circles ϕ_1 and ϕ_2 while holding volume V constant obtains:

$$\frac{\partial}{\partial t}\int_V \rho M d\mathbf{V} = \int_{S_1} \rho M v_1 dx dz - \int_{S_2} \rho M v_2 dx dz$$
$$- \int_V R\frac{\partial P}{\partial x} d\mathbf{V} - \int_V RF_x d\mathbf{V} \tag{4.4}$$

where S_1 equals the surface area of the "wall" at latitude circle ϕ_1; S_2 equals the surface area of the "wall" at latitude circle ϕ_2; and v_i equals meridional velocity directed positive into ϕ_i from the south.

Figure 4.1 illustrates the geometry. The only surface integrals obtained in (4.4) are those along the meridional boundary "walls." No contribution comes from the longitudinal direction because the domain encircles the earth, i.e., because of the

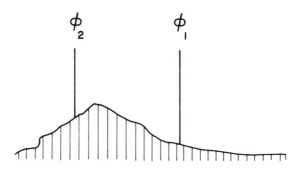

FIG. 4.1   The two boundary integrals in equation (4.4) have different surface areas due to the different circumference of the latitude circles and to topography.

periodic boundary condition. While obstructions cause a non-zero contribution for the pressure gradient term, they cannot do so for the zonal velocity gradient term. Finally, the top and bottom boundaries make no contribution because the vertical velocity vanishes there.

The time derivative term is approximately zero when averaged over a year's time. The time derivative can be ignored when dealing with long period problems of the general circulation. For a long-term average the friction term is cumulative and eventually becomes a large accumulation. In the short term, the time derivative must be included but the friction term can be dropped.

The first three terms on the right-hand side of (4.4) are responsible for bringing angular momentum in (or out) of the volume. Either of the first two right-hand side terms in (4.4) can be rewritten. For example, by applying a time average to the first term and using the hydrostatic equation to make a change of independent variable, one obtains:

$$\int_{S_1} \int_t \rho M v_1 \, dx \, dz \, dt = \frac{R^2 \Omega}{g} \int_0^{P_0} \int_x \int_t v_1 \, dt \, dx \, dP$$
$$+ \frac{R}{g} \int_0^{P_0} \int_x \int_t u v_1 \, dt \, dx \, dP$$

The flux has been split into two parts. One part is advection of planetary angular momentum (which includes a mass flux across the latitude circle). Using the shorthand notation developed in the Appendix, the right-hand side equals

$$\underbrace{\frac{R^2 \Omega L \tau P_0}{g} \widehat{[\overline{v_1}]}}_{(A)} + \underbrace{\frac{R L \tau P_0}{g} \widehat{[\overline{u v_1}]}}_{(B)} \tag{4.5}$$

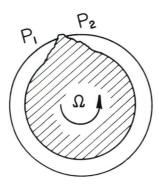

Fig. 4.2 Schematic diagram showing the mountain torque set up by differing pressures on the two sides of a mountain. The view is from above the north pole.

- *Term (A)* transfers (planetary) angular momentum through a net mass shift into or out of the volume, across latitude ϕ_1.
- *Term (B)* is a velocity correlation (discussed in the Appendix) that transfers relative angular momentum. Following the discussion in the Appendix, this flux of mass can be partitioned among several phenomena.

The third term on the right-hand side of equation (4.4) is the pressure torque term. If the zonal integral goes completely around the earth, then

$$\int \frac{\partial P}{\partial x} dx = 0$$

and the term vanishes. If, however, a mountain obstructs the circuit, then the integral takes the form:

$$\int \int R(P_2 - P_1) dy dz$$

for $P_1 \neq P_2$.

Figure 4.2 schematically illustrates the mountain torque. The integration is in the same direction as the earth's rotation. If pressure is greater on the west side than on the east side of the mountain ($P_2 > P_1$), then atmospheric westerlies are slowed down and the earth's rotation is sped up. Naturally, the far greater mass of the earth causes the change in motion of the atmosphere to be far greater than that of the earth. More precisely, it is said that *easterly* momentum is added to the atmosphere when $P_2 > P_1$. Figure 4.3 shows observed mountain torques.

The final term in (4.4) is the friction integral. It is useful to obtain an expression for "F". To use molecular friction coefficients would be to greatly underestimate the magnitude of the actual frictional drag on the air. Instead, an analogy to molecular friction known as turbulent viscosity is used whereby the microscopic-scale molecular friction formulation is expressed in terms of macroscopic motions. The analogy proceeds by using a "mean free path" of a "blob" of air that is similar to a "real" mean free path of a molecule. The analogy is appropriate because the blobs have pressure forces acting on them that may be analogous to the collisions of molecules. One chooses an *eddy* coefficient of viscosity κ_E and assumes the processes involved are the same. For example,

$$\kappa_E \nabla^2 \bar{u} = F_x$$

For *molecular* physics, the coefficient is a constant; here it is not. κ_E varies with the stability of the atmosphere, surface roughness, wind speed, etc. κ_E is very much bigger (a blob has billions of molecules) than the molecular coefficient, which refers to individual particles.

The friction integral is small in the free atmosphere, so the prime contribution is from the surface dissipation. Thus, the friction integral can be approximated in terms of stress tensors, τ

$$\int_V RF_x dV = \frac{-1}{r} \int_V \left(\frac{\partial \epsilon \tau_{xx}}{\partial x} + \frac{\partial \epsilon \tau_{xy}}{\partial y} + \frac{\partial \epsilon \tau_{xz}}{\partial z} \right.$$
$$\left. + \frac{\tau_{xy} \tan \phi}{r} + \frac{\tau_{xz}}{r} \right) dxdydz \tag{4.6}$$

where $\epsilon = r^2 \cos \phi$ and r equals the Earth's radius. The subscripts denote components of the tensors, not derivatives. In the x direction, for example

$$\kappa_E \nabla^2 \bar{u} = \nabla \cdot \mathcal{T}$$

where $\mathcal{T} = (\tau_{xx}, \tau_{xy}, \tau_{xz})$ and the "∇" operator above is three-dimensional. The stress tensor can be thought of as stresses applied to the sides of a small cube. The first subscript denotes the direction of the motion that is affected (in this case x, because momentum M depends upon u). The second subscript denotes the direction of the axis normal to the given face of the cube. Hence, τ_{xx} is the pressure force in the x direction, whereas τ_{xz} is a stress caused by vertical shear in the x direction. (See Figure 4.4). One expects the frictional interaction with the ground to exceed the internal atmospheric friction, so τ_{xz} will dominate in the surface boundary layer. If the horizontal components of $\mathcal{T}$ are ignored, then the vertical integral in (4.6) reduces to evaluating the stress tensor at the earth's surface.

$$\int_V RF_x dV \approx \frac{1}{r} \int_{x_0} \int_{y_0} \epsilon \tau_{x_0 z_0} dy_0 dx_0 \tag{4.7}$$

Stanton (1911) introduced the concept of a drag coefficient and Taylor (1916) suggested that stress be proportional to velocity squared. An approximate formula for the vertical stress tensor component may be defined as:

$$\tau_{x_0 z_0} = \mu \rho u_0 \sqrt{u_0^2 + v_0^2}$$

(4.8)

where $\mu \sim 4 \times 10^{-3}$ and has no units. Using (4.8) to approximate the friction force yields:

$$\int_V RF_x dV \approx \int_{x_0} \int_{y_0} R\mu\rho_0 u_0 \sqrt{u_0^2 + v_0^2} \; dy_0 dx_0$$

(4.9)

where the subscript zero refers to values measured at standard anemometer height. The surface stress is always dissipating atmospheric motion, but the right-hand side of (4.9) can be positive. Friction can increase M if the surface winds are easterly ($u_0 < 0$).

Frictional torques vary in sign with latitude since there are belts of surface westerlies and easterlies. Curiously, the measured mountain torques are often in the same direction as the frictional torque. Lorenz (1967, p. 52) points out that this correlation in sign is fortuitous. If the friction and mountain torques were of equal magnitude and uncorrelated in sign, then one could not deduce the total torque upon the atmosphere from the surface wind. Also, any theory that assumes a smooth spherical earth would not apply to the observed circulation because one could not parameterize the mountain torque in terms of friction. The correlation is not exact however, as Figure 4.3 indicates.

The frictional torque at most latitudes is much larger than the mountain torque. Oort and Peixoto (1983) also note that the seasonal variation of the mountain torque is a major part of the seasonal change of the total torque. Recent data from Wahr and Oort (1984) can be compared with Newton's estimates in Figure 4.3. The general patterns are similar in the studies; the largest differences are located near 30 N. Compared to Figure 4.3, Wahr and Oort find these mountain torque differences: during December through February there is no minimum at 50 N and the peak is at 30 N, not 20 N; during June through August they do not have a positive peak at 30 N, but a strong negative peak value there. Compared to Figure 4.3, Wahr and Oort find these frictional torque differences: during December through February the minimum at 37 N is about 10% less and the maximum at 17 S is about 20% less; during June through August the minimum at 47 N is about half that in the figure and the maximum at 17 N is more than twice as large.

Swinbank (1985) calculates the components of the momentum balance during January and July from the FGGE data. The FGGE data have generally larger mountain fluxes than those shown in Figure 4.3. However, the mountain fluxes are still smaller than needed for balance. Swinbank reports that the friction terms are more steady in time, while the majority of the total fluctuation is due to the mountain torques.

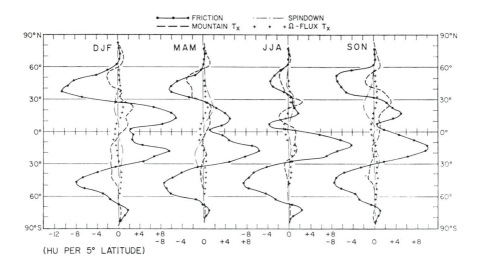

FIG. 4.3 Latitudinal variation of the contributors to the momentum balance during different seasons. The solid line is friction; the dashed line is the mountain torque. Other torques shown are the time rate of change of zonal momentum (dot-dashed line) and the flux of planetary momentum across the latitude circle ("plus" symbols). The units are in 10^{14} kg m^2 s^{-2} per $5°$. From *Meteor. Mono.*, Newton (1972), by permission of American Meteorological Society.

One might also wonder about the torque from small hills. According to Lorenz (1967), the hill torque may be, by implication, in the same direction as the mountain torque. The addition of "hill torque" may eliminate the factor of two difference noted by Lorenz between the transport of momentum (the surface integrals in eq. (4.4)) and what goes into the earth (the mountain plus friction terms).

4.2   Component kinetic energy equations

A natural extension of the angular momentum concepts presented in the previous section is the analysis of kinetic energy. The total kinetic energy can be partitioned into contributions by the zonal mean flow and by the zonally varying flow. Horizontal component expressions for the zonal average kinetic energy ($K_Z = \{K_Z^x, K_Z^y\}$) are derived first. Then the horizontal total kinetic energy component equations ($K_H = \{K^x, K^y\}$) are derived. The eddy kinetic energy ($K_E = \{K_E^x, K_E^y\}$) is simply the difference between the other energy components:

$$K^x = K_Z^x + K_E^x$$

$$K^y = K_Z^y + K_E^y \tag{4.10}$$

$$K_H = K_Z + K_E$$

The subscript Z refers to the kinetic energy of the zonal average flow. The subscript "E" refers to the kinetic energy of the deviations from the zonal mean flow, i.e., from the eddies. The zonal mean of an eddy quantity is zero. The energies here are defined with volume averages. The zonal average part of the volume average eliminates all cross products between eddy and zonal mean quantities (see Appendix; e.g., $u'[u]$ terms vanish).

$$
\begin{aligned}
K_H &= \int_V \left(\frac{[uu] + [vv]}{2} \right) \rho \, dx \, dy \, dz \\
&= \int_V \left(\frac{[u]^2}{2} + \frac{[v]^2}{2} \right) \rho \, dx \, dy \, dz + \int_V \left(\frac{[u'^2]}{2} + \frac{[v'^2]}{2} \right) \rho \, dx \, dy \, dz
\end{aligned}
\tag{4.11}
$$

The inclusion of density in (4.8) gives kinetic energy densities; the integrals are thus over the mass within a given volume.

4.2.1 Derivations

The equations for K_Z are derived first. The longitudinal (brackets) average of the horizontal equations of motion are taken. (This step parallels eqs. 2.19 and 2.20 in Holton, 1979, except using pressure as the vertical coordinate.) The u equation (in flux form) is multiplied by $[u]$ and the v equation (also in flux form) by $[v]$. Bringing the brackets term inside the time derivative obtains:

$$
\frac{\partial}{\partial t} \left(\frac{[u]^2}{2} \right) + \frac{[u]}{R} \frac{\partial R[uv]}{\partial y} + [u] \frac{\partial [u\omega]}{\partial P} - \frac{[u][uv] \tan \phi}{r}
$$
$$
- f[u][v] + [u][F_x] = 0
$$

$$
\frac{\partial}{\partial t} \left(\frac{[v]^2}{2} \right) + \frac{[v]}{R} \frac{\partial R[vv]}{\partial y} + [v] \frac{\partial [v\omega]}{\partial P} - \frac{[v][uu] \tan \phi}{r}
$$
$$
+ f[u][v] + [v][F_y] + g[v] \frac{\partial [z]}{\partial y} = 0
$$

The pressure gradient term integrates out of the $[u]$ equation if mountain torque is neglected. That assumption is made here on the basis that zonal mountain torque could be parameterized by modifying the F_x friction term. Also, $dy = r \, d\phi$ where ϕ is latitude and r is the earth's radius. Two terms may be combined:

$$
\frac{[u]}{R} \frac{\partial R[uv]}{\partial y} - \frac{[u][uv] \tan \phi}{r} \equiv \frac{[u]}{R^2} \left(\frac{\partial R^2 [uv]}{\partial y} \right)
$$

Integrating the K_Z^x and K_Z^y equations over the mass of the fluid gives:

$$\frac{\partial}{\partial t} \int \underbrace{\left(\frac{[u]^2}{2}\right) dm}_{(A)} + \int \underbrace{\left(\frac{[u]}{R}\right)}_{(B)} \underbrace{\left(\frac{1}{R}\frac{\partial R^2[uv]}{\partial y} + \frac{\partial R[u\omega]}{\partial P}\right) dm}_{(C)}$$

$$- \int f[u][v] dm + \int [u][F_x] dm = 0 \qquad (4.12)$$

The integrand (A) is K_Z^x as in (4.10).

$$\frac{\partial}{\partial t} \int \underbrace{\left(\frac{[v]^2}{2}\right) dm}_{(D)} + \int [v] \underbrace{\left(\frac{1}{R}\frac{\partial R[vv]}{\partial y} + \frac{\partial [v\omega]}{\partial P}\right) dm}_{(E)}$$

$$+ \int \frac{[v][uu]\tan\phi}{r} dm + \int g[v]\underbrace{\frac{\partial [z]}{\partial y}}_{(G)} dm \qquad (4.13)$$

$$+ \int f[u][v] dm + \int [v][F_y] dm = 0$$

The integrand (D) is K_Z^y as in (4.10). Parts (B) and (C) have an *angular* weighting by R. Term (E) does not have the R weighting as in (B) because the $\tan\phi$ term has not been incorporated into (E). Term (G) is a pressure work term.

The Coriolis term appears in both (4.12) and (4.13), but it is merely acting to exchange energy between those components. The Coriolis force terms cancel when (4.12) and (4.13) are summed. They must cancel since the rotation of the earth cannot be a source of energy to the atmosphere.

Upon integrating (4.12) over the mass of the atmosphere, the second term can be expanded:

$$\int \frac{[u]}{R}\left(\frac{1}{R}\frac{\partial R^2[uv]}{\partial y} + \frac{\partial R[u\omega]}{\partial P}\right) dm \equiv \int \left(\frac{1}{R}\frac{\partial R[u][uv]}{\partial y}\right. \qquad (4.14)$$

$$\left. + \frac{\partial [u][u\omega]}{\partial P}\right) dm - \int \left([Ruv]\frac{\partial}{\partial y}\left(\frac{[u]}{R}\right) + [Ru\omega]\frac{\partial}{\partial P}\left(\frac{[u]}{R}\right)\right) dm$$

If the integration is over the whole atmosphere (a closed system), then the first right-hand side integral in (4.14) is zero since the integrand contains two perfect differentials. The $1/R$ term vanishes since $dm = -gr R dp d\phi d\lambda$ where λ is longitude. The kinetic energy equations (4.12) and (4.13) can be interpreted using either side of the expression in (4.14) depending upon which is more convenient. The left-hand side integrand is convergence of relative angular momentum weighted by an angular

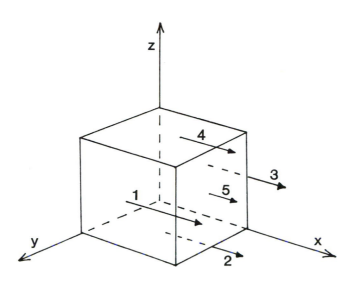

FIG. 4.4 Stress tensor notation. Stress applied along the top and bottom surfaces in the x direction, as indicated by arrows 4 and 2 respectively, contribute to τ_{xz}. Pressure force in the x direction is τ_{xx} and is labelled arrow 5. Stress tensors in the direction of arrows 1 and 3 contribute to τ_{xy}.

velocity. The right-hand side is the gradient of an angular velocity weighted by the momentum flux.

The observed vertical distribution of momentum flux looks schematically like Figure 4.5a. The convergence of the flux builds $[u]/R$ westerly momentum. Since this convergence is largest at an upper level, then the jet will be at an upper level (as in the observations; e.g., Figure 3.15). Figure 4.5 is a schematic illustration of the link between convergence and $[u]$. The last integral in (4.14) indicates that to change $[u]/R$, the eddy flux of angular momentum must go up or down the gradient of the angular velocity (up the gradient means from lower to higher values). The last integral in (4.14) equals zero at the maximum angular velocity.

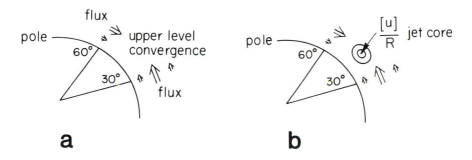

FIG. 4.5 Schematic diagrams of momentum flux and momentum convergence that maintains the mid-latitude jet streams.

Similarly, for meridional motion the second term in (4.13) can be expanded as

$$
\int [v] \left(\frac{1}{R}\frac{\partial R[vv]}{\partial y} + \frac{\partial [vw]}{\partial P} \right) dm \equiv \int \left(\frac{1}{R}\frac{\partial R[v][vv]}{\partial y} \right.
$$
$$
\left. + \frac{\partial [v][vw]}{\partial P} \right) dm - \int \left([vv]\frac{\partial [v]}{\partial y} + [vw]\frac{\partial [v]}{\partial P} \right) dm \tag{4.15}
$$

The interpretation here is essentially the same as for the $[u]/R$ term just discussed except that the factor of R is missing. A critical distinction here is that the left-hand sides of (4.14) and (4.15) are intuitively useful for understanding local contributions. The right-hand sides are tailored more for a global average description since the first integral on the right-hand side vanishes in that latter case.

The final form of the kinetic energy equation for the zonal mean motions is given by (4.16).

$$
\frac{\partial K_Z^x}{\partial t} = \int \left([Ruv]\frac{\partial}{\partial y}\left(\frac{[u]}{R}\right) + [Ruw]\frac{\partial}{\partial P}\left(\frac{[u]}{R}\right) \right) dm
$$
$$
+ \int f[u][v]dm - \int [u][F_x]dm = 0 \tag{4.16a}
$$

$$
\frac{\partial K_Z^y}{\partial t} = \int \left([vv]\frac{\partial [v]}{\partial y} + [Rvw]\frac{\partial [v]}{\partial P} \right) dm - \int \frac{[v][uu]\tan\phi}{r}dm
$$
$$
- \int g[v]\frac{\partial [z]}{\partial y}dm - \int f[u][v]dm - \int [v][F_y]dm = 0 \tag{4.16b}
$$

To derive the *total* kinetic energy equations one multiplies the u equation of motion (in advective form) by u and the v equation by v, bringing the terms inside the time integration and then taking the brackets average. The derivation is:

$$
\underbrace{\frac{\partial}{\partial t}\left(\frac{u^2}{2}\right) + u\frac{\partial}{\partial x}\left(\frac{u^2}{2}\right) \frac{v}{R}\frac{\partial}{\partial y}\left(\frac{Ru^2}{2}\right) + w\frac{\partial}{\partial P}\left(\frac{u^2}{2}\right)}_{(A)}
$$
$$
- \frac{u^2 v \tan\phi}{r} + gu\frac{\partial z}{\partial x} - fuv + uF_x = 0 \tag{4.17}
$$

Term (A) is rearranged by using the continuity equation in pressure coordinates and the chain rule:

$$
(A) = \frac{\partial}{\partial x}\left(\frac{uu^2}{2}\right) + \frac{1}{R}\frac{\partial}{\partial y}\left(\frac{Rvu^2}{2}\right) + \frac{\partial}{\partial P}\left(\frac{wu^2}{2}\right)
$$

The equation is now in flux form. The meridional component kinetic energy is similarly manipulated:

$$\frac{\partial}{\partial t}\left(\frac{v^2}{2}\right) + u\frac{\partial}{\partial x}\left(\frac{v^2}{2}\right) \underbrace{\frac{v}{R}\frac{\partial}{\partial y}\left(\frac{Rv^2}{2}\right) + \omega\frac{\partial}{\partial P}\left(\frac{v^2}{2}\right)}_{(B)}$$

(4.18)

$$+\ \frac{u^2 v \tan\phi}{r} + gv\frac{\partial z}{\partial y} + fuv + vF_y = 0$$

One notes that

$$(B) = \frac{\partial}{\partial x}\left(\frac{uv^2}{2}\right) + \frac{1}{R}\frac{\partial}{\partial y}\left(\frac{Rvv^2}{2}\right) + \frac{\partial}{\partial P}\left(\frac{\omega v^2}{2}\right)$$

Applying the zonal average yields:

$$\frac{\partial}{\partial t}\left[\frac{u^2}{2}\right] + \frac{1}{R}\frac{\partial}{\partial y}\left[\frac{Rvu^2}{2}\right] + \frac{\partial}{\partial P}\left[\frac{\omega u^2}{2}\right]$$

$$-\ \frac{[uuv]\tan\phi}{r} + g\left[u\frac{\partial z}{\partial x}\right] - f[uv] + [uF_x] = 0$$

$$\frac{\partial}{\partial t}\left[\frac{v^2}{2}\right] + \frac{1}{R}\frac{\partial}{\partial y}\left[\frac{Rvv^2}{2}\right] + \frac{\partial}{\partial P}\left[\frac{\omega v^2}{2}\right]$$

$$-\ \frac{[uuv]\tan\phi}{r} + g\left[v\frac{\partial z}{\partial y}\right] + f[uv] + [vF_y] = 0$$

Integrating over the entire mass of the atmosphere, the second and third terms vanish using $\omega = 0$ at the top and bottom on average. The first terms are just the rate of change of total kinetic energy

$$\frac{\partial K^x}{\partial t} = \int\left(\frac{[uuv]\tan\phi}{r} + f[uv]\right)dm - \int g\left[\frac{\partial z}{\partial x}\right] - \int[uF_x]dm = 0 \quad (4.19a)$$

$$\frac{\partial K^y}{\partial t} = -\int\left(\frac{[uuv]\tan\phi}{r} + f[uv]\right)dm - \int g\left[\frac{\partial z}{\partial y}\right] - \int[vF_y]dm = 0 \quad (4.19b)$$

The first integral is a conversion between K^x and K^y; it merely states that the Coriolis terms must cancel in the net summation; planetary rotation cannot be a source of energy.

The two equations in (4.19) indicate that the time rate of change of total kinetic energy is due to the work done by pressure forces and the work done by friction.

The component *eddy* kinetic energy equation is derived by subtracting (4.16) from (4.19). Component terms in (4.11) are combined in these steps.

$$\frac{\partial K_E}{\partial t} = -\int \left([Ruv]\frac{\partial}{\partial y}(\frac{[u]}{R}) + [Ruw]\frac{\partial}{\partial P}(\frac{[u]}{R}) + [vv]\frac{\partial [v]}{\partial y} + [vw]\frac{\partial [v]}{\partial P} \right.$$
$$\left. - \frac{[v][uu]\tan\phi}{r} \right) dm + \int g\left(\left[u'\frac{\partial z'}{\partial x}\right] + \left[v'\frac{\partial z'}{\partial y}\right] \right) dm$$
$$- \int ([u'F_x'] + [v'F_y'])dm = 0$$

The first integral converts kinetic energy between zonal and eddy components. The second integral is related to vertical overturnings. A closed system model is used in the section on available potential energy (§4.5.2) to elucidate these overturnings; they are conversions between kinetic and available potential energy. The last integral is frictional extraction.

The five parts of the first integral can be rearranged and combined. For example:

$$[Ruw]\frac{\partial}{\partial P}\left(\frac{[u]}{R}\right) = \underbrace{R[u][w]\frac{\partial}{\partial P}\left(\frac{[u]}{R}\right)}_{(A)} + \underbrace{R[u'w']\frac{\partial}{\partial P}\left(\frac{[u]}{R}\right)}_{(B)}$$

R is not a function of P, so it cancels out. Using the chain rule obtains:

$$(A) = [w]\frac{\partial}{\partial P}\left(\frac{[u]^2}{2}\right) = \underbrace{\frac{\partial}{\partial P}\left([w]\frac{[u]^2}{2}\right)}_{(C)} - \underbrace{\left(\frac{[u]^2}{2}\right)\frac{\partial[w]}{\partial P}}_{(D)}$$

Terms like (B) will contribute to the final K_E equation. Terms like (D) combine and cancel from the continuity equation. Terms like (C) combine to form:

$$\nabla_2 \cdot \left\{ \left(\frac{[u]^2 + [v]^2}{2}\right)[\mathbf{V}_2] \right\} \equiv (E)$$

where $\mathbf{V}_2 = (v, w)$ and $\nabla_2 = (\partial/\partial y, \partial/\partial P)$. Term (E) vanishes when integrated over the entire mass of the atmosphere. Term (E) is a two-dimensional perfect differential because longitude was removed from the problem by the brackets average.

When the eddy kinetic energy tendency is integrated over the entire mass of the atmosphere (dm), only squares of the primed quantities are left.

$$\frac{\partial K_E}{\partial t} = -\int \left(\underbrace{[Ru'v']\frac{\partial}{\partial y}(\frac{[u]}{R}) + [Ru'w']\frac{\partial}{\partial P}(\frac{[u]}{R})}_{(A)} \right.$$

$$
+ [v'v']\frac{\partial [v]}{\partial y} + [v'\omega']\frac{\partial [v]}{\partial P} - \underbrace{\frac{[v][u'u']}{r}}\tan\phi \Bigg) dm \tag{4.20}
$$

$$
\underbrace{\qquad\qquad\qquad\qquad\qquad\qquad\qquad}_{(B)}
$$

$$
- \int g\left(\underbrace{\left[u'\frac{\partial z'}{\partial x}\right] + \left[v'\frac{\partial z'}{\partial y}\right]}_{(C)}\right) dm - \int \underbrace{([u'F'_x] + [v'F'_y])}_{(D)}\, dm = 0
$$

4.2.2 Interpretation

The schematic features of the momentum budget are seen in the following observations. Figure 4.6 shows a calculation of the frictional torque exerted by the atmosphere on the surface of the earth, as determined from surface wind and pressure observations. Oort (1985) also notes that oceans exert a torque upon the solid earth in a manner similar to the atmospheric mountain torque; the dashed line in Figure 4.6 shows this "continental" torque. Negative continental torque is created by sea levels that slope upward from east to west. The atmospheric circulations impart frictional torque onto the oceans, which in turn transfer the torque to the solid earth. Consequently, the two curves in Figure 4.6 are quite similar. Both curves show the *atmosphere* gaining momentum in the tropics and losing momentum in the extratropics. To balance the gains, there must be large poleward transport of absolute angular momentum occurring in the subtropics of both hemispheres. For net balance by the earth, atmosphere, and oceans, the solid earth must transport angular momentum the opposite way. Oort (1985) proposes the controversial notion that the slipping of the crust along properly oriented fault lines could accomplish the task.

The time rate of change of kinetic energy is proportional to friction and to the convergence of momentum fluxes. From (4.14) and (4.15) the latter is proportional to the gradient of momentum transport weighted by the angular velocity. As detailed in the Appendix, many different physical phenomena contribute to a $[uv]$ momentum flux. Figure 4.7 illustrates the relative contributions by several classes of phenomena. The Appendix defines the notation.) The discussion of the observed momentum fluxes' contribution to zonal mean kinetic energy balance proceeds as follows. The fluxes by the mean meridional cells are discussed first, then the eddy fluxes, and finally both circulations are considered together.

4.2.2.1 Zonal mean cells transport

The contribution by the mean meridional cells has the opposite sign in the middle latitudes as it does in the tropics (Figure 4.7d). The sign reversal follows from the three-cell description of the mean meridional circulation. ($[u]$ is large and mainly westerly in the upper troposphere; at that altitude $[v]$ is poleward in the subtropics and equatorward in middle latitudes.) As mentioned in §3.3, the Hadley cell is much

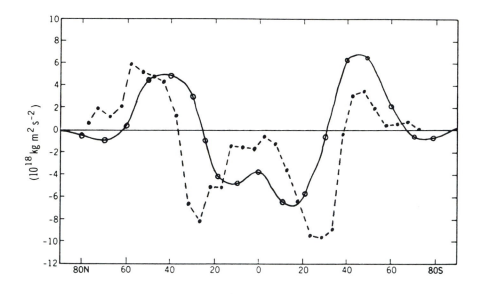

F**IG. 4.6**   The solid line is the net torque exerted by the atmosphere upon the earth, including mountain
torques and surface friction based on atmospheric observations. The dashed line is an estimate of the net
torque exerted by the oceans upon the solid earth—the continental torque. Diagram redrawn from Oort,
(1985).

stronger in the winter hemisphere. The poleward fluxes in the subtropics (Figure 4.7d)
reflect that seasonal change.

Figure 4.8 illustrates the vertical variation of the northward momentum flux
due to the mean meridional cells. The vertical variation is schematically deduced
in Figures 4.8a,b; confirming observations are shown in Figure 4.8c. The Northern
Hemisphere Hadley cell has $[v] < 0$ at low levels and $[v] > 0$ in the upper troposphere,
where $[u]$ is negative (easterly, shaded in Figure 4.8a). When a vertical average is
taken, there is some cancellation, but the low-level contribution is greater in the
tropics because $[u]$ is greater at low levels (Figure 3.14). Thus Figure 4.7d shows
northward (positive) transport. In the subtropics $[u]$ and $[v]$ are both positive in
the upper atmosphere and the vertical average transport is strongly poleward. The
Northern Hemisphere Ferrel cell has positive $[v]$ at low levels and negative $[v]$ above,
while $[u]$ is positive (westerly) throughout. Again there is some cancellation when a
vertical average is taken but the westerlies at the jet stream level are much stronger
than the low-level zonal wind. Hence, the vertical average is negative in midlatitudes
as seen in Figure 4.7d.

4.2.2.2   *Eddy transports*

(a) Momentum transport by the transient eddies (Figure 4.7b) is much greater than that
by the mean meridional circulations. Hence, eddies are the primary mechanism main-

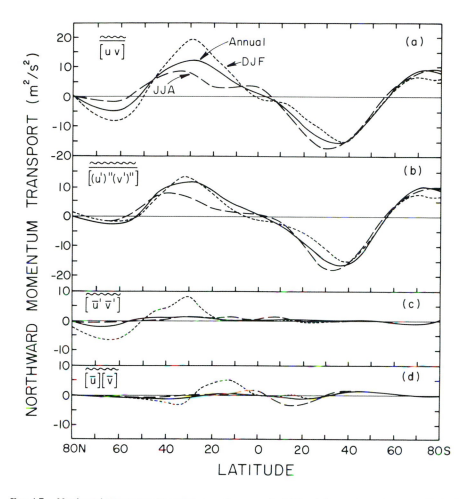

FIG. 4.7 Northward momentum transport: annual average (solid lines), June—August average (dashed lines), and December—February average (dotted lines). (a) All processes, (b) transient eddies, (c) stationary eddies, and (d) mean meridional cells. Figure redrawn from Oort and Peixoto (1983).

taining the zonal mean zonal flow. One could expect this. The velocity in the mean meridional cells ($[v]$) is an order of magnitude smaller than the typical v measured instantaneously at a point. So one anticipates that $[u][v] \ll [u'v']$, even though $[u]$ is large. (b) Transient eddy transports (Figure 4.7b) are less during summer, especially in the Northern Hemisphere. The eddy transports are greatest at about 30° latitude in both hemispheres. (c) The stationary wave flux (Figure 4.7c) is only significant in the Northern Hemisphere; it is quite large during winter. The standing wave pattern includes poleward transport in midlatitudes and equatorward transport at high latitudes. The former is seen in the characteristic shape of the long wave troughs. Those troughs are often oriented southwest to northeast, an orientation that creates poleward momentum flux (Figures 4.9, 5.9, 6.8, and 6.9). The equatorward momen-

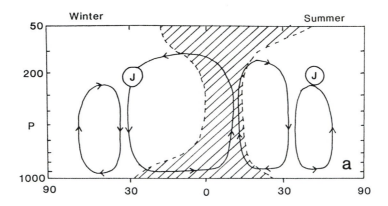

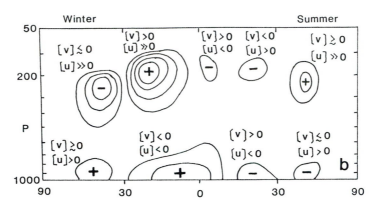

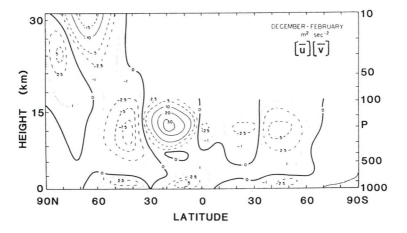

Fig. 4.8 Schematic illustration of the momentum transport by the mean meridional cells. (a) Locations of zonal mean easterly wind (hatching), circuits in the meridional plane of the Hadley and Ferrel cells (arrows) during summer and winter, and centers of the subtropical jets. (b) Local contributions of the mean meridional cells to the momentum flux. The vertical integral of data in (b) gives the values shown in Figure 4.7d. (c) Observed pattern from Newell et al. (1972).

tum flux at high latitudes can be understood from the typical life cycle of the frontal cyclones. As these storms grow, they move poleward as well as eastward. Eventually they often merge with the "semi-permanent" Aleutian and Icelandic Lows. At this point they decay, and one mechanism of decay (barotropic instability) has eddy momentum convergence. That convergence creates equatorward fluxes at high latitudes. (d) Classical theories about the general circulation (§1.3) completely neglected these eddy fluxes because eddies were not included. Those theories stated that the pressure forces, modified by the Coriolis effect, were the generators of zonal kinetic energy.

Even though the eddy velocities (u', v') have no zonal mean, the product of these velocities can have a nonzero zonal mean if the eddy has a special structure. Schematic diagrams (Figures 4.9 and 4.10) illustrate how the eddy momentum flux can have a zonal average contribution. For circular low and high pressure patterns (Figure 4.10), there is no net contribution to the zonal average eddy momentum flux even though the term may be locally large; the contributions to the zonal average on the east and west sides of the low (or high) cancel. The eddies must have asymmetry (Figure 4.9) to obtain a net zonal average eddy momentum flux. More precisely, there must be horizontal tilts of the trough and ridge axes. The horizontal tilts deform, in effect *rotate*, the location of maximum and minimum $u'v'$. A northwest to southeast tilt will give a southward flux $(u'v' < 0)$ in the middle latitudes by rotating the $u'v'$ pattern clockwise. A southwest to northeast tilt will give a northward flux $(u'v' > 0)$ by rotating the $u'v'$ pattern in Figure 4.10 counter clockwise. The direction of the flux is understood as an advection of eddy zonal momentum (u') by the eddy meridional velocity (v') in the meridional direction. Hence $u'v' > 0$ can occur with northward advection $(v' > 0)$ of positive zonal momentum $(u' > 0)$. Equivalently, southward advection $(v' < 0)$ of easterly momentum $(u' < 0)$ also is a northward flux of westerly zonal momentum. This interpretation follows simply because *the $u'v'$ terms in (4.20) originate from the meridional advection term in the zonal momentum equation.* Zonal averages are fundamental to the discussion here. If these terms were examined for *local* building or damping the jet, then one would need to retrieve the cross-product terms (e.g., $[u]v'$ type terms) that were neglected previously.

The eddy momentum fluxes, like the fluxes from mean meridional cells, are largest near the tropopause level. For the eddies, the maximum transport is located close to the maximum velocity found for the tropospheric jet streams described earlier. Figure 4.11 shows observed annual mean momentum fluxes by all circulations. The maximum transient eddy contribution to the total flux is about 8 to 10 times greater than that by meridional cells or standing eddies. Transient eddy fluxes are shown in Figure 4.12e and f. Inspection of Figures 4.8c, 4.11, and 4.12e and f shows that the Hadley cells reinforce the eddy fluxes but the Ferrel cells oppose the eddy fluxes in midlatitudes. This point is returned to in the theoretical explanation of the Ferrel cell given in §6.3.2.

At this stage some care is needed to account for the *angular* velocity being used. The cosine of latitude in R shifts the location of maximum $[u]/R$ poleward of the

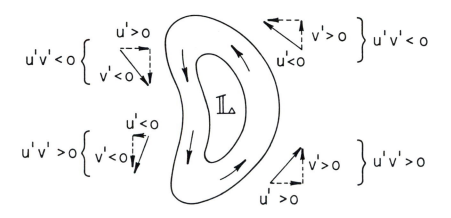

Fig. 4.9   Boomerang-shaped pressure pattern with associated geostrophic velocities. This shape has eddy momentum fluxes as indicated. The contributions to the zonal average eddy momentum flux on the east and west sides of a low (or high) are additive in contrast to Figure 4.10.

climatological jet stream position. Also, the cos ϕ is missing from the numerator of the eddy term displayed in Figure 4.11; correcting for that shifts the eddy momentum flux maximum slightly equatorward. With these adjustments, the momentum transport is directed up the gradient of the angular velocity, at least on the equatorward side of the jet. However, the match between the $[u]/R$ and momentum convergence fields is not exact.

Figure 4.12 shows other observations, based upon diagrams in Newell et al. (1970). R is not included. If one weights the momentum fluxes (Figures 4.12e and f) by $[u]/R$, then the maximum change of the mean kinetic energy will be near the 200 to 300 hPa layer and around 40 N and 30 to 50 S in the respective winter seasons. These resultant locations match the locations of the jet streams better than simply using the momentum convergence. These locations are poleward of where the eddy momentum convergence (a gradient of the quantity diagrammed) is greatest. There

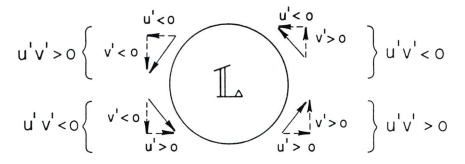

Fig. 4.10   For a circular low the east and west sides have eddy momentum fluxes that cancel when a zonal average is taken.

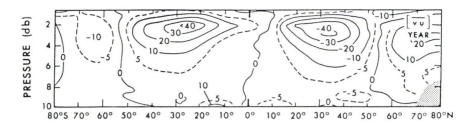

FIG. 4.11   Meridional cross-section of the zonal mean of the northward transport of eastward momentum by all circulations. Ten years of data (1963–1973) contributed to this annual mean. Redrawn from Oort and Peixóto, (1983).

are two matters to consider and §6.3 and §6.5 return to this subject. First, the eddy momentum and heat fluxes occur in a way that destroys thermal wind balance. To regain the balance, a Ferrel cell motion is set up that shifts the position of the jet equatorward from the location of maximum momentum convergence (§6.3). Second, Figure 4.10 shows that the subtropical jets seem to lie at the boundary between the Ferrel and Hadley cells. In §6.5 the linkage between jets and meridional cells is shown to more subtle. The full explanation of these mechanisms requires further background and that explanation is left to §6.3 and §6.5.

   The discussion so far has centered around the observed sources of the meridional flux of zonal momentum ($[uv]$). This flux is an important element in the maintenance of the zonal mean jet streams. The zonal and eddy kinetic energy can be changed by other processes. The eddy kinetic energy equation (4.20) groups the processes into four categories. The zonal mean kinetic energy equation can be similarly grouped. The four groups in (4.20) are specifically interpreted next.

- *Term (A)* contains fluxes of eddy momentum weighted by the zonal mean angular zonal velocity gradient. Term (A) is a conversion between different forms of kinetic energy; the same term but with opposite sign is found in (4.16).

- *Term (B)* contains fluxes of eddy momentum weighted by the mean meridional velocity gradient. Terms (A) and (B) together are sometimes labelled the "barotropic" energy conversion terms. Since $[v] \ll [u]$ and $[v'v'] \sim [u'v']$, (B) tends to be smaller than (A). The same term, but with opposite sign, occurs in (4.16).

- *Term (C)* is comprised of work done by pressure forces. The correlations of eddy velocity and eddy pressure gradient are sometimes categorized as "baroclinic" terms. The term is conversion between potential energy and eddy kinetic energy, so a related term is found in the available potential energy equation. To visualize the role played by this term, one must understand the concept of available potential energy. Thus, term (C) is examined in much greater detail in §4.5.2.

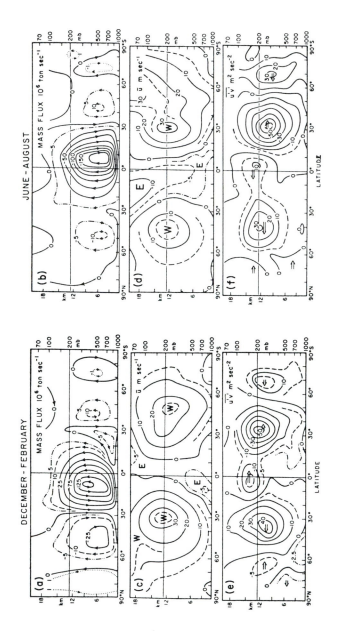

FIG. 4.12   Meridional cross-sections relating the Hadley and Ferrel cells (a) and (b), to the zonal mean zonal wind (c) and (d) and to the eddy momentum flux (e) and (f) during the two extreme seasons. From *Meteor. Mono.*, Newton (1972), by permission of American Meteorological Society.

- *Term (D)* includes the net loss of kinetic energy due to various frictional processes.

Terms (A) and (B) in (4.20) can either be a source or sink. The variation in the sign of these terms will be described in more detail when baroclinic and barotropic instability theory are considered over limited longitudinal segments. The pressure work terms (C in 4.20) can be a source or a sink as well. The frictional dissipation term (D) is always a sink. It is essentially a leakage of energy down in scale to turbulent length scales.

The subject of angular momentum is returned to in §6.2 when more recent theoretical ideas about the general circulation are discussed. The reader need not complete this chapter in order to follow the discussion in §6.2. This section briefly considered kinetic energy; the next section examines the role and definition of potential energy. After that, the linkage between kinetic and potential energy is considered.

4.3   Available potential energy

4.3.1   *Introduction*

Ultimately, differential solar heating sets up the atmospheric potential energy. Yet not all the potential energy is usable to drive atmospheric and oceanic motions; i.e., not all is convertible to kinetic energy. That which is convertible will be labeled "available potential energy" or APE.

It is appropriate to place the derivation and discussion of APE in a chapter dealing with observations, because APE is principally a diagnostic tool. While it is commonplace to interrogate the energy conversions that occur in theoretical models, the concept of available potential energy is derived here without the benefit of dynamical equations of motion. Instead, the derivations performed here use the equations: continuity, ideal gas, hydrostatic (or non-hydrostatic) and first law of thermodynamics.

Knowledge of the APE can not directly specify the type of circulation that would be expected. However, theoretical studies use APE implicitly to anticipate when a given flow will be (baroclinically) unstable to small disturbances, and APE enters into the equations used to specify the structure of the unstable modes. The problem is that several situations may be imagined that give the same APE but have quite different baroclinic instabilities. A simple example would be to imagine two states: both have the same temperature field but the rotation rate of the earth would be different. The APE would be the same in both cases, but the instability would be different. Theories in the quasi-geostrophic framework do not have a problem here, because temperature, rotation rate, and velocity must be linked; so that changing the rotation rate also changes the velocity field, and the quasi-geostrophic APE is not the same in the two states.

Dutton and Johnson (1967) attempted to link APE in a more general framework to the dynamics. Their desire was to use APE knowledge to forecast preferred atmospheric circulations. They base their analysis upon a "least action principle" that employs a variational calculus technique to find the solution to a fluid flow problem that minimizes an integral quantity. The integral is taken over the whole atmosphere and is thus subject to surface boundary conditions and is also a function of time (hence there is some hysteresis built into the procedure). The integrand being minimized is the difference between two quantities: a form of kinetic energy that includes the earth's angular momentum and a potential energy (made up of internal and gravitational parts). Unfortunately, they could not complete the solution, but they do discuss related problems and propose ways in which minimizing that integral could explain general features found in the laboratory simulations (§7.1.2) and observations of the general circulation. Therefore, the derivations made here are higher-level diagnostic tools to interpret the observations, but it should be apparent that energy conversions also bridge the observational and theoretical studies.

Expressions for APE can be derived in several ways. Margules formulated the original idea in a 1903 paper on the energetics of cyclonic storms. He idealized such storms by assuming that they are contained within a region bounded by vertical walls (i.e., a closed system). He thus ignored transports across the boundaries, though the transports are probably significant in the problem he was studying. Interestingly, he noted that sometimes the whole atmosphere could be treated as a closed system. This is how APE is generally considered today. It is of course possible to derive APE and kinetic energy equations for limited areas without further assumption; §4.5 does that.

The most well-known development is by Lorenz (1955). Lorenz's approach relies heavily on an assumption of hydrostatic balance. It is relatively easy to follow conceptually and a derivation follows shortly. Another approach is by Van Mieghem (1956, 1957). Contrary to Lorenz, Van Mieghem allows non-hydrostatic effects; but his method suffers from needing the velocity at the instant the reference state (of relative minimum potential energy) is reached. It requires knowledge of the reference state to calculate the reference state. It may be possible to find reference states by iterative means. A definitive paper on the subject is by Dutton and Johnson (1967; hereafter D and J). D and J derive expressions that are "more exact" than the "exact" expressions obtained by Lorenz because the hydrostatic assumption is not made. Analysis of their expressions provides insight into the generation and conversion of energy in the atmosphere. Conversion between kinetic and APE needs cross isobaric motion (non-zero horizontal pressure advection) *or* vertical advection of the non-hydrostatic part of the pressure field. Obviously, both processes are *non-geostrophic*, and Lorenz's equations include only the former. D and J even point out some pitfalls with using the approximate expressions also derived by Lorenz (and used by some scientists for diagnostic calculations with observed data). Nonetheless, the approximate expressions are more intuitively understandable, and are well worth discussing.

4.3.2 General comments

A simple model for APE and kinetic energy is the motion of a pendulum. At the top of its arc a pendulum is motionless, so its kinetic energy is zero. Since it is at its highest point, its potential energy is a maximum. As the pendulum swings through the bottom of its arc, it is moving at its fastest speed. Thus, the kinetic energy is greatest at the bottom of the arc and the potential energy is least (but not zero) there. The difference in potential energy between the top and bottom of the arc is that energy available to drive the motion of the pendulum; it is the available potential energy.

The second simple model consists of two fluids in a container separated by a thin wall as shown in Figure 4.13a. The fluids are immiscible and have different densities with $\rho_2 > \rho_1$. When the barrier is removed from a non-rotating system, the fluid accelerates until reaching the state indicated in Figure 4.13c. In Figure 4.13c the kinetic energy is a maximum, and has resulted from the lowering of the center of mass of the system. However, no further lowering can take place, so this is a minimum potential energy. Thus, the energetics are analogous to that for the pendulum example. The difference in the potential energy between the initial state and that in Figure 4.13c is the energy available for conversion to kinetic energy. In other words, the difference is the APE of the initial state. Margules used a similar model involving two ideal gases of differing potential temperature; Gill (1982; pp. 223–225) summarizes the results.

As pointed out by Dutton and Johnson (1967), when mass and specific entropy are conserved, then the sum of internal, (gravitational) potential, and kinetic energies is a constant. The pendulum example illustrates this conservation: as the fluid motion *evolves*, a minimum state of potential energy corresponds to a maximum in kinetic. Typically, the APE is derived as a difference between the current state and a "reference state" from which no energy is available to drive further atmospheric motions. An important point here is that the kinetic energy in these simple models results from the evolution of the fluid toward a reference state. The velocity obtained when the reference state is reached depends upon the initial state of the fluid. The reference state in these simple analogies could have almost any velocity, depending on the initial state. The reference state could be motionless. The pendulum could be placed at the bottom of the arc and with no motion. The two fluids could be motionless and set in the orientation of Figure 4.13c. *The motion discussed here comes from conversion of some potential energy present at the start*; this concept holds for creating atmospheric motions, but there is a caveat.

The reference state for the atmosphere *must* be motionless because of the crucial difference that the atmosphere is on a rotating planet. The state of minimum potential energy for the atmosphere will be where the pressure (P) and potential temperature (θ) surfaces have constant elevation. In the momentum equations, Coriolis terms result from the rotating coordinate frame; they would be unbalanced by any other term if the velocity were not zero. For example, one might suppose that the local time rate

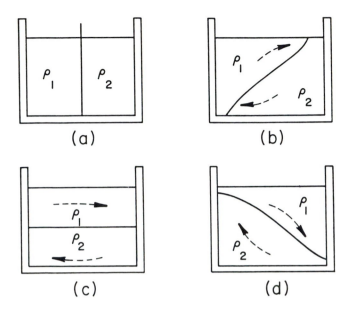

FIG. 4.13   A schematic illustration of available potential energy and its relationship to kinetic energy. (a) Initial state: a barrier separates immiscible fluids of differing density, ρ, where $\rho_2 > \rho_1$. The barrier is removed and gravity accelerates the flow (b) until maximum velocity (maximum kinetic energy) is reached (c). At (c) the potential energy is a minimum, but not zero. The change in potential energy between (a) and (c) is the available potential energy. (d) The current overshoots, the deceleration converts kinetic into potential energy.

of change of velocity would balance the Coriolis terms in the horizontal equations of motion. On a sphere, that would create an inertial oscillation whose frequency varies with latitude and whose meridional and zonal velocities would vary from solid body rotation. From the continuity equation, the vertical velocities would no longer be zero. Consequently, the resultant vertical advections would move the atmosphere away from the state of minimum potential energy.

While the atmospheric reference state of minimum potential energy must be at rest, there is no unique choice for the reference state. A wide range of possible vertical profiles of density could be chosen. The ultimate state of minimum potential energy would be where the entire atmosphere condensed out, but such a state is not more enlightening to use than, say, an isothermal atmosphere with a temperature of $250\,\mathrm{K}$.

Since the minimum reference state for the atmosphere is motionless, one may wonder how the inverse relationship between kinetic and available potential energy can still hold. Formally, it is possible to imagine atmospheric reference states that have non-zero horizontal velocities and yet still satisfy the definitions of APE to be used here. In §4.3.3 and §4.4 the APE is defined using a "reference pressure" (P_r) that is the average pressure on a θ surface. The P and θ lines could be parallel as in Figure 4.14. The horizontal pressure gradient balances the Coriolis terms (geostrophic balance). In Figure 4.14 the velocities are uniform; there are no vertical or horizontal

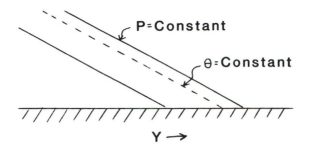

FIG. 4.14 Schematic meridional cross-section showing potential temperature (θ) surfaces coincident with pressure (P) surfaces. Since $P = P_r$ on each θ surface, then APE is a minimum from equation (4.26); $\epsilon = 0$. Since the isobaric surfaces have a meridional slope, this atmospheric state has zonal winds.

shears (solid body rotation on a sphere). Such a reference state might be defined by converting all the APE (estimated using a reference state at rest) into a kinetic energy for a globally uniform zonal velocity. Gill (1982; p. 225) caculates a globally uniform speed of about 30 m/s this way. If topography were included, for example, then shears would likely be necessary to find a stationary flow (§7.2.1). Alternatively, a reasonable reference state might be a steady-state solution to the primitive equations forced by a zonal distribution of heating (Van Mieghem, 1956, 1957). Figure 4.14 also shows that the pressure at every point on each θ surface equals P_r and thus the APE would be zero when the atmosphere reached the chosen reference state; $APE = 0$ in (4.22) and (4.26). One final caveat here is that the adjustment of the isentropes may be impossible with thermal wind balance specifying both temperature and density gradients (Fjørtoft, 1960).

The average amount of radiational heating does not increase the APE; instead, it is the *differential* heating that changes APE. The two-fluid example (Figure 4.13) illustrates this point. The motion will clearly be stronger if the density difference between the two fluids is greater; the motion will not be changed by changing the density of both fluids by the same amount. For the atmosphere, the main way to create APE is to cool the air where it is cold and/or heat it where it is hot. Clearly, any heating generates or destroys APE, depending upon whether the region was already warmer or cooler than the surroundings.

4.3.3 Derivation of available potential energy

The total potential energy (PE) in a column is the sum of gravitational potential (PG) and internal (I) energies.

$$PG = \int_0^\infty \rho g z \, dz = \int_0^{P_0} z \, dP$$

where the hydrostatic law has been used. P_0 is the pressure at $z = 0$. Using the chain rule and ideal gas law, this can be rewritten as

$$\int_0^{P_0} z\,dP = \int_0^{\infty} P\,dz = \int_0^{\infty} \rho RT\,dz$$

Internal energy is defined as:

$$I = \int_0^{\infty} \rho C_v T\,dz$$

Combining and using the hydrostatic law again gives

$$PG + I = \frac{C_p}{g} \int_0^{P_0} T\,dP \equiv PE \tag{4.21}$$

Hydrostatic balance is implicit here because the integral was converted from volume (using height limits) into pressure limits by the hydrostatic law. I is the internal energy in a vertical column.

Using $T = \theta P_{00}^{-\kappa} P^{\kappa}$ where P_{00} is 1000 hPa and $\kappa = R/C_p$,

$$PG + I = \left(\frac{C_p}{g P_{00}^{\kappa}}\right) \int_0^{P_0} \theta P^{\kappa}\,dP$$

$$= \left(\frac{C_p}{g(\kappa + 1) P_{00}^{\kappa}}\right) \int_0^{P_0^{\kappa+1}} \theta\,dP^{\kappa+1}$$

Integrating by parts,

$$PG + I = \frac{C_p}{g(\kappa + 1) P_{00}^{\kappa}} \left(\int_{@P=0}^{@P=P_0} d(\theta P^{\kappa+1}) - \int_{@P=0}^{@P=P_0} P^{\kappa+1}\,d\theta \right)$$

The integration limits refer to evaluating the integration variable at a level where pressure P equals some value.

Though it is not physically sensible, the following identity is useful.

$$\theta_0 \equiv \int_0^{\theta_0} d\theta$$

Essentially this is an integration underground, assuming $\theta = 0$ somewhere inside the earth. θ_0 is the potential temperature at the earth's surface where $P = P_0$. The definition

$$\int_0^{\theta_0} P_0^{\kappa+1}\,d\theta \equiv \theta_0 P_0^{\kappa+1}$$

is useful, where zero subscripts refer to the $P = P_0$ values. The identities assume P_0 is constant outside of the atmosphere (from $\theta = 0$ to $\theta = \theta_0$). Since $\theta = \infty$ at $P = 0$, then

$$
\begin{aligned}
PE = PG + I &= \left(\frac{C_p}{g(\kappa + 1)P_{00}^\kappa} \right) \Bigg\{ \left(\theta_0 P_0^{\kappa+1} = \int_0^{\theta_0} P^{\kappa+1} d\theta \right) \\
&+ \left(-\int_{@P=0}^{@P=P_0} P^{\kappa+1} d\theta = \int_{\theta_0}^{\infty} P^{\kappa+1} d\theta \right) \Bigg\} \\
&= \left(\frac{C_p}{g(\kappa + 1)P_{00}^\kappa} \right) \int_0^{\infty} P^{\kappa+1} d\theta
\end{aligned}
$$

This is equation (108) in Lorenz (1967). It is important to understand that the integrand varies in the atmosphere and this term does not contribute to potential energy "below" the atmosphere.

The minimum possible PE is that remaining after the atmosphere has been brought, parcel by parcel, adiabatically to the minimum possible state. That state is where the density, pressure, and potential temperature surfaces all coincide with the geopotential field. All of the variables are constant on concentric surfaces. This means, for example, that pressure will be constant on a constant potential temperature surface for the reference state. Equivalently, the pressure anywhere on a given θ surface will equal the average pressure on that θ surface. Mathematically,

$$
PE \Bigg)_{min} = \frac{C_p}{g(\kappa + 1)P_{00}^\kappa} \int_0^{\infty} (\overline{P}^\lambda)^{\kappa+1} d\theta
$$

where $\overline{P}^\lambda$ is the average pressure for a θ level. The overbar with superscript λ means a global "horizontal" average on a θ surface. Therefore, the total atmospheric available potential energy is:

$$
APE = \frac{C_p}{g(\kappa + 1)P_{00}^\kappa} \int_0^{\infty} \left(\overline{P^{\kappa+1}}^\lambda - (\overline{P}^\lambda)^{\kappa+1} \right) d\theta \tag{4.22}
$$

This is the so-called "exact" formula of Lorenz (1955).

The integrand term $\overline{P^{\kappa+1}}^\lambda$ is written using $\hat{P}$ and $\overline{P}^\lambda$ contributions where $\hat{P}$ are local deviations of P from $\overline{P}^\lambda$. Assuming that $\hat{P} \ll \overline{P}^\lambda$, one expands the resulting $\hat{P}/\overline{P}^\lambda$ powers as a Taylor series summation.

$$
\begin{aligned}
\overline{P^{\kappa+1}}^\lambda &= (\overline{P}^\lambda)^{\kappa+1} \overline{\left(1 + \frac{\hat{P}}{\overline{P}^\lambda} \right)^{\kappa+1}}^\lambda \\
&= (\overline{P}^\lambda)^{\kappa+1} \overline{\left(1 + (\kappa + 1)\left(\frac{\hat{P}}{\overline{P}^\lambda} \right) + \frac{(\kappa + 1)\kappa}{2} \left(\frac{\hat{P}}{\overline{P}^\lambda} \right)^2 + \cdots \right)}^\lambda
\end{aligned}
$$

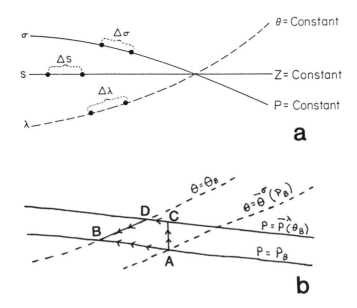

Fig. 4.15 Schematic diagrams to illustrate the geometry of the three isosurfaces used in the APE derivation. (a) Notation definition, (b) Two alternate paths to follow between points A and B that lead to equation (4.24).

Therefore, the integrand of (4.22) becomes

$$\overline{P^{\kappa+1}}^{\lambda} - (\overline{P}^{\lambda})^{\kappa+1} = (\overline{P}^{\lambda})^{\kappa+1}\left(\frac{(\kappa+1)\kappa}{2}\overline{\left(\frac{\hat{P}}{\overline{P}^{\lambda}}\right)^2}^{\lambda} + \cdots\right)$$

where $\overline{\hat{P}}^{\lambda} = 0$ by definition. Lorenz finds that $(\hat{P}/\overline{P}^{\lambda})^2 = 0.075$ and the next term $(\hat{P}/\overline{P}^{\lambda})^3 = -0.019$ from observations. When observed values are plugged in, the second term left in the series is only about 6% of the first term and is neglected. Van Mieghem (1956), who used a more rigorous approach, found that this approximation only resulted in about a 10% error. Thus, an approximate formula is derived:

$$APE = \left(\frac{\kappa C_p}{2gP_{00}^{\kappa}}\right)\int_0^{\infty}(\overline{P}^{\lambda})^{1+\kappa}\overline{\left(\frac{\hat{P}}{\overline{P}^{\lambda}}\right)^2}^{\lambda}\,d\theta \qquad (4.23)$$

where the overbar indicates an average over a potential temperature surface.

One can reformulate (4.23) into a temperature version. Figure 4.15a will define the notation; dS is a horizontal distance. Figure 4.15b shows two paths between points A and B. Points A and B are on the isobaric surface $P = P_B$. Points C and D lie on the isobaric surface $P = \overline{P}^{\lambda}(\theta_B)$ corresponding to the average pressure of the θ

surface (θ_B) passing through the point B. Point C and A have the same location S. The change in θ following each path is as follows. For the path $A \Rightarrow B$:

$$\Delta\theta_1 = \int_A^B \frac{\partial\theta}{\partial\sigma}d\sigma = \theta_B - \bar{\theta}^\sigma(P_B)$$

For the path $A \Rightarrow C \Rightarrow D \Rightarrow B$:

$$\Delta\theta_2 = \int_A^C \frac{\partial P}{\partial z}\frac{\partial\theta}{\partial P}dz + \int_C^D \frac{\partial P}{\partial\sigma}\frac{\partial\theta}{\partial P}d\sigma + \int_D^B \frac{\partial\theta}{\partial\lambda}d\lambda$$

The last integral vanishes because the path is on an isentropic surface. Since the P and θ surfaces are approximately horizontal, then the static stability is approximated by it's average on a pressure surface.

$$\Delta\theta_2 = \overline{\frac{\partial\theta}{\partial P}}^\sigma \left(\int_A^C \frac{\partial P}{\partial z}dz + \int_C^D \frac{\partial P}{\partial\sigma}d\sigma\right)$$

The second integral vanishes since the path is on an isobaric surface. The changes in θ are equal, leaving

$$\theta_B - \bar{\theta}^\sigma(P_B) = -\overline{\frac{\partial\theta}{\partial P}}^\sigma \left\{P_B - \bar{P}^\lambda(\theta_B)\right\}$$

Defining the notation $\tilde{\theta}$ for the deviation of θ from its mean on a P surface, then

$$\tilde{\theta} = -\overline{\frac{\partial\theta}{\partial P}}^\sigma \hat{P} \tag{4.24}$$

$\Im$ designates the integral in (4.23). Substituting (4.24) into $\Im$ yields:

$$\Im \cong \int_0^\infty (\bar{P}^\lambda)^{\kappa+1}\left\{\overline{\frac{\tilde{\theta}^2}{(\bar{P}^\lambda)^2\left(\overline{\frac{\partial\theta}{\partial P}}^\sigma\right)^2}}^\lambda\right\}d\theta$$

Some of the averages over λ are approximated by averages over σ. The average over P of static stability is replaced by an average over a θ surface to get

$$\Im \cong \int (\bar{P}^\sigma)^{\kappa-1}\overline{(\tilde{\theta})^2}^\sigma \left(\overline{\frac{\partial\theta}{\partial P}}\right)^{-1\sigma}\overline{\frac{\partial P}{\partial\theta}}^\lambda d\theta$$

Noting that

$$\overline{\frac{\partial P}{\partial\theta}}^\lambda d\theta = dP$$

then

$$APE \cong \frac{C_p \kappa}{2g P_{00}^\kappa} \int_0^{P_0} (\overline{\theta^\sigma})^2 (\overline{P^\sigma})^{\kappa-1} \left(-\overline{\frac{\partial \theta}{\partial P}}^\sigma \right)^{-1} \overline{\left(\frac{\tilde{\theta}}{\overline{\theta^\sigma}} \right)^2}^\sigma dP$$

From the Poisson relation and the hydrostatic relation,

$$\frac{\partial \theta}{\partial P} = \left(\frac{P_{00}}{P} \right)^\kappa \frac{\partial T}{\partial P} - \frac{T \kappa P_{00}^\kappa}{P^{\kappa+1}}$$

$$= \frac{\theta}{T} \frac{\partial T}{\partial P} - \frac{\kappa \theta}{P} = -\frac{\theta}{T \rho g} \frac{\partial T}{\partial z} - \frac{\kappa \theta}{P}$$

Since the adiabatic lapse rate $\gamma_d = g/C_p$ and defining $\gamma = -\partial T/\partial z$, then

$$\overline{\frac{\partial \theta}{\partial P}}^\sigma = -\overline{\frac{\kappa \theta}{P} \left(1 - \frac{P}{\rho T R} \frac{\gamma}{\gamma_d} \right)}^\sigma = -\overline{\frac{\kappa \theta}{P} \left(\frac{\gamma_d - \gamma}{\gamma_d} \right)}^\sigma$$

from the ideal gas law. Using Poisson's equation for $\tilde{\theta}$ and θ obtains, after some rearranging,

$$APE \cong \frac{1}{2} \int_0^{P_0} \overline{T}^\sigma (\gamma_d - \gamma)^{-1} \overline{\left(\frac{\tilde{T}}{\overline{T}^\sigma} \right)^2}^\sigma dP \tag{4.25}$$

This is the approximate formula (110) in Lorenz (1967).

Equation (4.23) shows that pressure departures on a theta surface or, from (4.25), temperature departures on a pressure surface are needed in order to have any usable potential energy. In general terms, when hot areas are being heated and cold areas cooled, APE is generated since that builds departures of θ and P from their mean values. The need for departures is evident in observations already shown. In the summer the solar heating is more uniformly distributed with latitude than during winter (Figure 3.8). Alternatively, it could be said that the solar heating of the tropics is about the same all year, but the polar regions have much greater long-wave cooling during winter. Stronger horizontal temperature gradients in midlatitudes are set up during winter, so there is accordingly more APE. In turn, it is not surprising that the winter hemisphere has more vigorous circulations.

As stated previously, only a small fraction of the atmosphere's reservoir of potential energy is available to drive atmospheric motions. Indeed, Lorenz derived (4.25) in order to estimate that fraction, since total potential energy is easily expressed in terms of temperature. The fraction is on the order of 0.1 to 0.2%. In Table 4.1, the amount of available potential energy (APE) and total potential energy (PE) are tabulated by month and hemisphere. The global average APE found by Dutton and Johnson (1967) agrees well with the estimate 4.5×10^6 J m^2 by Price (1975). While the total potential energy is constant on a global average, it is slightly larger in the

Table 4.1 Selected monthly values of PE, APE, and standard deviation of the daily values of APE ($= \sigma_{APE}$).

	Northern Hemisphere		
Month	**APE** (10^6 Jm^2)	σ_{APE} (10^5 Jm^2)	**PE** (10^9 Jm^2)
January	4.34	7.83	2.51
March	4.25	4.5	2.53
May	2.72	4.99	2.56
July	1.57	3.2	2.58
September	2.3	4.52	2.57
November	4.15	7.87	2.53
Average:	3.3		

	Southern Hemisphere		
Month	**APE** (10^6 Jm^2)	σ_{APE} (10^5 Jm^2)	**PE** (10^9 Jm^2)
January	2.76	3.84	2.56
March	2.75	4.14	2.56
May	4.18	4.55	2.53
July	4.41	5.29	2.52
September	4.44	7.26	2.52
November	3.15	5.52	2.54
Average:	3.7		

	Entire Globe		
Month	**APE** (10^6 Jm^2)	σ_{APE} (10^5 Jm^2)	**PE** (10^9 Jm^2)
January	3.86	5.64	2.53
March	3.7	3.75	2.54
May	3.65	2.64	2.54
July	3.46	2.97	2.54
September	3.74	4.77	2.54
November	3.83	6.18	2.53
Average:	3.79		

Source: Data from Dutton and Johnson (1967).

summer hemisphere than in the winter. This arises because the hemisphere is warmer in summer than in winter. However, the available potential energy has the opposite, and *proportionally*, much larger variation. APE is much larger in winter than in summer. Again, this occurs because the temperature (and other) gradients are smaller in summer than in winter. The emphasis here is that the strong cooling near the winter pole creates the sharper gradients that cause greater APE in winter and ultimately, greater cyclonic storm activity during that season.

4.4 Limited volume energy generation equations

Greater insight into APE is gained by examining the processes that create and destroy it. Because latitude and height variations (and soon longitude variations) are used in this book, it is best to derive appropriate equations for a limited area. Several authors have derived limited area APE budget equations: Smith (1969), Johnson (1970), Vincent and Chang (1973), Smith et al. (1977), and Edmon (1978). The derivation given in Smith (1969) is used for a fixed volume in isobaric coordinates. Edmon (1978) pointed out a deficiency of this system in that two of the resulting terms are very large and nearly compensating. Edmon also showed how this problem could be avoided and that change is incorporated into the discussion. Retain the hydrostatic assumption and rewrite (4.21):

$$PE = C_p \int_0^\infty \int_S T\rho dSdz = C_p \int_M TdM$$

which also defines the mass of the atmosphere, M. Using Poisson's relation,

$$PE = C_p \int_M \left\{ \frac{\theta P^\kappa}{P_{00}^\kappa} \right\} dM$$

and the minimum is given by

$$(PE)_{min} = C_p \int_M \left\{ \frac{\theta P_r^\kappa}{P_{00}^\kappa} \right\} dM$$

Dutton and Johnson (1967) show that an indefinite integral defines P_r as the global average P on a θ surface.

$$P_r = \frac{1}{S} \int_S \int_{\theta_{top}}^\theta \frac{\partial P}{\partial \theta} d\theta dS$$

Therefore, $APE = PE - PE)_{min}$ and

$$APE = C_p \int_M \left\{ \left(\frac{P^\kappa - P_r^\kappa}{P^\kappa} \right) T \right\} dM \tag{4.26}$$

where $\kappa = R/C_p$. One might have expected σ to be used, but it is more convenient to use S in the definition of P_r since $\partial S/\partial t = 0$ is assumed later.

Using the hydrostatic relation, the APE is partitioned into limited volumes, each with mass M_j.

$$APE = \sum_j A_j = \sum_j \left\{ C_p \int_{M_j} \left(\frac{P^\kappa - P_r^\kappa}{P^\kappa} \right) TdM \right\}$$

$$= \sum_j \left\{ \frac{C_p}{g} \int_S \int_{P_2}^{P_1} \left(\frac{P^\kappa - P_r^\kappa}{P^\kappa} \right) TdPdS \right\}$$

Taking a local time derivative and using Leibnitz's rule,

$$
\frac{\partial A_j}{\partial t} = \frac{C_p}{g} \int_S \int_{P_2}^{P_1} \left\{ \frac{\partial T}{\partial t} - P_r^\kappa \frac{\partial}{\partial t}\left(\frac{T}{P^\kappa}\right) - \frac{T}{P^\kappa}\frac{\partial P_r^\kappa}{\partial t} \right\} dP dS
$$
$$
+ \frac{C_p}{g} \int_S \left\{ \left(\frac{P_1^\kappa - P_r^\kappa}{P_1^\kappa}\right) T_1 \frac{\partial P_1}{\partial t} - \left(\frac{P_2^\kappa - P_r^\kappa}{P_2^\kappa}\right) T_2 \frac{\partial P_2}{\partial t} \right\} dP dS
$$

where it is assumed that S is not a function of time and $P_2 < P_1$. If one were to integrate between fixed pressure levels (away from the earth's surface, for example), then the boundary integral would vanish, but it is retained for generality. Substituting (θ/P_{00}^κ) for (T/P^κ) and expanding the local time derivative using the total derivative definition and the chain rule obtains:

$$
\frac{\partial A_j}{\partial t} = \frac{C_p}{g} \int_S \int_{P_2}^{P_1} \left\{ \frac{dT}{dt} - \mathbf{V}_P \cdot \nabla_P T - \omega \frac{\partial T}{\partial P} \right.
$$
$$
- \left(\frac{P_r}{P_{00}}\right)^\kappa \left(\frac{d\theta}{dt} - \mathbf{V}_P \cdot \nabla_P \theta - \omega \frac{\partial \theta}{\partial P}\right) \tag{4.27}
$$
$$
\left. - \left(\frac{\theta}{P_{00}^\kappa}\right)^\kappa \left(\frac{dP_r^\kappa}{dt} - \mathbf{V}_P \cdot \nabla_P P_r^\kappa - \omega \frac{\partial P_r^\kappa}{\partial P}\right) \right\} + boundary\ integral
$$

where the subscript "P" refers to components on a constant pressure surface. The next three steps are as follows. The first law of thermodynamics is:

$$
\frac{dT}{dt} = -\frac{P}{C_v}\frac{d\alpha}{dt} + \frac{q}{C_v}
$$

where α is specific volume ($\equiv 1/\rho$) and q is the heating rate per unit mass. Plugging in the ideal gas law and rearranging yields

$$
C_p \frac{dT}{dt} = \frac{RT}{P}\frac{dP}{dt} + q = \omega\alpha + q. \tag{4.28}
$$

This expression is substituted for the dT/dt term in (4.27). In the second step, the advection terms in (4.27) are written in flux form by application of the continuity equation. The third step expands two θ terms in (4.27). The total derivative is written as

$$
\left(\frac{P_r}{P_{00}}\right)^\kappa \frac{d\theta}{dt} = P_r^\kappa \frac{d}{dt}\left(\frac{T}{P^\kappa}\right) = \frac{P_r^\kappa}{P^\kappa}\frac{dT}{dt} - \left(\frac{RP_r^\kappa T}{C_p P^{\kappa+1}}\right)\omega
$$
$$
= \frac{P_r^\kappa}{P^\kappa}\left(\frac{dT}{dt} - \frac{\omega\alpha}{C_p}\right) = \left(\frac{P_r}{P}\right)^\kappa \frac{q}{C_p}
$$

The last step uses the first law of thermodynamics again. The horizontal advection of θ term is written as

$$\left(\frac{P_r}{P_{00}}\right)^\kappa \nabla_P \cdot (\theta \mathbf{V}_P) = P_r^\kappa \nabla_P \cdot \left(\frac{T\mathbf{V}_P}{P^\kappa}\right) = \left(\frac{P_r}{P}\right)^\kappa \nabla_P \cdot (T\mathbf{V}_P)$$

The result of these three steps is

$$\frac{\partial A_j}{\partial t} = \frac{1}{g} \int_S \int_{P_2}^{P_1} \{\omega\alpha + \epsilon q\} dP dS \tag{4.29}$$

$$- \frac{C_p}{g} \int_S \int_{P_2}^{P_1} \left\{\epsilon \nabla_P \cdot (T\mathbf{V}_P) + \frac{\partial}{\partial P}(\omega T) - \left(\frac{P_r^\kappa}{P_{00}}\right) \frac{\partial \omega\theta}{\partial P}\right\} dP dS$$

$$- \int_S \int_{P_2}^{P_1} \left\{\left(\frac{\theta}{P_{00}^\kappa}\right)^\kappa \left(\frac{dP_r^\kappa}{dt} - \mathbf{V}_P \cdot \nabla_P P_r^\kappa - \omega \frac{\partial P_r^\kappa}{\partial P}\right)\right\} dP dS$$

$$+ boundary\ integral$$

where

$$\epsilon = \frac{P^\kappa - P_r^\kappa}{P^\kappa}$$

is sometimes called the "efficiency factor."

The next steps combine the vertical advection terms. These terms are collectively labelled "W".

$$W = \frac{\partial(\omega T)}{\partial P} - \left(\frac{P_r}{P_{00}}\right)^\kappa \frac{\partial(\omega\theta)}{\partial P} - \frac{\theta}{P_{00}^\kappa}\left(\omega \frac{\partial P_r^\kappa}{\partial P}\right)$$

$$= \frac{\partial(\omega T)}{\partial P} - \frac{1}{P_{00}^\kappa} \frac{\partial(\theta\omega P_r^\kappa)}{\partial P}$$

Substituting in Poisson's relation and combining terms yields

$$W = \frac{\partial(\omega T)}{\partial P} - \frac{\partial}{\partial P}\left(\frac{T\omega P_r^\kappa}{P^\kappa}\right) = \frac{\partial}{\partial P}\left\{\left(\frac{P^\kappa - P_r^\kappa}{P^\kappa}\right)\omega T\right\}$$

The chain rule applied to the following heat flux divergence gives

$$\nabla_P \cdot \left(\epsilon T\mathbf{V}_P\right) = \nabla_P \cdot (T\mathbf{V}_P) - \left(\frac{P_r}{P}\right)^\kappa \nabla_P \cdot (T\mathbf{V}_P)$$

$$- \frac{T\mathbf{V}_P}{P^\kappa} \cdot \left(\nabla_P P_r^\kappa\right)$$

This combines with terms in (4.29), leaving

$$\frac{\partial A_j}{\partial t} = \frac{1}{g} \int_S \int_{P_2}^{P_1} \{\omega\alpha + \epsilon q\} dPdS$$

$$- \frac{C_p}{g} \int_S \int_{P_2}^{P_1} \left\{ \nabla_P \cdot (\epsilon T \mathbf{V}_P) \right.$$

$$+ \frac{\partial}{\partial P}(\epsilon\omega T) + \frac{T}{P^\kappa} \frac{dP_r^\kappa}{dt} \right\} dPdS$$

$$+ \; boundary \; terms.$$

At this point, the derivation by Smith (1969) and Smith et al. (1977) diverges from that by Edmon (1978). The next step chosen by Smith (1969) eliminates the integrand involving the total derivative of the reference pressure. Egger (1976) later pointed out that the term must be retained. Smith concurred, and therefore the time rate of change of APE obtained by Smith et al. (1977) is:

$$\frac{\partial A_j}{\partial t} = \underbrace{\frac{1}{g} \int_S \int_{P_2}^{P_1} \{\epsilon q\} dPdS}_{(A)} + \underbrace{\frac{1}{g} \int_S \int_{P_2}^{P_1} \{\omega\alpha\} dPdS}_{(B)}$$

$$- \underbrace{\frac{1}{g} \int_S \int_{P_2}^{P_1} \left\{ \nabla_P \cdot (\epsilon C_p T \mathbf{V}_P) + \frac{\partial}{\partial P}(\epsilon C_p \omega T) \right\} dPdS}_{(C)} \qquad (4.30)$$

$$+ \underbrace{\frac{1}{g} \int_S \left\{ C_p \epsilon_1 T_1 \frac{\partial P_1}{\partial t} - C_p \epsilon_2 T_2 \frac{\partial P_2}{\partial t} \right\} dS}_{(D)}$$

$$- \underbrace{\frac{C_p}{g} \int_S \int_{P_2}^{P_1} \frac{T}{P^\kappa} \frac{dP_r^\kappa}{dt} dPdS}_{(E)}$$

The generation of APE for a limited volume, (4.30), is derived by Smith (1969). While this formulation is illustrative, it does present some problems. One problem is that the conversion between kinetic and available potential energy is ambiguous due to pressure work terms that are observed to be large. This discrepancy between terms (B) in (4.30) and (4.34) will be identified in (4.36). The second problem is that term (C) in (4.30) requires knowledge of the derivative of P_r at the boundaries. Given that P_r is not defined outside the limited area, the evaluation of this term is difficult in practice. The third problem is that term (B) and the vertical derivative part of term (E) are both large but nearly cancel.

All three of these problems can be avoided. The term involving the total derivative of P_r can be expanded.

$$(E) = -\frac{C_p}{g} \int_S \int_{P_2}^{P_1} \frac{T}{P^\kappa} \left\{ \underbrace{\frac{\partial P_r^\kappa}{\partial t}}_{(I)} + \underbrace{\mathbf{V}_P \cdot \nabla_P P_r^\kappa}_{(H)} + \underbrace{\omega \frac{\partial P_r^\kappa}{\partial P}}_{(G)} \right\} dPdS$$

The vertical derivative term (G) can be combined with (B) in (4.30). Term (B) is rewritten using the ideal gas law and the chain rule.

$$(B) = \frac{1}{g} \int_S \int_{P_2}^{P_1} \{\omega\alpha\} dPdS = \frac{C_p}{g} \int_S \int_{P_2}^{P_1} \left\{ \frac{T\omega}{P^\kappa} \right\} \frac{\partial P^\kappa}{\partial P} dPdS$$

Terms (B) and (G) combine by collecting derivatives, applying the chain rule, and introducing the ideal gas law. The result of these three steps is

$$(B) + (G) = \frac{C_p}{g} \int_S \int_{P_2}^{P_1} \left\{ \frac{T\omega}{P^\kappa} \right\} \left(\frac{\partial P^\kappa}{\partial P} - \frac{\partial P_r^\kappa}{\partial P} \right) dPdS$$

$$= \frac{C_p}{g} \int_S \int_{P_2}^{P_1} \frac{T\omega}{P^\kappa} \left\{ \epsilon \frac{\partial P^\kappa}{\partial P} + P^\kappa \frac{\partial \epsilon}{\partial P} \right\} dPdS$$

$$= \frac{1}{g} \underbrace{\int_S \int_{P_2}^{P_1} \epsilon\omega\alpha dPdS}_{(B)} + \frac{C_p}{g} \underbrace{\int_S \int_{P_2}^{P_1} T\omega \frac{\partial \epsilon}{\partial P} dPdS}_{(J)}$$

Term (B) is very similar to (B) in (4.30). Term (I) is written as a local derivative of ϵ. Terms (H) and (J) are written

$$(H) + (J) = \frac{C_p}{g} \int_S \int_{P_2}^{P_1} \left\{ T\mathbf{V}_P \cdot \nabla_P \epsilon + T\omega \frac{\partial \epsilon}{\partial P} \right\} dPdS$$

Using the chain rule, the terms (H) and (J) combine with (C) to obtain

$$\frac{\partial A_j}{\partial t} = \frac{1}{g} \underbrace{\int_S \int_{P_2}^{P_1} \{\epsilon q\} dPdS}_{(A)} + \frac{1}{g} \underbrace{\int_S \int_{P_2}^{P_1} \{\epsilon\omega\alpha\} dPdS}_{(B)}$$

$$- \frac{C_p}{g} \underbrace{\int_S \int_{P_2}^{P_1} \epsilon \left\{ \nabla_P \cdot (T\mathbf{V}_P) + \frac{\partial}{\partial P}(\omega T) \right\} dPdS}_{(C)} \qquad (4.31)$$

$$+ \frac{C_p}{g} \int_S \left\{ \epsilon_1 T_1 \frac{\partial P_1}{\partial t} - \epsilon_2 T_2 \frac{\partial P_2}{\partial t} \right\} dS - \frac{C_p}{g} \int_S \int_{P_2}^{P_1} T \frac{\partial \epsilon}{\partial t} dP dS$$

$$\underbrace{\phantom{+ \frac{C_p}{g} \int_S \left\{ \epsilon_1 T_1 \frac{\partial P_1}{\partial t} - \epsilon_2 T_2 \frac{\partial P_2}{\partial t} \right\} dS}}_{(D)} \qquad \underbrace{\phantom{\frac{C_p}{g} \int_S \int_{P_2}^{P_1} T \frac{\partial \epsilon}{\partial t} dP dS}}_{(E)}$$

The improved equation for the time rate of change of APE, (4.31), is derived by Edmon (1978). Terms (A) and (D) are the same as in (4.30); terms (B), (C), and (E) have their counterpart in (4.30). No longer are there spatial derivatives of ϵ. In addition, the original formulation had two very large terms (terms (B) and (G)) which nearly cancelled. Term (B) in (4.31) is much smaller than term (B) in (4.30) because of the small size of ϵ.

Before discussing (4.31) it is useful to derive a complementary kinetic energy equation. The horizontal wind kinetic energy in a limited volume is

$$K_j \equiv \frac{1}{g} \int_S \int_{P_2}^{P_1} \left\{ \frac{(\mathbf{V}_P)^2}{2} \right\} dP dS \qquad (4.32)$$

where $P_2 < P_1$ as before.

The dot product of $\mathbf{V}_P$ with the vector momentum equation gives

$$\frac{\partial}{\partial t} \left\{ \frac{(\mathbf{V}_P)^2}{2} \right\} + \mathbf{V}_P \cdot \nabla \left\{ \frac{(\mathbf{V}_P)^2}{2} \right\} + \omega \frac{\partial}{\partial P} \left\{ \frac{(\mathbf{V}_P)^2}{2} \right\} + \mathbf{V}_P \cdot \nabla_P \Phi$$

$$+ \mathbf{V}_P \cdot \{\Omega \times \mathbf{V}\}_P + \mathbf{V}_P \cdot \{\mathbf{G} \times \mathbf{V}\}_P + \mathbf{V}_P \cdot \mathbf{F} = 0$$

where $\mathbf{G}$ includes geometric terms such as $(u \tan \phi)/r$. The cross-products vanish. The subscript "P" is dropped as needed for clarity.

Rewriting,

$$\frac{\partial}{\partial t} \left\{ \frac{\mathbf{V}^2}{2} \right\} = -\mathbf{V} \cdot \nabla_P \Phi - \mathbf{V}_P \cdot \mathbf{F} - \nabla_P \cdot \left\{ \mathbf{V} \frac{\mathbf{V}^2}{2} \right\} - \frac{\partial}{\partial P} \left\{ \frac{\mathbf{V}^2 \omega}{2} \right\} \qquad (4.33)$$

where the continuity equation in pressure coordinates has been used to express the advection terms in flux form. The local time derivative of (4.32) is taken next.

$$\frac{\partial K_j}{\partial t} = \frac{1}{g} \int_S \int_{P_2}^{P_1} \frac{\partial}{\partial t} \left\{ \frac{\mathbf{V}^2}{2} \right\} dP dS + \frac{1}{g} \int_S \left\{ \frac{\mathbf{V}_1^2}{2} \frac{\partial P_1}{\partial t} - \frac{\mathbf{V}_2^2}{2} \frac{\partial P_2}{\partial t} \right\} dS$$

Substitution from (4.33) obtains

$$\frac{\partial K_j}{\partial t} = -\frac{1}{g} \int_S \int_{P_2}^{P_1} \mathbf{V}_P \cdot \mathbf{F} dP dS - \frac{1}{g} \int_S \int_{P_2}^{P_1} \mathbf{V}_P \cdot \nabla_P \Phi dP dS$$

$$\underbrace{\phantom{-\frac{1}{g} \int_S \int_{P_2}^{P_1} \mathbf{V}_P \cdot \mathbf{F} dP dS}}_{(A)} \qquad \underbrace{\phantom{-\frac{1}{g} \int_S \int_{P_2}^{P_1} \mathbf{V}_P \cdot \nabla_P \Phi dP dS}}_{(B)}$$

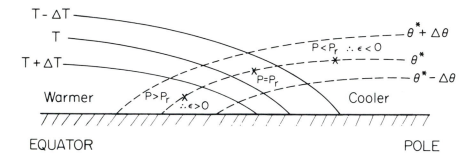

FIG. 4.16   A meridional cross-section of schematic isopleths of temperature, T, and potential temperature, θ. Because the polar regions are cooler than the tropics, the isopleths bend as shown. If P_r is the average pressure on the isopleth $\theta = \theta^*$, then the pressure on that θ surface will be higher in the tropics ($\epsilon > 0$ there) and lower in the polar regions ($\epsilon < 0$ there). Vertical coordinate is P.

$$- \frac{1}{g} \int_S \int_{P_2}^{P_1} \left\{ \nabla_P \cdot \left\{ \mathbf{V}_P \frac{\mathbf{V}_P^2}{2} \right\} + \frac{\partial}{\partial P} \left\{ \frac{\mathbf{V}_P^2 \omega}{2} \right\} \right\} dP dS \qquad (4.34)$$

$$\underbrace{\phantom{- \frac{1}{g} \int_S \int_{P_2}^{P_1} \left\{ \nabla_P \cdot \left\{ \mathbf{V}_P \frac{\mathbf{V}_P^2}{2} \right\} + \frac{\partial}{\partial P} \left\{ \frac{\mathbf{V}_P^2 \omega}{2} \right\} \right\} dP dS}}_{(C)}$$

$$+ \frac{1}{g} \int_S \left\{ \frac{\mathbf{V}_1^2}{2} \frac{\partial P_1}{\partial t} - \frac{\mathbf{V}_2^2}{2} \frac{\partial P_2}{\partial t} \right\} dS$$

$$\underbrace{\phantom{+ \frac{1}{g} \int_S \left\{ \frac{\mathbf{V}_1^2}{2} \frac{\partial P_1}{\partial t} - \frac{\mathbf{V}_2^2}{2} \frac{\partial P_2}{\partial t} \right\} dS}}_{(D)}$$

As with the APE generation equation, if the vertical boundaries are fixed, then the last surface integral vanishes. It is retained if the boundaries are variable, for example, when using the earth's surface or the tropopause.

4.5   Interpretation of the limited volume energy equations

This section interprets the terms on the right-hand sides of (4.30),(4.31) and (4.34).

4.5.1   Term (A)

This term appears once in each equation and represents generation or destruction by various diabatic processes. The APE equation is considered first.

Relation (4.31) is more precise than the "approximate" formulas given by Lorenz. The term is not simply equal to the integrated diabatic heating; it also includes the efficiency factor. The factor ϵ can have either sign, hence introducing heat ($q > 0$) where $\epsilon < 0$ could have the effect of destroying APE and so forth. In general, regions that are hotter than the reference state will have higher pressure than P_r and the positive departures imply $\epsilon > 0$. The converse relationship exists for cold regions, where $\epsilon < 0$. The reason for this pattern is made clear in Figure 4.16, which shows a schematic meridional cross section of T and θ.

The pattern of ϵ can be deduced from Figure 4.17, reproduced from Dutton and Johnson. The efficiency factor plotted by D and J is $\epsilon(\partial Z/\partial\theta)$. For large scale motions, $(\partial Z/\partial\theta)$ is positive, monotonic, and about three times larger in the troposphere. The data shown are along a single meridian, whose location corresponds to the long wave trough (e.g., §5.4; Figure 5.9). Nonetheless, the general pattern is similar to a zonal average of ϵ. (Newell et al. , 1970, show seasonal and zonal average patterns of ϵ; the main difference from D and J is that they find smaller values in the tropical tropopause.) In the troposphere, generally positive values occur in tropical regions and negative values elsewhere. The largest magnitudes are near the tropopause. Near the surface the values are quite small, and outside the tropics there is some question regarding the sign. In the stratosphere the pattern is reversed and is considerably smaller in amplitude due to the $(\partial Z/\partial\theta)$ factor. The reversal of sign across the tropopause reflects the reversal of the horizontal temperature gradient. By inspection, heating at low latitudes and cooling at high latitudes is an effective way to generate APE; in a very broad sense, this is what happens in the atmosphere.

D and J partition the APE generation term into five prime contributors:

4.5.1.1   Solar (shortwave) radiation

Short wave radiation clearly produces zonal APE by the absorption of energy in low latitudes. Seasonal differences imply strong latitudinal contrast in the winter hemisphere, while in summer the radiation reaching the top of the atmosphere is rather uniform with latitude. So, there is strong APE generation in winter. In summer there still may be some generation, but it is more due to a modulation by latitudinally varying atmospheric constituents, primarily clouds. The clear subtropics versus the cloudy polar regions can generate some APE during summer due to the variation of the albedo (§3.1; Figure 3.7c).

4.5.1.2   Outgoing (longwave) infrared radiation

Does longwave radiation create or destroy APE? If the efficiency factor is ignored, terrestrial radiation might appear to destroy APE because the emission is somewhat greater at low latitudes (§3.1; Figures 3.6 and 3.8) than in the polar regions. However, ϵ changes the interpretation. The emission at the high latitudes is strong at the levels just below the tropopause—where ϵ is large and negative. This situation makes the longwave (IR) radiation a generator of APE as well. Since the atmospheric state determines the efficiency factor, then it could be said that the atmosphere maintains a structure that can generate APE by both radiative processes.

4.5.1.3   Sensible heat transfer

On a zonal average, the surface heat flux is directed into the atmosphere over the tropics (the surface warms the air above). Over the tropics $\epsilon > 0$, hence this mechanism generates APE. To understand the situation in midlatitudes, some care is needed. From Figure 4.16, one might expect ϵ at the surface to be positive in the tropics,

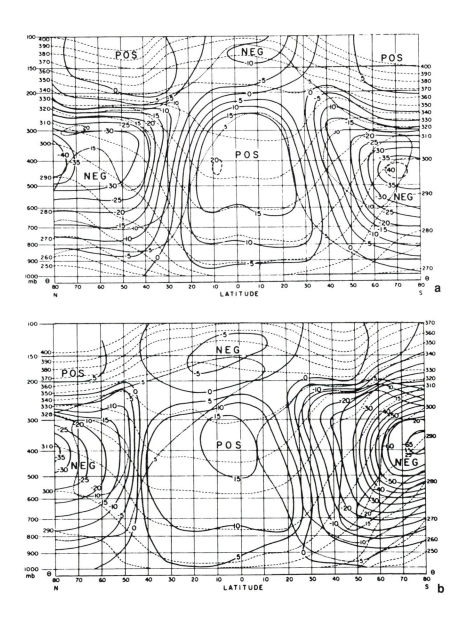

FIG. 4.17 Meridional cross-sections along 75 W longitude of the efficiency factor (solid contours), the standard deviation of ϵ (dotted contours), and isentropes (dashed contours). (a) January 1958 (b) July 1958. From Dutton and Johnson, (1967).

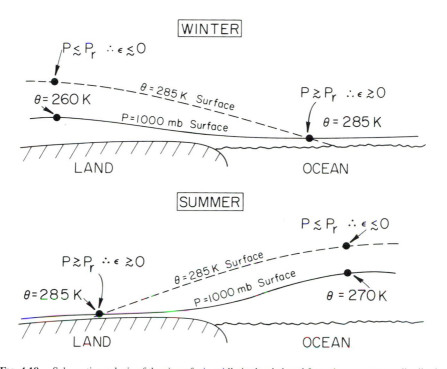

FIG. 4.18 Schematic analysis of the sign of ϵ in midlatitudes deduced from the temperature distribution observed on the 1000 hPa isobaric surface (on which $T - \theta$). The temperature is lower over the continents in winter and the oceans in summer. The 1000 hPa surface is indicated by a solid line; the dashed line is an isentropic surface deduced from the fact that θ must increase with height. As indicated, the sign of ϵ is only approximate because at a particular latitude, P_r may be greater than the highest pressure shown in either figure or lower than the lowest pressure marked. However, from the reasoning in Figure 4.16 one may anticipate $P \simeq P_r$ in midlatitudes.

negative in the polar regions, and indeterminate in midlatitudes. To determine the sign of ϵ for the Northern Hemisphere, one must emphasize differences between continental and oceanic regions. In §5.4, observations show that the time mean surface pressure over the oceans is higher than over the land areas during summer. During winter the time mean surface pressure is lower over the oceans. Observations (§5.2) also show that surface air over the continents is warmer than over the oceans during summer. During winter the air over the oceans is warmer. As shown in Figure 4.18, the *time mean* pressure and temperature distributions reinforce each other insofar as they tend to alter the sign of ϵ. Those distributions tend to cause ϵ to be negative over cold areas and positive over warm areas. The sign of ϵ therefore seems positively correlated with the sign of heating due to sensible heat transfer. Therefore, this diabatic process is positive for the climatological oceanic lows in winter and for thermal lows over continents in summer (because these surfaces are usually warmer than the air above). Conversely, destruction of APE should occur over oceans in summer and continents in winter.

The situation near oceanic western boundary currents (WBCs; e.g., the Gulf Stream) requires careful analysis. Behind a developing surface low, very cold air is drawn off the continent and over warm water. Very large surface heat fluxes occur. These fluxes are routinely far greater than the surface fluxes discussed in the preceding paragraph. Since the heat fluxes occur in the cold air mass, Lorenz's approximate formulas imply destruction of APE because $T' < 0$ and $Q' > 0$. However, since the more precise formulas of D and J and (4.31) retain the efficiency factor, the surface heat fluxes can in fact *generate* APE. The efficiency factor reverses the sign because θ surfaces vary widely in altitude. Figure 4.19 shows θ surfaces at a time when explosive cyclogenesis is occurring. The coldest air is *not* where the heat fluxes are greatest. Consequently, ϵ is still likely to be negative. (If P_r is calculated only using the domain shown in the figure, then $\epsilon < 0$ where the largest heat fluxes occur.) One reason why the air is not coldest where the heat fluxes are greatest is simply because the air is very rapidly warmed by the ocean; the local time rate of change of temperature is near zero during the passage of the continental air over the WBC (e.g., Grotjahn and Wang, 1989).

The sensible (and latent) heat fluxes near a WBC are localized phenomena. Consequently, the fluxes make a direct generation of *eddy APE* and help explain the observed fact that the western boundary regions of the oceans are areas of strong cyclogenesis. (Observations are shown in §5.8; Figure 5.19.) Grotjahn and Wang (1989) discuss this point at length and show how latent heating is a direct source of eddy APE near the Kuro Shio WBC. Grotjahn and Lai (1991) discuss the physical mechanism by which the sensible heat fluxes can enhance the development even though they occur in the cold air sector of the storm; basically, those fluxes enhance eddy instability by reducing the static stability.

4.5.1.4 *Latent heat*

On a large scale, much latent heat is released near the equator in the rising branch of the Hadley-type cells (§3.5; Figure 3.27). This heating occurs primarily at mid- and upper- troposphere levels where ϵ is large and positive; thus, APE is generated. Latent heat release is a complex problem on the smaller scale of individual storms.

Latent heat is the primary driving mechanism of tropical disturbances. The "CISK" mechanism described in §6.7 can explain the maintenance of Hadley-type circulations. CISK can also be applied to tropical disturbances like hurricanes. The basic idea is that large scale convergence occurs in a region of cyclonic vorticity; the release of latent heat causes air to rise, stretching vortex tubes in the vertical; the stretching increases the vorticity leading to more convergence, etc.

Latent heating could either generate or destroy APE locally for midlatitude frontal cyclones. For midlatitude cyclones, an isentropic surface rises considerably from the warm sector into the cold. The θ surface has low pressure on the cold side relative to the higher pressure on the warm side; the pattern is seen in Figure 4.19. If latent heat is released in the warm sector, APE is generated. That occurs for devel-

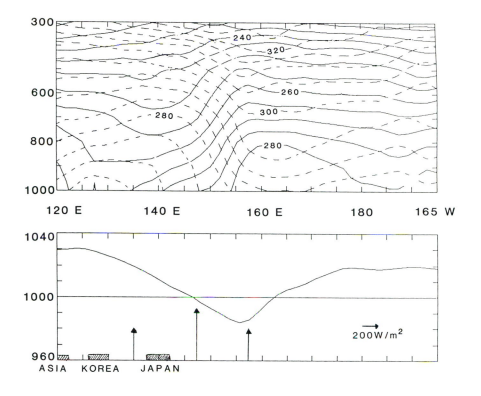

FIG. 4.19   Cross-sections along 37.5 N latitude. Upper chart shows observations of T (solid contours) and ϵ (dashed contours) for an explosively developing surface low to the east of Japan. Lower chart shows observed sea level pressure. ECMWF FGGE data for 00 GMT on 19 January 1979 are used; Grotjahn and Wang (1989) study this storm. Arrows indicate magnitude of surface latent heat flux using U.S. Navy data. Sensible heat fluxes are smaller but similarly distributed in space. The surface fluxes are largest where ϵ is likely to be positive.

oping storms, many of which form over the WBCs (§5.8). For a decaying midlatitude cyclone (an "occluded" low) the situation is more complex. The center of the surface low has migrated into the cold air; but along the eastern and poleward sides of the low is warm air drawn poleward leading to the thickness ridge by which an occluded front is identified. Condensation in the warm tongue of the thickness ridge generates APE. The pattern shown in Figure 4.20 illustrates the motion of the warm sector air found in trajectories calculated by Grotjahn (1987). In contrast, surface convergence caused by cross-isobaric motion around the surface low pressure center may cause clouds to form in the cold air sector; the thin arrows in Figure 4.20 show this motion. The rising cold air destroys eddy APE. Hoskins et al. (1985) assume a latent heating rate of 2.5 K/day to show that convection within a cut-off low could eliminate an oceanic low in 1 to 2 days. However, Shutts (1987) finds that including radiative cooling counteracts the latent heating and lengthens the estimated decay time to 5 to 10 days.

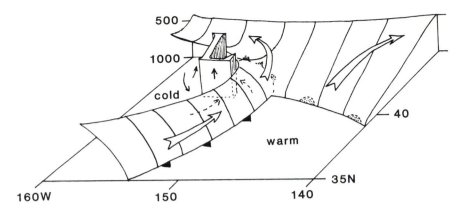

FIG. 4.20   Three-dimensional schematic view of a decaying frontal cyclone. The frontal surfaces are shown in perspective view; the "valley" in the frontal surface to the northeast of the low center overlies the location of the surface "occluded" front. The elongated "L" indicates the trough center throughout the lower troposphere; the trough axis is nearly vertical, indicating that the low is centered in the cold air mass. The open arrows show motion relative to the storm in the warm air mass. The thin arrows show motion in the cold air that rises due to surface frictional convergence.

4.5.1.5   *Frictional dissipation*

For frictional retardation of the winds to affect APE, then the loss of energy by friction must go into heating. Frictional heating has to be a small effect. While friction is an ancient way to start a fire, it is hard to boil soup by stirring it. If the kinetic energy in a 10 m/s wind is converted into internal energy, then the temperature of the air increases by almost 10^{-1} K. The frictional temperature change is small compared to the radiative relaxation rate of 2 K/day and local surface heat fluxes that often exceed 10 K/day.

In D and J this term depends upon temperature and static stability, so that weaker stability enhances the term. If one assumes that this term is mainly due to frictional losses in the boundary layer, then the frictional processes could turn out to generate APE. This conclusion is based on having positive values of ϵ near the ground over much of the earth. But, the values of ϵ are small near the surface and possibly negative outside the tropics, according to D and J. Newell et al. (1970) find the diabatic friction term to be small as well, but positive over nearly all latitudes at the surface.

In the free atmosphere (away from the ground) one might expect dissipation to be greatest near the jet streams due to the strong shears there. D and J argue that the turbulent dissipation losses in the vicinity of the midlatitude jets would destroy APE. They base their argument upon some observed atmospheric average dissipation values from Kung (1966). Kung found that the dissipation in the free atmosphere is three times larger in sum than that in the boundary layer. D and J conclude that friction is a net destroyer of atmospheric APE. Another frictional effect is caused by topographic lee waves. The importance of lee waves to the atmospheric energy budget is uncertain. Typical values of lee wave drag have been estimated to be 50 to 100 W/m^2 (Lilly and Kennedy, 1973); these values are comparable to the surface stress.

In summary, for the zonal average circulation, the five diabatic processes generate APE, except possibly friction.

Moving on to kinetic energy (KE), the only diabatic effect in (4.34) is friction. Of course, friction will only act to destroy KE. All dissipative processes are lumped into the friction term "**F**". Friction is important in the boundary layer. Friction is also important higher up, in the free atmosphere, in conjunction with lee waves, as mentioned earlier. Friction could also include the contributions by mountain torques as discussed in the opening section of this chapter.

4.5.2 Term (B)

This term reflects a conversion between APE and KE. A simple derivation can illustrate the connection. From hydrostatic balance,

$$\omega\alpha = -\omega\frac{\partial\Phi}{\partial P} \tag{4.35}$$

Clearly, (4.35) involves shifting mass in the vertical, a feature that is consistent with the simple models discussed in §4.3.2.

Expanding the right side of (4.35) by using the chain rule followed by the continuity equation yields

$$\omega\alpha = -\frac{\partial\omega\Phi}{\partial P} + \Phi\frac{\partial\omega}{\partial P} = -\frac{\partial\omega\Phi}{\partial P} - \Phi(\nabla_P\cdot\mathbf{V}_P)$$

$$= -\frac{\partial\omega\Phi}{\partial P} - \nabla_P\cdot(\Phi\mathbf{V}_P) + \mathbf{V}_P\cdot\nabla\Phi$$

Thus

$$\frac{1}{g}\underbrace{\int_S\int_{P_1}^{P_2}\omega\alpha\,dP\,dS}_{-(B)\text{ from (4.30)}} = \frac{1}{g}\underbrace{\int_S\int_{P_1}^{P_2}\mathbf{V}_P\cdot\nabla\Phi\,dP\,dS}_{(B)\text{ from (4.34)}}$$

$$-\frac{1}{g}\underbrace{\int_S\int_{P_1}^{P_2}\left\{\nabla_P\cdot(\Phi\mathbf{V}_P) + \frac{\partial}{\partial P}(\omega\Phi)\right\}dP\,dS}_{\text{boundary terms}} \tag{4.36}$$

The integration limits have been reversed from before, but $P_2 < P_1$ as before. The left-hand side of (4.36) is the negative of the conversion in (4.30); therefore (4.36) is positive for conversion from A_E to K_E.

By comparing (4.34) with (4.30), the efficiency factor is ignored and the discussion simplified. The two terms (B) do not exactly match when expressed in this form. The third integral in (4.36) represents work done by pressure forces upon the

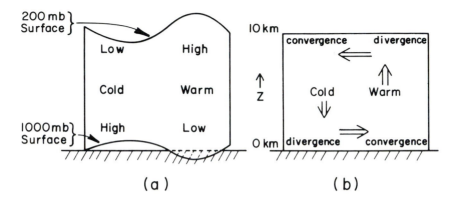

FIG. 4.21   Two views of a *developing* cyclone drawn in cross-section through the trough and ridge. (a) The top and bottom show the deformation of the 200 and 1000 hPa surfaces, regions of high and low pressure, and cold and warm temperature. (b) The divergence pattern, the resultant ageostrophic circulation and the vertical advection of temperature.

boundaries. If the boundaries are ignored (as in a closed system) then there are equal and opposite terms in (4.30) and (4.34). Accordingly, the terms are designated as conversion between APE and KE. In reality, the boundary terms are not necessarily zero, and some care must be taken when referring to either term (B) in (4.30) or (4.34) as *the* conversion. Indeed, observations show the boundary pressure work terms to be quite large (Kornegay and Vincent, 1976). However, this problem may be avoided if one uses the efficiency factor weighted conversion term (B) found in (4.31).

When dealing with a *closed system*, the boundary terms vanish in (4.36). The closed system analogy succinctly shows how the conversion mechanism works. It is sufficient to show that warm lows and cold highs create kinetic energy by reducing the potential energy.

A typical developing cyclone has cold (dry) air flowing equatorward behind the surface low center and warm (moist) air flowing poleward ahead of it. The surface low is accompanied by warm air above it and the trailing high by cold upper-level air. Figure 4.21 illustrates the process as a closed system viewed with a longitudinal cross section.

The motion field implies that the center of gravity is being lowered. Kinetic energy is being produced as the air is accelerated from the highs into the lows. Also, the secondary circulation set up by the divergence field is reinforced by the pressure gradients. Warm (less dense) air is rising and cold (more dense) air is sinking.

Schematically,

$$\int\int \mathbf{V}_P \cdot \nabla_P \Phi dP = -\underbrace{\int\int \Phi \nabla_P \cdot \mathbf{V}_P dP dS}_{(CKA)} + \underbrace{\int\int \nabla_P \cdot (\mathbf{V}_P \Phi)\, dP dS}_{(D)}$$

CKA is the conversion from available potential to kinetic energy. For a closed system D=0 since there will be no flux across a boundary. The continuity equation and chain rule rewrite CKA.

$$\text{CKA} = -\underbrace{\int\int\omega\frac{\partial\Phi}{\partial P}dPdS}_{(AC)} + \underbrace{\int\int\frac{\partial}{\partial P}(\omega\Phi)\,dPdS}_{(E)}$$

E vanishes for a closed system. From (4.35), (AC) equals term (B) in (4.30) after substituting in the hydrostatic law. To illustrate, the vertical (pressure) velocity approximately equals its vertical average, such that

$$\text{CKA} \approx -\int_S \overbrace{\omega}\int_{P_1}^{P_2}\left(\frac{\partial\Phi}{\partial P}\right)dPdS$$

The horizontal surface integrals over S are divided into separate integrals over the surface low (indicated by subscript "L") and the surface high (subscript "H"). The $\overline{(\,)}^\sigma$ indicates the surface integration on a constant pressure surface. Thus,

$$\text{CKA} \approx -\overline{\widehat{\omega_H}}^\sigma S_H\left(\overline{\Phi_{200}}^\sigma - \overline{\Phi_{1000}}^\sigma\right)_H - \overline{\widehat{\omega_L}}^\sigma S_L\left(\overline{\Phi_{200}}^\sigma - \overline{\Phi_{1000}}^\sigma\right)_L$$

For a closed system,

$$\overline{\widehat{\omega_H}}^\sigma S_H = -\overline{\widehat{\omega_L}}^\sigma S_L \equiv \delta$$

Thus

$$\text{CKA} \approx -\delta\left(\underbrace{\left(\overline{\Phi_{200}}^\sigma - \overline{\Phi_{1000}}^\sigma\right)_H}_{g\cdot\text{TH}} - \underbrace{\left(\overline{\Phi_{200}}^\sigma - \overline{\Phi_{1000}}^\sigma\right)_L}_{g\cdot\text{TL}}\right)$$

Where TH and TL are the thicknesses of the 1000 to 200 hPa layers in their respective surface high and surface low pressure regions. Finally,

$$\text{CKA} \approx -\delta g(TH - TL)$$

Since $\omega > 0$ for downward motion, and that occurs over the surface high, then $\delta > 0$. Thus the sign of CKA equals the sign of TL $-$ TH.

For the developing cyclone shown in Figure 4.21, TL > TH, so CKA > 0. That is, kinetic energy is growing at the expense of the potential energy.

For the decaying cyclone shown in Figure 4.22, TL < TH, hence CKA < 0. This occurs because the cyclone has larger amplitude at 200 hPa than it does at 1000 hPa. Since thickness is proportional to the mean virtual temperature in the layer, then cold air above a decaying surface low leads to smaller thickness there. The cyclone is

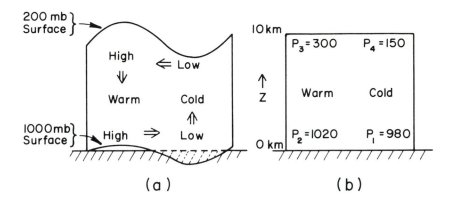

FIG. 4.22   Views of a *decaying* cyclone. (a) Similar to Figure 4.21a with the inclusion of the circulation driven by the surface friction. (b) Approximate pressures at the top (10 km elevation) and bottom. The implied circulation at high levels is up the pressure gradient; the vertical advection of temperature is the opposite of Figure 4.21.

decaying by converting kinetic energy into potential energy. Putting it another way, the center of mass of the system is being raised in Figure 4.22.

Observed analogues of the closed system schematics shown in Figures 4.21 and 4.22 are the following. Figure 4.21 can represent a typical east-west oriented cross section of a developing cyclone. In contrast, Figure 4.22 can be thought of as a north-south oriented cross section through a decaying storm. For example, the surface low in Figure 4.22 might be the climatological "Icelandic low," and the warm high pressure might be the "Bermuda high." The secondary circulation of Figure 4.22 would therefore be the same as a local "Ferrel cell." The connection between the eddy circulations and the Ferrel cell is fundamental; it is just what is observed, as will be shown in §6.3 and §6.5 (e.g., Figure 6.20).

The APE to KE conversion can be related to the density weighted heat flux. In order to do so, it helps to recall the simple models of §4.3.2: the conversion may take place by shifting the center of mass in the vertical. Shifting the entire mass of the atmosphere uniformly up or down does not change the APE, consequently, the vertical flux of specific volume ($\omega\alpha$) must be non-uniform. The vertical motion or the specific volume or both must be horizontally varying. In quasi-geostrophic theory, the dependent variables are linearized about a vertically varying state (called the "static state") that describes the bulk of the vertical variation of the variables.

$$P = P_s(1 + \mu\nu P')$$

$$\rho = \rho_s(1 + \mu\rho') \tag{4.37}$$

$$\theta = \theta_s(1 + \mu\theta')$$

Subscript "s" indicates the static state; these quantities only vary with height. The prime denotes the nondimensional state that may vary in space and time; this state might be called the "quasi-geostrophic" state. μ is a nondimensionalizing factor that is quite small; it is proportional to the Rossby number squared (Grotjahn, 1979). $\nu = D/H$ is an aspect ratio between a scaling length D, used for vertical derivatives, and the scale height H. The α in the energy conversion uses the quasi-geostrophic part of the total specific volume.

$$\alpha' = \alpha - \alpha_s = \frac{1}{\rho_s(1+\mu\rho')} - \frac{1}{\rho_s} \simeq \frac{1}{\rho_s}(1-\mu\rho') - \frac{1}{\rho_s} = -\frac{\mu\rho'}{\rho_s} \qquad (4.38)$$

where the smallness of μ has been exploited. The next step is to use vertical velocity (W) instead of pressure velocity (ω). Hydrostatic balance for the static state and (4.38) combine to transform term (B) in (4.30) as follows.

$$\frac{1}{g}\omega'\alpha' = \frac{1}{g}\frac{dP}{dt}\alpha' \simeq \frac{1}{g}W'\frac{\partial P_s}{\partial z}\alpha' = $$
$$= -\rho_s W'\alpha' = \mu W'\rho' \qquad (4.39)$$

Equation (4.39) is consistent with the discussion of the simple models in §4.3.2 and the schematic diagrams in Figures 4.21 and 4.22. Raising the center of mass should lead to greater APE. Air that is heavier than the static state has $\rho' > 0$; lighter than average air has $\rho' < 0$. Therefore, when heavier air moves upwards or when lighter than average air moves downward, then $W'\rho' > 0$ and the center of mass is raised.

The quasi-geostrophic density is not commonly examined. Instead, it is common practice to express the conversion in terms of a vertical heat flux. The derivation proceeds as follows. From Poisson's equation and the ideal gas law,

$$C_p \ln\theta = C_v \ln P - C_p \ln\rho - C_p \ln(RP_{00}^{-\kappa}) \qquad (4.40)$$

Substituting (4.37) into (4.40) yields

$$\ln\left(\theta_s(1+\mu\theta')\right) = \gamma\ln\left(P_s(1+\mu\nu P')\right) - \ln\left(\rho_s(1+\mu\rho')\right) - C_p\ln(RP_{00}^{-\kappa}) \quad (4.41)$$

where $\gamma = C_v/C_p$. Since $\mu \ll 1$, then (4.41) reduces to

$$\theta' = \gamma\nu P' - \rho' \qquad (4.42)$$

The hydrostatic law is used to derive an expression relating quasi-geostrophic pressure and potential temperature. The hydrostatic law is

$$0 = g + \frac{\left(\ln\left(P_s(1+\mu P')\right)\right)_z}{\ln\left(\rho_s(1+\mu\rho')\right)} \qquad (4.43)$$

Using the chain rule and the hydrostatic balance of the static state obtains

$$0 = g - \left(\frac{g}{1 + \mu\rho'}\right) + \frac{\mu\nu(P_s P')_z}{\rho_s(1 + \mu\rho')} \tag{4.44}$$

where the z subscript denotes differentiation with respect to height. The scale height H is assumed to be constant, where

$$P_s = H\rho_s g \tag{4.45}$$

Then substitution of (4.45) into (4.44) gives

$$0 = g\left(1 - \frac{1}{1 + \mu\rho'}\right) + \frac{gH\mu\nu(\rho_s P')_z}{\rho_s(1 + \mu\rho')} \tag{4.46}$$

The term in the large brackets simplifies. Multiplying (4.46) by $(1 + \mu\rho')$ obtains

$$0 = g\mu\rho' + \frac{gH\mu\nu}{\rho_s}(\rho_s P')_z \tag{4.47}$$

The chain rule is used on (4.47) and the common $g\mu$ factor is divided out.

$$0 = \rho' + H\nu(P')_z + H\nu P'(\ln \rho_s)_z \tag{4.48}$$

Using (4.40) applied to the static state for $\ln \rho_s$ and (4.42) for ρ' obtains

$$0 = \gamma\nu P' - \theta' + H\nu(P')_z + H\nu P'\left(\gamma(\ln P_s)_z - (\ln \theta_s)_z\right) \tag{4.49}$$

The atmosphere in the quasi-geostrophic formulation is assumed to be weakly statically stable so that the $(\ln \theta_s)_z$ term may be neglected.

Rearranging,

$$\theta' = (P')_{z'} + \gamma\nu P' + H\nu\gamma P'(\ln P_s)_z \tag{4.50}$$

where the z' denotes a nondimensional derivative. The assumption that the scale height is a constant means that $P_s = P_0 \exp(-z/H)$ where P_0 is the pressure at $z = 0$. Substitution of the definition of P_s obtains

$$\theta' = (P')_{z'} \tag{4.51}$$

to lowest order in the Rossby number.

Returning to (4.48) and solving for ρ' obtains

$$-\rho' = (P')_{z'} + H\nu P'(\ln \rho_s)_z \tag{4.52}$$

Using (4.45) and the hydrostatic law gives

$$-\rho' = (P')_{z'} - P'\nu \tag{4.53}$$

Using (4.42)

$$-\rho' = (P')_{z'} - \frac{1}{\gamma}(\theta' + \rho') \tag{4.54}$$

From (4.51)

$$-\rho' = \theta' \tag{4.55}$$

to lowest order in the Rossby number. Finally, substituting (4.55) into (4.39) yields

$$\frac{1}{g}\omega\alpha \simeq -\mu W'\theta' \tag{4.56}$$

to lowest order in the Rossby number.

The vertical heat flux is also consistent with the simple models shown in §4.3.2 and the eddy conversion diagramed in Figures 4.21 and 4.22. Air that is warmer than the quasi-geostrophic average has $\theta' > 0$; colder than average air has $\theta' < 0$. Warm air rising or cold air sinking has $W'\theta' > 0$, making the conversion negative, a result that is consistent with the loss of APE.

The heat flux form of the energy conversion can be visualized using the schematic diagrams in Figure 4.23. The quasi-geostrophic system has quasi-horizontal motion. The horizontal velocities in the advection terms are geostrophic as in Figure 4.23a, but vertical motion is also allowed. The surfaces of P and θ are parallel for the static state, as in Figure 4.23b. Motion along the θ surface in Figure 4.23b does not change the center of mass, hence the APE is unchanged. In Figure 4.23c, the motion is along θ surface, but the pressure is decreasing as the parcel moves from point C to D. Consequently, the parcel goes from a higher to a lower temperature. Relative to the environment, warm air is rising, but the APE is not changed. APE is unchanged because nothing has changed: there is no change in the θ pattern, so the mass field (P pattern) is not changed, and thus the center of mass is unchanged. In Figure 4.23d the motion follows a pressure surface. In going from points E to F, the air parcel sinks and proceeds from a higher to a lower θ; warm air is brought downward because the temperature at point F increases. In Figure 4.23d *the motion changes the distribution of θ because the air parcels move adiabatically*. Since the θ field is changed, then so must the P field. Bringing the warm air downward raises the center of mass, converting some KE to APE.

4.5.3   Term (C)

These integrals are boundary terms, essentially due to divergence of mass in or out of the limited area. When integrated over the entire mass of the atmosphere these

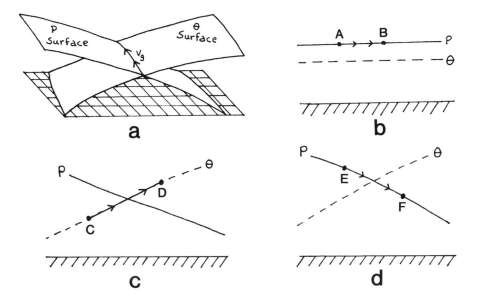

FIG. 4.23   (a) Geostrophic, adiabatic winds lie along the intersection of θ and P surfaces. (b), (c), (d) These diagrams show alternative air parcel motions that may occur in *quasi*-geostrophic balance. In (b) and (c) there is no change of APE. In (d) the APE is changed.

terms vanish; but they are significant if only a limited region is considered, or if the atmosphere is divided into different types of phenomena. For example, when K_j is subdivided into zonal and eddy parts, this term becomes the internal conversion between the eddy and zonal components of K_j. That conversion was seen earlier in this chapter (§4.2.1). Terms (A) and (B) in the eddy kinetic energy equation (4.20) have their counterparts in the zonal mean kinetic energy equation (4.16); the terms have opposite sign in the two equations, indicating a conversion between the two forms of energy. A similar statement holds for the available potential energy. Though it has not been derived in this chapter, term (C) in (4.31) can lead to conversion terms between zonal mean and eddy A_j if the limited volumes are so partitioned.

4.5.4   *Term (D)*

These integrals are also boundary contributions, but of a different sort—they are a function of the integration limits. If the boundaries are constant pressure surfaces, then these terms vanish. If not, then these terms can be generators or destructors of energy. For example, if the integration ranges from the surface to 500 hPa, then only the surface term contributes here. If the surface pressure *increases*, then $\partial P_1/\partial t > 0$ and K_j is increasing. Since kinetic energy *density* is being used, simply adding mass without changing the speed means more K_j within the volume. Of course, if adding the mass also decreases the speed, then other terms are changed in (4.34).

4.5.5 *Term (E)*

This term was only found for the available potential energy formulas. It arises from local changes in the efficiency factor for a given temperature profile. The local rate of change of ϵ is given by the local rate of change of the reference pressure. When (4.30) is used, this term removes the part of the diabatic heating that is uniformly distributed. (Part of term (E) combined with (B) in Eqn. 4.30 to get (B) in Eqn. 4.31.) In (4.31) the term is more subtle: if the reference pressure is systematically increasing, then term (E) increases A_j because ϵ is proportional to minus P_r. Increasing P_r implies increasing mass in the domain; term (E) points out that when all else is equal, the increasing mass increases the potential energy; hence the APE is increased proportionally.

4.6 Eddy versus zonal average energetics

Expressions (4.31) and (4.34) involve averaging. As shown in the Appendix, averages may be partitioned several ways, depending upon what phenomena are to be isolated. The most frequent partition is between the zonal average and the deviation from the zonal average. Lorenz (1955) was the first to partition the energetics this way. When a zonal average is made around a complete latitude circle, there can be conversion between zonal APE and zonal KE, similarly between eddy APE and eddy KE. After such a partitioning, no term remains for conversion between the eddy component of one energy type and the zonal component of a different energy type. Instead, conversion between eddy and zonal APE and between eddy and zonal KE remain. Generation and destruction by diabatic processes become partitioned, too. The resulting equations (4.57) express the energy balances in symbolic form; Newell et al. (1974) give further details. The eddy and zonal mean equations for available potential energy ("A") and kinetic energy ("K") are:

$$\frac{\partial A_Z}{\partial t} = GZ - CZ - CA$$

$$\frac{\partial A_E}{\partial t} = GE - CE + CA$$

$$\frac{\partial K_Z}{\partial t} = -DZ + CZ - CK \tag{4.57}$$

$$\frac{\partial K_E}{\partial t} = -DE + CE + CK$$

The notation used for the right-hand side terms is as follows: G refers to generation; D to destruction; E to eddy; Z to zonal; CK to conversion between KE components; and CA to conversion between APE components.

The discussion of the observed energy conversions draws upon diagrams published by Newell et al. (1974); they use approximate relations and integrate them over the entire mass of the atmosphere.

$$A_Z = C_p \int [\epsilon][\overline{T}]dM \qquad (4.58)$$

$$A_E = \frac{C_p}{2} \int \gamma[(\overline{T'})^2 + \overline{(T'')^2}]dM \qquad (4.59)$$

$$K_Z = \frac{1}{2} \int [\overline{u}]^2 + [\overline{v}]^2 dM \qquad (4.60)$$

$$K_E = \frac{1}{2} \int [(\overline{u'})^2 + (\overline{v'})^2] + [\overline{(u'')^2} + \overline{(v'')^2}]dM \qquad (4.61)$$

where

$$\gamma = -\left(\frac{\theta}{T}\right)\left(\frac{P_r}{P}\right)^\kappa \left(\frac{\kappa}{P_r}\frac{\partial P_r}{\partial \theta}\right)$$

is a parameter primarily related to static stability. In the definition of γ, reference state variables are evaluated at the level $\theta = [\theta]$. The definitions for A_Z and A_E above use approximate formulas similar to (4.25). As in §4.3.3, deviations of temperature from an average on a pressure surface are used.

The conversions between forms of energy used by Newell et al. (1974) are

$$GZ = \int [\epsilon][\overline{Q}]dM \qquad (4.62)$$

$$GE = \int \gamma[\overline{Q' \, T'}]dM \qquad (4.63)$$

$$CZ = -\int (\overline{\omega})_\sigma (\overline{\alpha})_\sigma dM \qquad (4.64)$$

$$CE = -\int [\overline{\omega' \, \alpha'} + \overline{\omega'' \alpha''}]dM \qquad (4.65)$$

$$CK = -\int [\overline{u' \, v'} + \overline{u'' v''}]\frac{\cos\phi}{r}\frac{\partial}{\partial\phi}\left(\frac{[\overline{u}]}{\cos\phi}\right)dM \qquad (4.66)$$
$$\quad -\int [\overline{u' \, \omega'} + \overline{u'' \omega''}]\frac{\partial}{\partial P}([\overline{u}])dM$$

$$CA = -C_p \int \left(\frac{\theta}{T}\right)[\overline{v' \, T'} + \overline{v'' T''}]\frac{1}{r}\frac{\partial}{\partial\phi}\left(\frac{[\epsilon][\overline{T}]}{[\theta]}\right)dM \qquad (4.67)$$
$$\quad -C_p \int \left(\frac{\theta}{T}\right)[\overline{\omega' \, T'} + \overline{\omega'' T''}]\frac{\partial}{\partial P}\left(\frac{[\epsilon][\overline{T}]}{[\theta]}\right)dM$$

$$DZ = -\int ([\overline{u}][\overline{F_x}] + [\overline{v}][\overline{F_y}])dM \qquad (4.68)$$

$$DE = -\int \left[\overline{u' \, F'_x} + \overline{v' \, F'_y} + \overline{u'' F''_x} + \overline{v'' F''_y}\right]dM \qquad (4.69)$$

where r is the earth's radius, $dM = r^2(\cos\phi/g)d\phi d\lambda dP$, λ is longitude, ϕ latitude. The subscript σ identifies deviations on a pressure surface.

These conversions differ slightly from those given by Lorenz (1967), particularly conversion CA. In addition, the terms involving $[v]$ are already neglected in (4.66). These equations differ from those derived in §4.4 because the decomposition is different. Previously, energy was decomposed into zonal mean and eddy contributions. Newell et al. (1974) choose a decomposition into zonal *and* time mean (4.58) and (4.60) versus all eddies plus *transient* zonal mean fields (4.59) and (4.61). Consequently, a variety of time deviation terms appear in equations (4.62) through (4.69). The plots shown later will also neglect the vertical flux terms written in (4.66) and (4.67).

The definition of CZ can be expressed in an approximate form by following steps similar to those used to derive (4.36):

$$CZ = -\int [\omega][\alpha]dM = \int g[\omega]\frac{\partial[Z]}{\partial P} = -\int g[Z]\frac{\partial[\omega]}{\partial P}$$

$$= \int [Z]\left[\frac{\partial u}{\partial x} + \frac{\partial v}{\partial y}\right]dM = -g\int [v]\frac{\partial[Z]}{\partial y} \approx g\int f[v][u_g]dM \tag{4.70}$$

where the g subscript denotes a geostrophic wind.

The generation and destruction terms have been discussed at length in §4.5.1.

$$GZ, \; GE \sim \int [QT]dM$$

while

$$DZ, \; DE \sim \int [\mathbf{V}_2 \cdot \mathbf{F}_2]dM$$

where Q is a heating rate per unit mass. These diabatic terms are split into appropriate $[s][s]$ and $[s's']$ type terms for Z and E components, respectively.

The γ above behaves like the efficiency factor. The ϵ in (4.31) is present in two parts. Effectively, the sign change for ϵ is buried in the T' factors included in (4.59) since these quantities are similar to deviations of T on a pressure surface. Hence, the sign of T' is similar to the sign of ϵ by reasoning similar to that shown in Figure 4.16. The static stability part included by Dutton and Johnson (1967) is included in the γ definition. For the zonal mean latent heating in the tropics, clearly $GZ > 0$ because heat is released where the atmosphere is warm ($T_\sigma > 0; Q > 0$). At high latitudes the air is cold ($T_\sigma < 0$), so the radiative cooling ($Q < 0$) leads to positive GZ there.

The CK and CA terms refer to conversions between the zonal and eddy components of the KE and APE, respectively. The conversions CA and CK were discussed before in connection with (4.20). Both integrands in each of CK and CA, (4.66) and (4.67), have two contributions. The first term is from standing eddies while the second is from transient phenomena.

For potential energy, the conversion is proportional to heat fluxes; primarily (or at least for quasi-geostrophic theory) this means horizontal heat fluxes. Using a thermal wind relationship to replace the meridional gradient of temperature with vertical shear gives

$$CA \sim \int \frac{\partial [u]}{\partial z} [v'T'] dM \qquad\qquad A_Z \to A_E$$

The transfer ($A_Z \to A_E$) is an eddy correlation *weighted* by the zonal mean flow *vertical* shear. Dynamically, the heat flux distorts the temperature pattern from a zonal orientation into a wavy configuration when v' and T' are positively correlated. The $\partial [u]/\partial z$ factor essentially measures the amount of zonal temperature gradient that is available to be distorted. In short, the rate of conversion depends upon the efficiency of the motions that transfer the energy and the magnitude of the available energy, respectively.

For kinetic energy, the conversion is proportional to the Reynolds stress (momentum flux).

$$CK \sim -\int \frac{\partial [u]}{\partial y} [u'v'] dM \sim \int [u] \frac{\partial [u'v']}{\partial y} dM \qquad\qquad K_Z \to K_E$$

The transfer between K_Z and K_E is weighted by the *horizontal* shear in the zonal-mean flow. The second form is based on (4.14).

It is emphasized above that both the CA and CK conversions can be related to the shear in the flow. Earlier in this chapter (§4.5.2) the baroclinic conversion was identified as being the term CE, and was illustrated schematically (Figure 4.21). In actuality, the baroclinic conversion is identified with CA as well. It is easy to see in the quasi-geostrophic system that the creation of the height field deviations (increasing A_E) simultaneously leads to greater K_E from the geostrophic relationship. The simultaneity is identified later in this section as being an "express bus route" for the energy conversion. The reader should notice that the baroclinic instability mechanism (CA) requires the vertical tilt to be oriented *against* the vertical shear in order for the eddy to grow. The trough and ridge axes tilt upstream with height as illustrated by the lower charts in Figure 4.24. In a similar vein, in order for an eddy to grow by barotropic instability (CK), the eddy horizontal tilts must be oriented against the horizontal shear. The upper charts in Figure 4.24 show the trough and ridge axes tilting upstream relative to the horizontal shear; this orientation creates a divergence of eddy momentum flux along the axis of the jet leading to $CK > 0$ there. The relationships depicted in Figure 4.24 powerfully illustrate that the baroclinic and barotropic instability mechanisms are special cases of a continuous spectrum of three-dimensional fluid flow shear instability.

The presence of an eddy in a flow with horizontal shear often creates distortions that lead to the development of energy conversions. In Figure 4.25a a zonal mean flow and a circular cyclone are combined at the start. In Figure 4.25b the cyclone has weaker flow than the mean flow and is distorted in a simple way by the zonal mean flow. In comparison with Figure 4.9, the eddy fluxes in Figure 4.25b lead to a

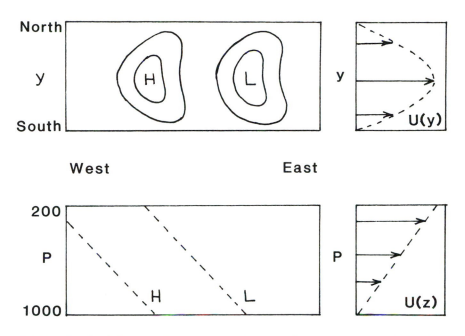

FIG. 4.24   Trough and ridge axes of extratropical cyclones must tilt against the shear in order for the eddy to gain energy from the mean flow. (a) Surface chart of high and low pressure pattern showing horizontal tilts of trough and ridge axes and jet maximum winds in center of domain (arrows). The eddy momentum flux divergence is centered on the jet, in contrast with the pattern shown in Figure 4.9. (b) East-west cross-section showing trough and ridge axes (dashed lines) and zonal wind increasing with height (arrows). Baroclinic conversion, identified with term CA has axes tilted upstream with height. Compare this figure with Figure 4.21.

loss of K_E; so the distortion of the eddy by the mean flow weakens the eddy in this case. In Figure 4.25c the mean flow and cyclone have comparable flow speeds. In this case the nonlinear advection may lead to a trough having the southeast to northwest tilt indicated in the figure. The eddy momentum flux $[u'v']$ is southward, leading to the destruction of the jet on its northern side and the enhancement of the jet on its southern side. The situation depicted in Figure 4.25c may also lead to a migration of the cyclone to the northern side of the jet, where the jet has cyclonic vorticity, too.

   The CZ and CE terms are conversions between KE and APE for either eddies or for the zonal mean fields.

$$CZ \sim \int [\omega][\alpha] dM \qquad CE \sim \int [\omega'\alpha'] dM$$

Recalling (4.70), the Hadley and Ferrel cells contribute to CZ. Even though the mean meridional cells are a tiny fraction of K_Z, they are the entire conversion term between A_Z and K_Z. This situation reflects the basic facts that (a) the atmosphere is largely in geostrophic balance, but (b) this conversion between energy forms needs ageostrophic winds. The CZ and CE conversions simply indicate lowering (or raising) of the center

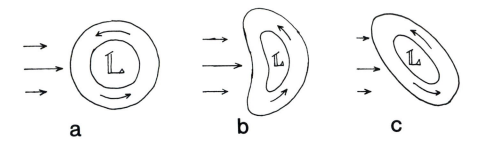

FIG. 4.25   Possible interactions between a mean flow with a jet and a cyclonic low pressure center in the horizontal plane. (a) Initial combination of a jet mean flow (arrows) and a circular cyclone (contours). (b) Case where the eddy flows are much weaker than the jet flows. Differential advection by the jet distorts the cyclone shape. From Figure 4.9, the cyclone develops eddy momentum fluxes (double-shafted arrows) whose convergence causes the eddy to feed its energy into the jet. (c) Case where the eddy and jet flows are comparable. Differential nonlinear advection leads to eddy momentum fluxes that build the jet on the south side and weaken the jet on the north side. This situation may lead to a northward migration of the cyclone toward the side of the jet with cyclonic shear vorticity.

of mass in order to increase (or decrease) the kinetic energy. In terms of the Hadley cell, overturning has warm air rising and cooler air sinking—the center of mass is lowered and A_Z is flowing into K_Z.

The volume averages of these different components are considered first. Figure 4.26a has three different estimates of the annual average energy budget. It should be immediately apparent that there is considerable disagreement about some of the quantities. In Figure 4.26b two estimates are based upon grid point data generated by two sophisticated objective analysis schemes. The two analysis schemes started with the same, unusually good, observations. Yet, the numeric values of the conversions differ widely.

The "box" diagram has been criticized in several ways. First, the prevalence of thermal wind balance links the two forms of energy as mentioned above. As demonstrated in Figure 4.27, when APE is generated by building departures of T and P from their global averages, then the geostrophic wind is simultaneously increased as well. Hence increasing A_Z simultaneously increases K_Z. Second, the atmosphere does not "recognize" the boxes as distinct reservoirs. A conceptual model might be parallel bus routes running through the different forms of energy. Some buses make stops along the way (at A_E, say) and discharge or take on passengers (i.e., lose or gain energy). Other buses run straight through ($A_Z \rightarrow A_E \rightarrow K_E$) like express bus lines. Thermal wind balance requires the express bus route. It is, nonetheless *mathematically* correct to partition the atmosphere in the manner of Figure 4.26, and it does provide some useful insight into the general circulation.

The box diagrams described here partition the problem into zonal averages and deviations, an approach labelled ZAD for short. An alternative to ZAD is to partition the atmosphere into transformed mean and eddy parts, whereby the transformed mean has the contributions by the eddies to the zonal mean removed from the usual zonal

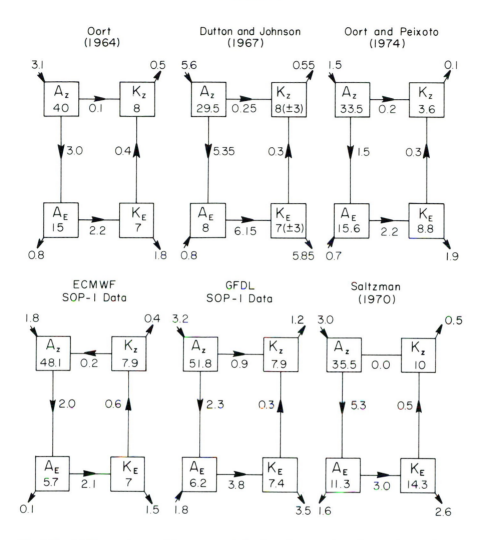

FIG. 4.26   (a) Three estimates of the energy cycle for the entire atmosphere. Some estimates of energy conversion differ by more than a factor of three; there is some question about the sign of the eddy available potential energy (A_E) generation. The $A_E \rightarrow K_E$ is inferred on the basis of energy balance. (b) Two estimates of the energy cycle (ECMWF and GFDL) based upon different objective analyses of the same observations. SOP–1 is the period 1 January to 5 March 1979. Data from Kung and Tanaka (1983). For comparison, related wintertime mean values computed by Saltzman (1970) are also shown.

mean. One example of eddy motion that contributes to a zonal mean is Stokes drift (Stokes, 1847). Plumb (1983) argues that the eddy forcing of the Ferrel cell described later (§6.3.2) is another example. The transformed Eulerian mean (TEM) approach removes such a Stokes drift from the zonal mean being used.

   TEM is developed in several articles (e.g., Andrews and McIntyre, 1976) and is relevant to the discussion of Eliassen-Palm fluxes in §7.4. Plumb (1983) compares

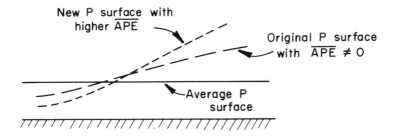

FIG. 4.27   As temperature is increased where it is warm (on the right) and decreased on the left, thickness changes increase the tilt of an isobaric surface. The increased tilt increases A_Z. From geostrophic (thermal wind) balance the zonal mean zonal wind is increased, increasing K_Z at the same time.

energetics calculated by ZAD and TEM formulations. The energy conversions in TEM (indicated by $\star$ superscripts here) have similar counterparts in ZAD. In TEM, the $(A_Z \rightarrow A_E)^\star$ conversion is excluded. Instead, horizontal heat fluxes in the ZAD formulation of $(A_Z \rightarrow A_E)$ are cast into both the $(A_E \rightarrow K_E)^\star$ and $(K_Z \rightarrow K_E)^\star$ conversions in TEM. The recasting of the eddy heat flux is not too surprising, given the connection it has to Stokes drift explained in §7.4.1.

TEM gains some benefits over ZAD for certain situations discussed by Plumb (1983) and others. But TEM loses some interpretive advantages enjoyed by ZAD: energy conversions can no longer be linked to unique types of eddy structural changes (such as horizontal versus vertical axis tilts) and energy must sometimes flow "outside" the TEM box diagram in order to go from one form to another. On the other hand, ZAD may create a misleading impression about energy conversions when applied to an intransient mountain lee wave, while TEM avoids that problem. Plumb (1983) also discusses the difference between ZAD and TEM interpretations of baroclinic, midlatitude disturbances. ZAD has other limitations—for example, the "express bus route" caused by thermal wind balance. TEM does not avoid the express bus route problem either. TEM reveals the problem since the energy conversion runs the opposite way: $(K_E \rightarrow A_E)^\star$ for baroclinic growth. A simultaneity is revealed that *neither* approach recognizes explicitly. Additional analysis of TEM versus ZAD can be found in Hayashi (1987). The writer concludes that ZAD is a straightforward and common tool for understanding the baroclinic waves and is sufficient for the purposes of this book.

Of course, other partitions could be devised: time mean versus transient, short versus long waves versus zonal mean. Using the time mean flow is logical since adiabatic, zonal-mean flow is not stationary when topography is present. Instead, the major mountain ranges (and land-sea contrasts) create a meandering time mean flow (§7.2). The only restrictions are that the partitions be orthogonal and physically meaningful.

One anticipates that the kinetic energy must be dissipated by friction in the net, and that it should remove energy from both K_E and K_Z. One expects differential solar heating to generate available potential energy in the net. Some of the heat energy is radiated back to space and lost in that way. But before that can happen, atmospheric motions are set up that redistribute much of the energy (Figure 3.6). Therefore, the basic path followed is: energy is input into A, some fraction is converted to K and then lost by dissipation. A_Z should *primarily* reflect the differential solar heating that is spread out around each latitude circle by the relatively rapid rotation of the earth, but one is not so sure about A_E.

A more detailed picture of the energy cycle can be deduced from information in §4.4. (1) Heating at low latitudes and cooling at high latitudes is an effective way to generate A_Z. However, whether A_E is created or destroyed by the sum of diabatic processes is a matter of some debate. (2) From the radiation balance discussion there must be a transport of heat from warm to cold areas. This is an effective way to convert A_Z to K_Z in the mean meridional cells and A_Z to A_E through hydrodynamic instability of the large scale gradients and through land-ocean contrasts. (3) The momentum transport by the eddies (§§4.1 and 4.2) is largely up the gradient of the zonal average zonal wind, suggesting a conversion from K_E to K_Z. On the equatorial side of the subtropical jets the flux is up the gradient; the direction is less clear on the polar side. (4) Friction destroys both K_Z and K_E. (5) These four factors lead to the conclusion that there must be conversion from A_E to K_E simply because there is no other source of K_E.

When calculating the various parts of the energy box diagrams, it is best to avoid direct calculation of items that require quantities that are not well measured. In particular, one tries to avoid conversions that involve the vertical velocity field (such as the $A_E \rightarrow K_E$). Dutton and Johnson, for example, calculate the $A_Z \rightarrow K_Z$ and $K_Z \rightarrow K_E$ conversions from observations and infer the others as necessary to satisfy a balance. The $A_Z \rightarrow K_Z$ conversion might seem to involve vertical motion, but (4.70) shows that it is approximated by the meridional wind and the geostrophic zonal wind. The $K_Z \rightarrow K_E$ conversion involves mainly correlations of the horizontal eddy velocities. One may wonder why the $A_Z \rightarrow A_E$ conversion is not calculated instead. The problem is that the $A_Z \rightarrow A_E$ conversion depends upon both meridional and vertical heat fluxes; it is not clear that one is much less than the other in this conversion.

The diabatic generation terms are only partly derived from observations. D and J accept the dissipation values given by Kung (1966). Kung's values are two or three times larger than the values used by Oort (1964). In order to maintain balance, the generation values must be larger for D and J than for Oort. In terms of zonal and eddy components, D and J make an estimate of the A_Z generation that is less than the total dissipation, so they deduce a net generation of A_E. On the other hand, the generation of A_Z was greater than the dissipation calculated by Oort, and that implies a net destruction of A_E as indicated in the Oort (1964) data. Apparently, this matter is still unsettled. For example, Kung and Tanaka (1983) calculate the conversions for the special observing periods of the Global Weather Experiment (GWE). Figure 4.26b shows data for the period 1 January to 5 March 1979. Unusually frequent observations were made during this part of GWE. Kung and Tanaka use the Geophysical Fluid Dynamics Laboratory (GFDL) and the European Centre for Medium-Range Weather Forecasts (ECMWF) data to make their calculations. Though the GFDL and ECMWF data are based upon virtually identical observations, the energy box diagrams are rather different for the different data sets. The A_E to K_E conversion is twice as large in the GFDL data as in the ECMWF data. The discrepancy causes the A_E diabatic generation to be positive for the GFDL but negative (and small) for the ECMWF data. The differences between the GFDL and ECMWF "observational" data sets arise from the differences in the forecast models and objective analysis techniques used by GFDL and ECMWF to convert the data at unevenly-spaced stations into data at regularly-spaced grid points. For comparison purposes, average wintertime values computed by Saltzman are also included.

While there is disagreement as to the magnitude, all researchers find that A_Z is much greater than A_E. Deviations of temperature tend to be oriented in zonal bands: the dominant variation in the temperature field is a pole-to-equator temperature gradient (e.g., Figure 5.4). In contrast, K_Z and K_E have about equal magnitudes.

The meridional distribution of some of the energy conversions completes the discussion of energetics. Definitions (4.62) through (4.69) are valid for integrals over the entire mass of the atmosphere. The following diagrams (Figures 4.28 through 4.30 and 4.32) present the zonal average values of the integrands in (4.62), (4.70), (4.67) and (4.66).

The A_Z generation by diabatic processes is shown in Figure 4.28 from Newell et al. (1970). Newell et al. calculate this quantity from radiation, precipitation, and boundary layer flux estimates. The estimate of $[Q]$ is tenuous, at best. For example, the net heating in a vertical column contributed by latent heating might be estimated from surface precipitation data. However, it would still be unclear how that component of Q is vertically distributed. An alternative is to calculate Q as a residual after calculating all the other terms in the θ conservation equation. Hoskins et al. (1989) estimate Q by this technique; a significant difference is found in middle latitudes. Hoskins et al. find heating from the surface to 400 hPa in middle latitudes while Newell et al. find cooling there. The midlatitude precipitation maximum (Figure 3.27) can explain the heating found by Hoskins et al. The lack of heating in Newell et al. data has critical impact upon theoretical diagnostics of the meridional circulation, as will be discussed in §6.3.2.

Maximum values of GZ occur in the upper troposphere near the equator and in the middle to upper troposphere in high latitudes. The efficiency factor at the latter location is five times the factor at the former. The former is principally due to the latent heat release in the rising branch of the Hadley cells. The latter is mainly due to radiational cooling. The sensible heat transport at the surface seems to make only a weak input into A_Z. While the sensible heat input has comparable magnitude to the other diabatic heating rates, it is introduced into the boundary layer where ϵ is very small. Hence, the change in A_Z by this process tends to be small.

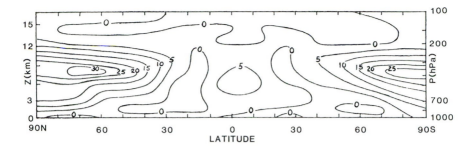

Fig. **4.28** Zonal mean cross-section of the diabatic generation of A_Z using the form of equation (4.62) that incorporates ϵ. The units are 10^{-4} J kg^{-1} s^{-1}. Figure based on data in Newall et al. (1970, 1974).

Five ways to generate A_Z (latent heat, friction, sensible heat, and short- and longwave radiation) are described in §4.5.1. Just two, latent heating in the tropics and long wave cooling in the high latitude upper troposphere, are easily identified in Figure 4.28. What about the other three? One should not conclude that sensible heating is unimportant because the efficiency factor is strongly modulating the APE generation. Strong heating occurs at the ground, but it is mainly used to evaporate water (which is transported away). The air that is actually warmed up becomes buoyantly unstable leading to convection (a smaller scale phenomenon than those incorporated into the figure).

Oort and Peixóto (1983) make a similar decomposition as shown here. They present vertical profiles (global horizontal averages) and meridional curves (zonal and vertical averages) of the conversions in Figures 4.29, 4.30, and 4.32. Comparison with Newell et al. (1974) is difficult, but attempted below.

Figure 4.29 shows the A_Z to K_Z conversion. From (4.70), this conversion can be approximated as the advection of planetary vorticity by using the mean meridional velocity and geostrophic zonal wind and this figure exploits that approximation. That being the case, the maximum values up near the tropopause are associated with the positive, larger $[u_g]$ there. Indeed, the Ferrel cell (which has quite weak meridional circulation compared with the Hadley) is visible primarily because of the large, positive $[u_g]$ of the midlatitude jet streams. The thermally direct Hadley cells tend to create zonal average kinetic energy, since $[v] > 0$ at high altitudes. By similar reasoning, the thermally indirect Ferrel cells tend to destroy it. There are two easily seen seasonal variations: (a) the dominance of the winter Hadley cell and (b) greater seasonal change (outside the tropics) in the Northern compared with the Southern Hemisphere. Oort and Peixoto (1983) find much larger values of this conversion in the lowest 150 hPa, and larger negative values near 40 N in winter. Newell et al. (1970) have a similar pattern but values are half as large in the Southern Hemisphere.

Newell et al. (1974) present diagrams for the A_Z to A_E conversion, reproduced here as Figure 4.30. This conversion is labelled a baroclinic instability

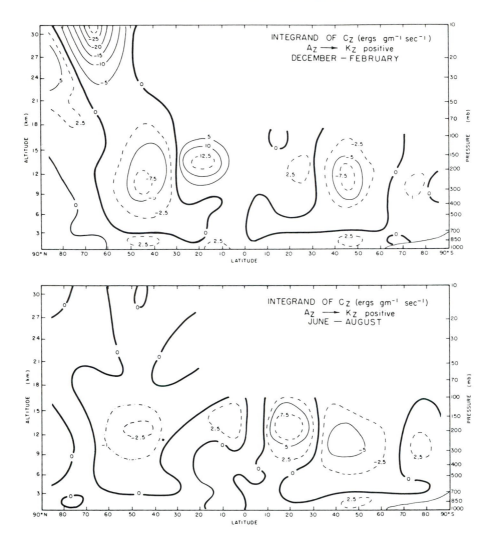

Fig. 4.29 Zonal mean cross-sections of conversion CZ from (4.62). (a) December through February (b) June through August. The units are 10^{-4} J kg^{-1} s^{-1}. From Newell et al. (1974).

process, and it is primarily created by the midlatitude transient cyclonic eddies. The conversion is proportional to the meridional transport of sensible heat, which is observed to be large in the lower troposphere ($\sim$850 hPa) and lower stratosphere ($\sim$200 hPa). The meridional heat flux is shown in Figure 4.31. As mentioned earlier, the conversion depends upon the matching of the eddy heat flux and the meridional temperature gradient. From a thermal wind argument, one anticipates the meridional temperature gradient to reverse sign above the jet stream core. This reversal is responsible for the tendency of the lower stratospheric val-

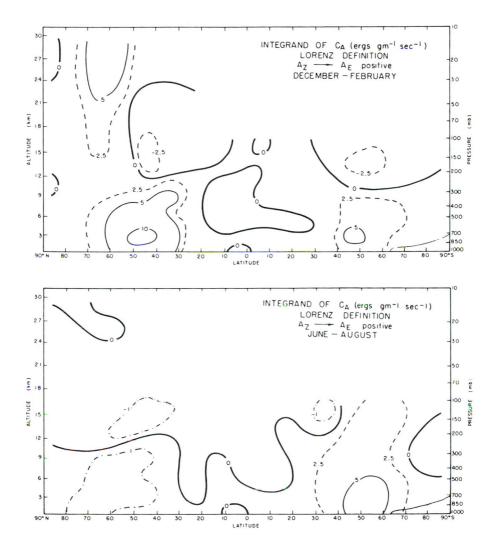

FIG. 4.30 Similar to Figure 4.29 except for the conversion CA in Lorenz (1967) and ignoring vertical fluxes. The units are 10^{-4} J kg^{-1} s^{-1}. From Newell et al. (1974).

ues to have opposite sign to the troposphere values at midlatitudes. Of course, where the stratospheric polar night jet is strong (Figure 3.15), the temperature gradient does not reverse, creating larger positive values in the winter stratosphere. Newell et al. (1974) ignore vertical heat fluxes, which implies that they will underestimate this conversion. Figure 4.30 uses the Lorenz (1967) formulas, which generally give larger values than (4.67). Curves in Oort and Peixoto (1983) show similar properties except for having higher positive values in the upper troposphere. Newell et al. (1970) values seem smaller than Oort and Peixoto's, but are still 2

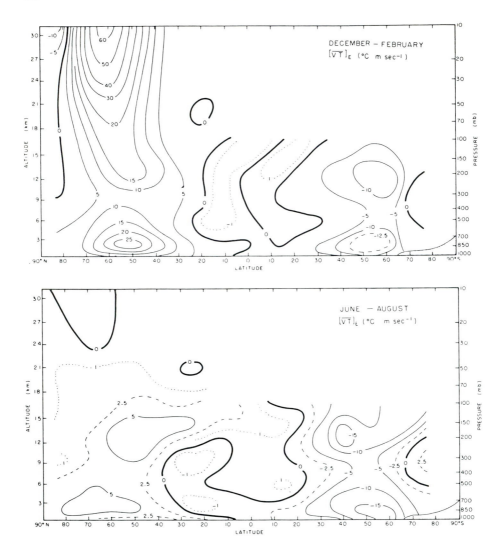

FIG. 4.31   Meridional cross-sections of zonal mean heat flux by eddies. (a) December through February. (b) June through August. The units are K m s^{-1}. From Newell et al. (1974).

to 6 times larger (depending on location) than those in Figure 4.30 in the stratosphere.

As stated before, the A_Z to A_E "baroclinic" conversion is partly proportional to the meridional heat flux. It has become popular in the observational literature to show figures of the heat flux rather than figures of the available potential energy conversion. First, while there is a superficial similarity between the heat flux and APE conversion (Figure 4.30 versus 4.31), there are also large differences, especially in the stratosphere, due to the sign reversal of the mean flow vertical shear. Second,

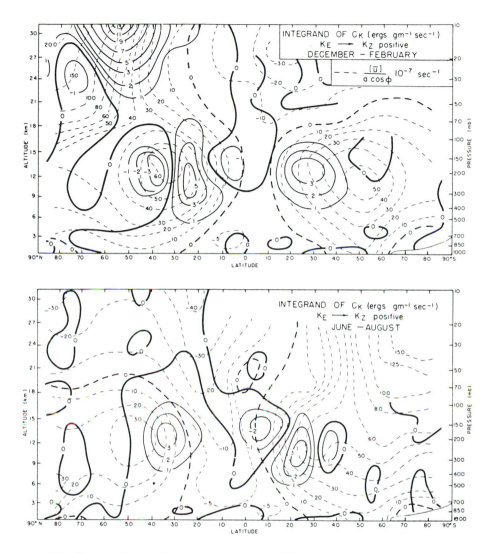

FIG. 4.32   Similar to Figure 4.29 except for conversion CK in equation (4.66) and ignoring vertical fluxes and all terms multiplied by $[v]$. Solid lines contour CK, while dashed lines contour $[u]/R$. The units are 10^{-4} J kg^{-1} s^{-1}. From Newell et al. (1974).

the very large stratospheric heat fluxes in the neighborhood of the polar night jet are much less important in the APE conversion due to the mass weighting. The integrals in (4.67) are over mass, so if height is the vertical coordinate, then the values shown in Figure 4.30 must be multiplied by density. (Indeed, density would multiply the data shown in all the conversion figures.) Third, the strongest tropospheric heat fluxes are centered 5 to 10 degrees poleward of the regions of strongest APE conversion.

Finally, the K_E to K_Z conversion is given in Figure 4.32. This conversion is labelled a "barotropic" conversion (positive values mean barotropic damping of eddies). As stated above, it matches the eddy momentum flux to the gradient of the zonal average zonal wind. (Newell et al. (1970) ignore the vertical fluxes.) The conversions tend to be largest near the tropopause, again in apparent association with the jet streams. K_Z tends to be created equatorward of the jet stream axis with some tendency to destroy it on the poleward side in winter. Examining zonal averages makes this result misleading. The point is discussed in more detail when considering sector averages (§7.3.2). This conversion can also include transport by mean meridional cells; with their inclusion, the maximum conversion is very close to the jet stream location. Generally, the pattern described above is consistent with Oort and Peixóto (1983), except that Newell et al. (1974) find stronger negative values near 35 S in winter (Figure 4.30b).

In summary, the ultimate direction of the energy cycle is as follows. Potential energy is generated by solar radiation; some fraction of that energy generates A_Z (along with other diabatic processes); some of that energy is converted to K_Z and K_E, where it is ultimately lost by frictional processes. These energy conversions are embedded in a cycle of energy whose path is low latitude net input, poleward transport and high latitude net output, and whose components are larger quantities of energy than the energy conversions of the box diagram. The size differential can be seen by noting that $\gamma \ll 1$, causing the integrand in (4.67) to be much smaller than the heat flux. The situation is similar to using a bank that charges a fee for each check you write. To transfer funds from the location of excess funds (your account!) to the location of deficit funds (the originator of your bill) requires an additional cost (the bank fee). The fee is (typically) a small fraction of the amount of the check. To accomplish the transport of heat (Figure 3.9) implied by the net radiation (Figure 3.8c), some energy is used to feed the circulations that handle the transport.

Figure 4.33 summarizes schematically the general properties of the energy flow in the atmosphere. The arrows in this figure reveal a pattern of generally poleward transport, including a complicated loop in the middle latitudes. Solar radiation (1a) enters the top of the atmosphere, with more arriving in tropical latitudes than in the higher latitudes, when averaged over a year. The net radiation (1b) is positive in the tropics and subtropics but negative in the high latitudes. A large fraction of the net radiation at the subtropical surface goes to evaporate water, which is transported (4) in two directions. The heat transport is handled by the Hadley cells in the tropics and the frontal cyclones in midlatitudes. Latent heat of condensation (2a) helps drive the Hadley cells. The Hadley cells (5a), in conjunction with the Ferrel cells (5b), focus the westerlies into subtropical jets whose vertical shear is a source of energy (6c) for frontal cyclones (by baroclinic instability). The "express bus route" baroclinic conversion includes poleward and vertical heat fluxes (3a) by the cyclones. Additional transport is accomplished by the oceans, whereupon the heat is put into the base of the atmosphere (3b) or is radiated back to space as part of the net terrestrial radiation

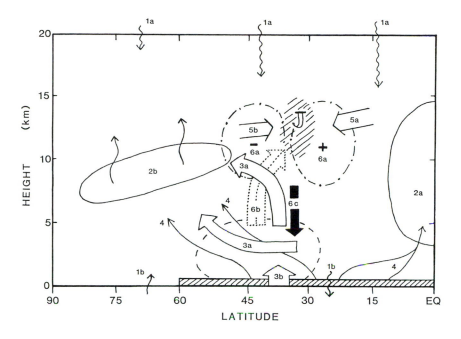

FIG. 4.33   Schematic diagram of the main elements affecting energy flow in the atmosphere. Radiation (short wavelength arrows for solar radiation; longer wavelength for terrestrial) are divided into (1a) solar radiation reaching the top of the atmosphere; and (1b) net radiation at the earth's surface. Diabatic processes generating A_Z include latent heating in the tropics (2a) and longwave radiative cooling from cloud tops (2b) in high latitudes. Open arrows show sensible heat transport by eddies (3a) and surface heat flux from the ocean (3b). Thin arrows show moisture transport (4), while hatching indicates areas of notable frictional energy loss. Double-shafted arrows indicate $A_Z \rightarrow K_Z$ conversion building a subtropical jet between the Hadley cell transport (5a) and the Ferrel cell transport (5b). The vertically pointing arrow (3a) shows $A_E \rightarrow K_E$ baroclinic growth. Momentum fluxes (6a) lead to eddy barotropic decay that builds westerlies by momentum convergence that is subsequently fed upon by new eddies (6c). Though the eddies transport heat poleward (3a), the Eliassen Palm fluxes (6b) show that eddies also transfer amplitude vertically and then decay by feeding some energy back to the zonal mean jet as well.

(1b). Other contributions to the higher latitude net terrestrial radiation include that from clouds (2b) that form in air holding the moisture and heat transported poleward by cyclones. The cyclone heat transport (3a) is strongest during the early and mature stages of the storms. During their decay stages the cyclones have momentum fluxes (6a), the flux convergence helps rebuild the midlatitude westerlies, allowing new storms to form and transport more heat, repeating the cycle again. The eddy heat (3a) and momentum fluxes (6a) oppose each other creating Ferrel cell transports (5b). The Eliassen-Palm flux (6b) shows the eddy amplitude shift from low to high elevations and the eventual feedback to build the jet stream (§7.4). Some kinetic energy is lost by friction (hatching), mainly at the earth's surface and where the shears are large, as near the jets.

5

OBSERVED NONZONAL FIELDS

The problem cannot be presented for solution with any real hope of success until observations have disclosed to us sufficient evidence of the true structure of the general circulation of the atmosphere and its variations to enable us to make an effective mental picture of the atmospheric problem, just as Kepler's co-ordination [with Tycho Brahe] made it possible to form a working mental picture of the solar system.

Shaw, 1926

So, naturalists observe, a flea
Hath smaller fleas that on him prey;
And these have smaller still to bite 'em;
And so proceed *ad infinitum*.

Swift, 1733

This chapter presents observed fields without applying a zonal average. Time averaging is retained. The additional degree of freedom requires some other restriction to keep the discussion manageable. Focus is placed upon certain levels of particular importance. Usually, one level is sufficient to show the pattern in the troposphere, away from the surface. A different set of balances influences the surface pattern than those influencing the free atmosphere, so it is useful to analyze the surface distribution separately.

The observed fields in this chapter reexamine the quantities whose zonal averages were shown in Chapters 3 and 4. The zonal mean fields provide a good, general, simplified sense of the atmospheric structure and major energetic properties. That is why the zonal averages were treated separately. That is also one reason why the classic books (e.g., Lorenz, 1967) focus exclusively upon zonal means. However, those averages are constructed from fields that have significant zonal variation. Features like frontal cyclones, extratropical long waves, and the Asian monsoon are fundamental elements of the general circulation. In order to comprehend fully such fundamental elements, the longitudinal structure must be shown. In this book the nonzonal fields

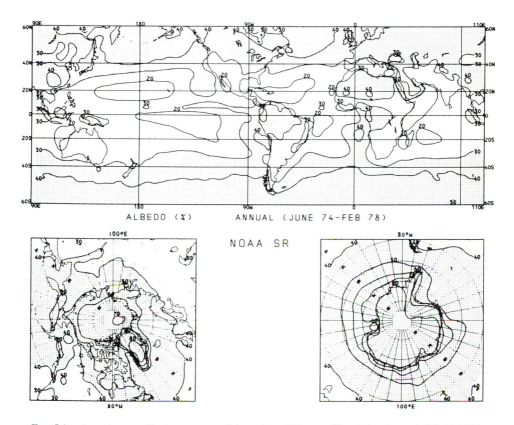

ALBEDO (%)   ANNUAL (JUNE 74-FEB 78)

NOAA SR

FIG. **5.1**   Annual mean albedo as measured by polar orbiting satellite during the period June 1974 through February 1978. Areas having albedo greater than 30% are shaded. The figure may be compared with Figure 5.13. From Winston et al. (1979).

are presented separately so as to build upon the base of understanding reached by examining the zonal means (and their interconnections) first.

5.1   Radiation fields

The longitudinal variation of annual average planetary albedo is shown in Figure 5.1 for the months of June 1974 through February 1978. Winston et al. (1979) obtain an annual average from this 45-month period by first making seasonal averages, then combining the four seasonal averages to form the annual average. The albedo, being a measure of the reflectance, is low in regions where the absorption of incoming solar radiation is high. The albedo is largest for areas that are cloud-covered, snow-covered, or desert. The areas where clouds are largely responsible for high albedos might be estimated by comparison with Figure 5.13. The data in Figure 5.1 are direct measurements by a polar orbiting satellite. Where polar regions are in darkness during winter, the albedos are averages only over months with daylight.

The following are general conclusions about the albedo distribution. (1) The general pattern consists of low values in the subtropics with high values in the mid-latitudes (associated with the traveling cyclone's clouds), in the tropics (associated with the deep convective clouds), and near the poles (due to snow cover as well as some cloudiness). The latitudinal variation was described in §3.1 (Figure 3.7c). (2) The major bright regions in the subtropics are due to stratiform cloudiness in the eastern oceans and bright desert areas (for example, the Sahara). (3) The land-ocean differences are mainly as follows. The high tropical albedos favor the land areas in summer due to preferential development of convective clouds over tropical lands. The land areas in winter tend to have higher albedos in middle and high latitudes due to snow cover. (4) The albedo has more longitudinal variation in the tropics and subtropics than at higher latitudes. On a time average, the net effect from the traveling midlatitude cyclones is fairly uniformly spread along a latitude circle.

The solar radiation that is not reflected is absorbed. Figure 5.2a shows the amount absorbed. Regions where the annual mean absorbed radiation is less than 250 Wm^{-2} are shaded. The absorption decreases rapidly toward the poles as one expects from the zonal mean absorption pattern (Figure 3.6). The absorption is rather similar to the albedo distribution with one main exception. The incoming solar radiation has a strong dependence upon latitude, so the absorbed radiation strongly decreases towards higher latitudes. The highest values of absorbed radiation are over the central Pacific and eastern Indian oceans.

For energy balance there must be longwave radiation back to space. The annual average longwave radiation is shown in Figure 5.2b. Radiative fluxes less than 250 Wm^{-2} are shaded. The bulk of the longwave radiation is emitted from the surface of the earth (if the sky is clear) or from the top of a cloud (where it is cloudy). The Stefan-Boltzmann law of blackbody emission states that the irradiance is proportional to the temperature of the object raised to the fourth power. For deep convective clouds, the cloud top is very high, very cold, and therefore emits comparatively little radiation. The regions of deep tropical convection (ICZ) show up quite well as shaded regions in this figure. The stratiform clouds to the west of Africa, North and South America have tops that are quite low (below 800 hPa), and so the emission is moderately high. Ice-covered areas such as Greenland and Antarctica are very cold at the surface and so have very low values of longwave emission. The contour intervals used in Figures 5.2a and 5.2b are not the same. In general, the gradient of the longwave emission is much less than that for the solar absorption. The change of gradient between absorbed and emitted was shown earlier, in Figure 3.6. Here, the differences between the magnitudes of emission and absorption imply zonal as well as meridional heat transports.

By subtracting the outgoing radiation (Figure 5.2b) from the incoming radiation (Figure 5.2a) one obtains the local net radiative balance. The net radiation is shown in Figure 5.2c. Shaded areas denote regions where the outgoing radiation exceeds the incoming radiation. Clearly, one expects the middle and high latitudes to be shaded (Figure 3.6.); but a deficit of incoming radiation exists at several other locations. The

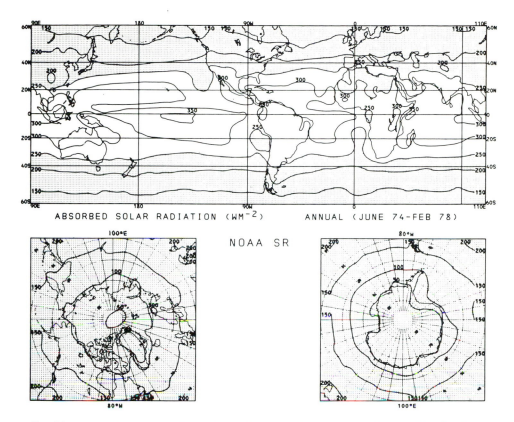

ABSORBED SOLAR RADIATION (WM⁻²)   ANNUAL (JUNE 74-FEB 78)

NOAA SR

FIG. 5.2a   Annual mean solar radiation absorbed by the earth, atmosphere, and ocean. Values less than 250 W m⁻² are shaded. The figure may be compared with Figure 3.6. Tropical and cloud–free regions generally absorb more radiation. From Winston et al. (1979).

Saharan and Arabian deserts have a net deficit because they emit longwave radiation at very high temperatures. They absorb less solar radiation than they emit due to the high albedo of their surfaces. Figure 5.2c implies that the high temperatures of these deserts are maintained by transports of heat into these desert regions. That heat transport is aided by air that is adiabatically warmed as it sinks and flows from subtropical surface highs toward surface lows. Later on in this chapter one sees the upper level convergent horizontal winds across northern Africa (Figure 5.25) that accompany the sinking. Not all deserts show a deficit; the Kalahari and the Australian Outback are two counterexamples. The persistent stratiform clouds located off the west coasts of Africa, North and South America also have a deficit of net radiation for a similar reason as the Sahara.

A key question in climate studies is the net effect of clouds upon radiative balance. The high albedo of clouds reflects incoming solar radiation; increasing cloudiness would cool the earth by this effect. However, clouds are very efficient at absorbing infrared radiation emitted by the earth; increasing cloudiness would warm the earth

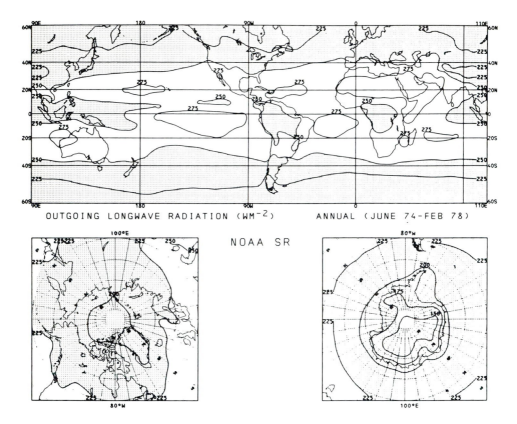

OUTGOING LONGWAVE RADIATION (WM^{-2}) ANNUAL (JUNE 74-FEB 78)

NOAA SR

FIG. 5.2b   Annual mean longwave emission by the earth, atmosphere, and ocean combined. This figure may be compared with Figures 3.6 and 5.1. Emission is greatest when coming from the earth's surface in the relatively cloud-free subtropics. Comparing with Plate 2, low tropical values of emission occur where clouds tops are frequent and high. Values less than 250 W m^{-2} shaded. From Winston et al. (1979).

by inhibiting radiative cooling of the surface. Which effect dominates? Only very recently have measurements from the Earth Radiation Budget Experiment (ERBE) confirmed that, on average, the albedo change has greater effect (e.g., Ohring, 1990). ERBE measurements show, on average, a net 14 W m^{-2} net loss of radiation due to clouds. A review by Arking (1991) places the net loss between 17 and 27 Wm^{-2}. Harrison et al. (1990) find smaller values of net cooling (14 to 21 Wm^{-2} on the annual average); they find near cancellation of the two effects in winter, so that the summer cloud cooling is quite strong. Clouds are thus reducing the seasonal change in net radiative heating.

5.2   Temperature fields

The first large-scale maps of surface air temperature were probably made by von Humboldt in the early 1800s; the maps were not officially published until 48 years after

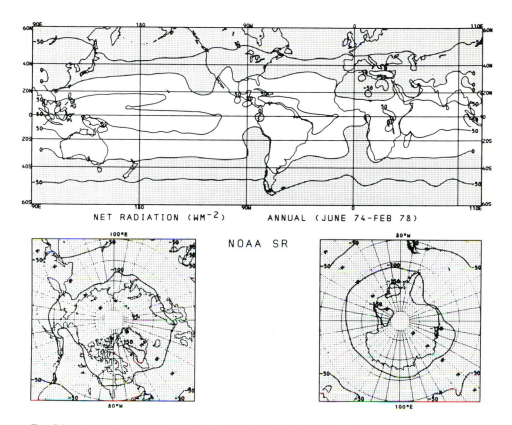

FIG. 5.2c   Annual mean net radiation budget is shown. Shading denotes negative values. Net radiation is the difference that results from Figure 5.2a minus Figure 5.2b. Integrated over the whole planet, the net radiation should equal zero. From Winston et al. (1979).

his death (Hildebrandsson and Teisserenc de Bort, 1907). Though highly smoothed, the isotherms drawn by von Humboldt showed cooler temperatures over eastern Asia and North America, with warmest temperatures over Europe. Other early maps were made by Kupffer (1829; annual mean) and Dove (1852; monthly means).

The surface air temperature pattern is indicated in Figure 5.3 for January and July. The surface air temperatures are based upon measurements at shelters located 1.5 to 2.0 m above the ground. As discussed in Chapter 3, the orientation of the temperature gradient is generally meridional; the major zonal asymmetries occur near the boundaries of continents. The midlatitude and polar land areas are usually colder in winter than the oceans at the same latitude. During summer the land areas are warmer. The seasonal change in temperature is estimated by taking the difference between Figures 5.3a and b. Obviously the surface temperature has much greater seasonal variation over the continents than over the oceans. Additionally, the seasonal variation is greater over the Northern Hemisphere.

The surface air temperature is strongly influenced by several local effects. Many

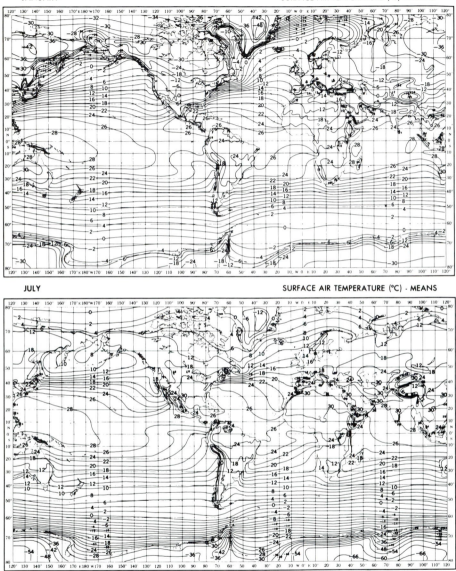

FIG. 5.3   Surface air temperature in °C during (a) January and (b) July over the globe. From Tanner (1990).

of the contours in Figure 5.3 outline regions of high elevation (e.g., Tibet). Surface type strongly influences the air temperature (e.g., cold eastern ocean boundary currents like the California current or cold glacier-covered regions like Antarctica). To gain a more comprehensive picture of the thermal structure of much of the troposphere, the temperature at a level above the atmospheric boundary layer is more appropriate.

The 700 hPa level is a representative level for the tropospheric temperature field. A lower troposphere level is chosen for several reasons. First, the eddy heat fluxes are largest near the 850 to 700 hPa levels (Figure 4.31). Second, one finds almost the same temperature pattern at the 700 and 850 hPa standard levels. Third, using a thermal wind argument, this temperature can be related to the surface and 500 hPa geopotential height patterns. Figure 5.4 depicts the average temperature fields at 700 hPa for 10 years of winter and summer months. This figure also provides information about the horizontal and vertical transient eddy heat fluxes. Hoskins et al. (1989) compile this figure from analyses archived at ECMWF.

While isotherms in the Southern Hemisphere line up strongly with latitude, both hemispheres have notable zonal asymmetry, mainly in zonal wave numbers 1, 2, and 3. As will be shown, this long wave pattern reappears in the geopotential and other fields. In the Northern Hemisphere winter, the main thermal troughs are centered near 80 W and 130 E, with a weaker trough over eastern Europe (roughly 30 E). The horizontal temperature gradient is strongest near the bases of the two major troughs in the Northern Hemisphere and to the south of Africa in the Southern Hemisphere. In summer the troughs are less prominent and the gradient is weaker. In the Northern Hemisphere the thermal troughs are located over the eastern sides of the continents because cold air forms over the central continent and is advected eastward as well as equatorward behind frontal cyclones. The situation in the Southern Hemisphere is a little different because only South America extends into middle latitudes. In the Southern Hemisphere weak thermal troughs exist downwind (east) of the continents; a weak wavenumber 3 pattern can be found in Figure 5.4d. While the time mean pattern is quite zonal in the Southern Hemisphere, long waves are prominent on any given day. In the Southern Hemisphere these long waves propagate more readily (due to a comparative lack of topography; §7.2.1) and therefore contribute less to a time average.

The heat flux is mainly poleward as one would expect. However, significant zonal components exist, directing the flux mainly down the local gradient of temperature (from warm air to cold). The eddy heat fluxes thereby reduce the temperature gradient. The largest heat fluxes occur along the east side of each thermal trough, or where the horizontal temperature gradients are strongest. The latter situation occurs in the region off southern Africa (Figure 5.4d). These maximum eddy heat fluxes are mainly associated with the mature stage of transient baroclinic eddies. The heat fluxes and their relationship to observed properties of the atmosphere and ocean are considered further in this chapter (§5.7; §5.8). More will be said about the heat flux geography in the discussion of eddy lifecycle theories (§7.3.2).

Most of the general features discussed above are found in studies using other data sets. Lau and Oort (1981) compare analyses by GFDL and the U.S. NMC at 850 hPa for nine winters. While the NMC and GFDL analyses differ by as much as 4°C (over the subtropical oceans) the same three-trough pattern is found over the Northern Hemisphere. A similar study was carried out by Karoly and Oort

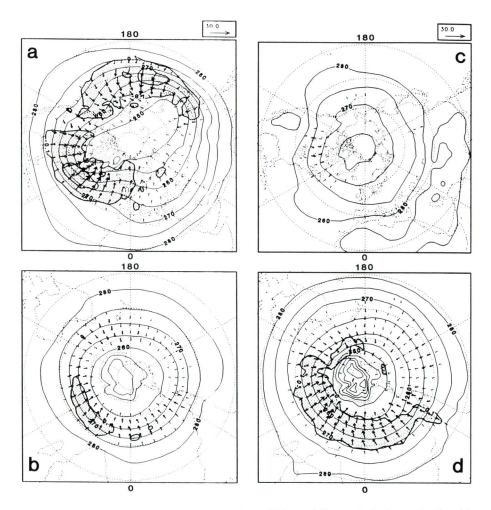

FIG. 5.4   Temperature at 700 hPa is contoured using a 5 K interval. Vectors depict the zonal and meridional components of transient eddy heat flux ($\overline{V'T'}$). The vector length is proportional to heat flux magnitude, where the vector length scale of $30\,\mathrm{K\,m\,s^{-1}}$ is indicated at the top of the figure. Ten years of
seasonal ECMWF final analyses are averaged. (a) Northern Hemisphere, December–February, (b) Southern Hemisphere, December–February, (c) Northern Hemisphere, June–August, (d) Southern Hemisphere,
June–August. From Hoskins et al. (1989).

(1987), except that they analyzed ten Southern Hemisphere winters and used GFDL
and Australian WMC data. Both analyses in Karoly and Oort (1987) show several
weak thermal troughs. Contours of the meridional component of eddy heat flux in
Karoly and Oort (1987) agree with Figure 5.4d: maxima are southeast of Africa
and east of South America, while minima are south of Australia and west of South
America.

5.3   Velocity fields

Halley (1686) made the earliest known large-scale chart of the general circulation winds. Halley's chart corresponds remarkably well with modern charts. A side-by-side comparison is found in Shaw (1926; his Figures 101 and 102) or can be made by comparing Figures 1.1 and 5.5 herein.

The time mean wind vectors at 1000 hPa illustrate the surface wind pattern. The arrow length is proportional to the vector wind speed in Figure 5.5. Since vectors are added, areas with short arrows on the chart are not necessarily regions of light winds. For example, equally strong winds that blow half the time from the east and half the time from the west would have zero magnitude on this chart. However, if the wind direction is approximately steady and the wind speed is normally distributed, then the vector average used here approximates (but underestimates) the time mean wind speed. Consequently, large arrows in Figure 5.5 more often indicate persistence of wind direction. The regions with large arrows in the tropics correspond particularly well with persistent winds, as can be seen in Figure 14.1 in Palmén and Newton (1969). These trade winds were so named because sailing sea traders could count upon the winds to blow in a certain direction. The lower branch of the Hadley cells should show up in Figure 5.5; one expects the equatorward motion to have a westward component. Such simple motion is indeed seen over the tropical Pacific and Atlantic oceans. But the "Hadley" cells are localized and highly distorted. They are seen as the equatorward flow from the large, subtropical, anticyclonic gyres. The equatorward sides of the subtropical highs provide the bulk of the low level "Hadley" cell motion, and these highs (each indicated by an "H") are centered over the oceans. The identifiable thermally direct ("Hadley") circulations include the summer monsoon over the Indian Ocean in July and the convergence over southwestern Brazil during January. The so-called intertropical convergence zone (or ICZ) is where one such gyre meets its counterpart from the opposite hemisphere. Even on these time mean maps the ICZ, indicated by open dash lines, meanders.

One cause of the meander is the geographical distribution of the seasonal change. As noted in §3.3, the winter hemisphere Hadley cell predominates. Over Brazil and from the Indian Ocean eastward to the western Pacific, the seasonal change is quite strong and consistent with the seasonal changes discussed in Chapter 3 (e.g., shift of the rising branch of the dominant winter Hadley cell across the equator). Elsewhere, particularly over the eastern Pacific, the ICZ position varies little with season. The low-level convergence zone thus meanders over a wide range of latitudes. The convergence is concentrated in certain regions, particularly land areas, as well.

In satellite film loops, upper-level clouds appear to stream eastward and poleward from areas of relatively more intense ICZ convection. The cloud bands often extend into both hemispheres. The film loops create the impression that the high-level return flow from the "Hadley" cells is being focused into narrow bands of stronger flow. The clouds are persistent enough to appear on time mean charts as well as instan-

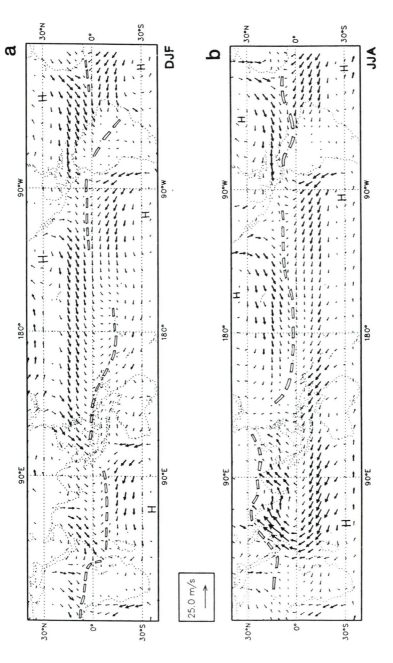

Fig. 5.5 Seasonal vector mean winds at 1000 hPa over 10 years of ECMWF final analyses. (a) December–February and (b) June–August. Time means of the instantaneous vector winds are plotted; see text. Vector length is proportional wind speed; vector length scale of 25 m/s is plotted at left center of chart. The open dashed lines indicate areas of low-level convergence and as such indicate the approximate time-mean locations of the ICZ. Redrawn from Hoskins et al. (1989).

170

taneous images discussed later (§5.5). From diagrams of the mean meridional cells (Figure 3.17), one expects cross-equatorial, high-altitude flow from the summer into the winter hemisphere. That ageostrophic cross-equatorial flow is difficult to identify in charts of upper-level winds.

Strong cross-equatorial flow is occasionally seen in the GWE data. For example, Bengtsson et al. (1982) remark upon several intense episodes of interhemispheric exchange during GWE. Figure 5.6 gives an example of the actual trajectories followed by air parcels over Brazil and the Atlantic during three days in January 1979. The low-level convergence occurs primarily over central Brazil. These trajectories reveal much cross-equatorial southward flow at low levels (e.g., trajectory **A**). Strong rising motion occurs along the line of low-level convergence. Above, in the upper troposphere, the air spreads out. The corresponding infrared satellite image mosaic gives the impression that two bands of high-level clouds are streaming out from the deep convection over Brazil—one towards the Northern and one towards the Southern Hemisphere. But contrary to the impression given by Figures 3.17 or 5.6b, very few trajectories actually cross back into the Northern Hemisphere. Instead, most parcels on the north side of the convection follow a path similar to trajectory **D** in Figure 5.6a. For example, parcels **B** and **C** are eventually captured by the midlatitude westerlies of the *Southern* Hemisphere. It is possible that the wind fields used in creating Figure 5.6a are significantly in error. The vertical motion field in the original ECMWF analyses has been criticized by several researchers (it is almost one-third the size of the vertical motions in the corresponding GFDL analyses, for example). But the ECMWF vertical motion was not used here; instead, a more accurate kinematic vertical motion using a variational calculus correction (O'Brien, 1970) has been employed. It is also possible that the geographic area shown in Figure 5.6 is not representative of the remainder of the tropics. Nonetheless, it seems clear from examining cloud motions and crude trajectories that the high-level tropical winds have complex structure.

Contours of wind speed and vectors of ageostrophic wind are given in Figure 5.7. (The geostrophic wind direction may be deduced from the time-mean pressure fields.) The largest velocities are occurring at high levels in the troposphere. To represent the jet-stream flow, the 200 hPa level has been chosen for the time-mean maps. The discussion here focuses upon three subjects: the jets, the ageostrophic winds, and the fidelity of the data shown. Because of the projection being used, it is not feasible to identify cross-equatorial motion in Figure 5.7.

The strongest winds in the Northern Hemisphere are near the eastern boundaries of the continents. The thermal trough in the middle troposphere and the long wave, time-mean pressure troughs are centered near these locations. The bands of maximum wind are identified with the so-called subtropical jet streams. Considering the Northern Hemisphere first, two separate maxima are found. The largest velocities are off the east coast of Asia, averaging in excess of $70\,\mathrm{ms}^{-1}$! The maximum off eastern North America is less in the time mean. Curiously enough, the pattern is rather similar in the Southern Hemisphere: a dominant jet maximum covers nearly the same

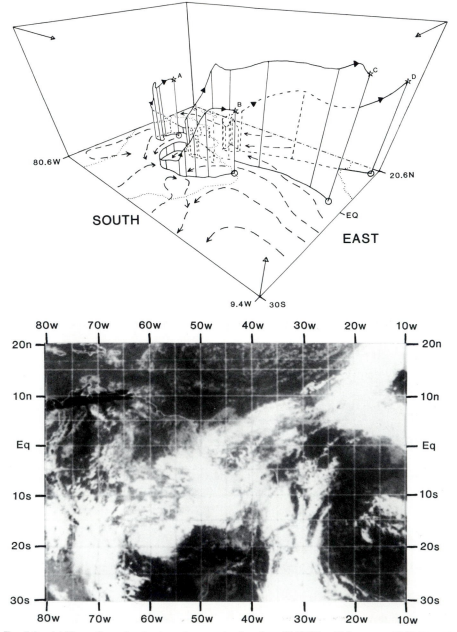

Fig. 5.6   (a) Three-dimensional trajectories over nine days from 00 GMT on 11 January to 00 GMT on 20 January 1979. ECMWF GWE data are used except for vertical velocities that are calculated by a variational, kinematic scheme. The star symbol shows the three-dimensional location of a parcel at 00 GMT on 20 January; the circle symbol shows the geographic location of the parcel on the earth at the same time. Vertical bars connect the three-dimensional path and its projection upon the earth at 24-hour intervals. Streamlines show the flow field at 850 hPa at 00 GMT on 15 January. See text for discussion. (b) Infrared image of the same domain shown in part (a). The satellite picture is a composite of polar orbiter images on 16 January 1979 when the satellite crosses the equator at 15:15 local time. Satellite image courtesy of NOAA.

longitudinal range as in the Northern Hemisphere and a weaker jet occurs at roughly the same longitudes, as well. The similar longitudinal locations arise because the jet maxima are linked with the areas of heaviest tropical rainfall (§6.5). The largest average velocities in the Southern Hemisphere just exceed $50 \, \text{ms}^{-1}$.

The zonal mean observations also contain a tropical easterly jet. From Figure 3.15, the largest easterly winds are in the stratosphere. However, easterly wind maxima are identifiable at $200 \, \text{hPa}$ and are usually centered between east Africa and Indonesia. At $200 \, \text{hPa}$, these two easterly jet maxima exceed $20 \, \text{ms}^{-1}$ on a summer-long average.

An alternative view of the jet streams is presented in Plate 1. Plate 1 shows the average conditions during January 1979 and during July 1979. The plate figures use a three-dimensional contour of the wind speed: the white, cloud-like masses enclose regions where the wind speed exceeds $35 \, \text{ms}^{-1}$. The entire globe is shown, as can be determined by the geography mapped onto the bottom of each box. (The topographic relief uses fake shadows; the surface as drawn is flat.) The vertical dimension in each color figure is greatly exaggerated to show the vertical structure better; each box is $20 \, \text{km}$ high. Contours of the sea-level pressure are also included ($12 \, \text{hPa}$ interval; $1000 \, \text{hPa}$ contour is dashed). Shadows are drawn on the bottom of each box to give a sense of the elevation of each jet; they are drawn as if a light source had been placed just north of the north edge of the box at roughly $50 \, \text{km}$ elevation; hence the shadows appear to be slightly south of where the jets actually are located. (This light source location was chosen to make the fake topographic relief shadows more easily understood.)

During January (upper figure in Plate 1) the subtropical jets predominate in both hemispheres. A tiny piece of the polar night jet is found above Siberia. The shadows clearly show how the northern subtropical jets start at a low latitude (e.g., near 15 N over the west African coast) and end in the middle latitudes on the southern side of the Aleutian and Icelandic lows. Though it is summer, a strong subtropical jet extends across the southern Indian ocean and halfway across the south Pacific (directly above the dashed contour). During July (lower figure in Plate 1) the northern subtropical jets have withered away, but the easterly jet is just visible (small volume centered near India). A similar easterly velocity maximum is seen west of India in Figure 5.7c. The southern subtropical jet is further north than before and the strongest wind speeds are found over Australia. Further south is a second, globe-girdling band where high winds from the polar night jet extend into the troposphere. The jet tube pattern in the lower color figure is deceptively smooth because the Southern Hemisphere long wave pattern is so weak on a time average. On individual days the three-dimensional subtropical jet tubes are broken into individual pieces that are located above the surface fronts, particularly where the fronts are equatorward of the surface lows.

Vectors of ageostrophic velocity are also given in Figure 5.7. These vectors reach significant magnitudes. (The largest appear to be about $15 \, \text{ms}^{-1}$, off the east coast of Asia. More typical maximum values are about $5 \, \text{ms}^{-1}$). The ageostrophic vectors

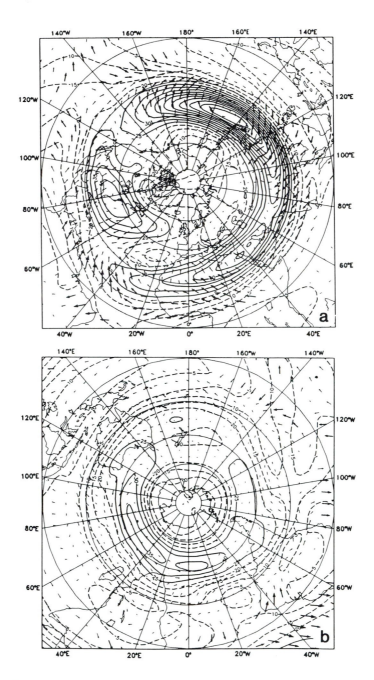

FIG. 5.7   Time-mean winds at 200 hPa. Contours of wind speed use a 5 m/s interval; arrows depict vector ageostrophic velocity where the largest arrows near 30 N, 100 E in (a) have magnitudes of 10 to 15 m/s. (a) Northern Hemisphere, December–February, (b) Southern Hemisphere, December–February,

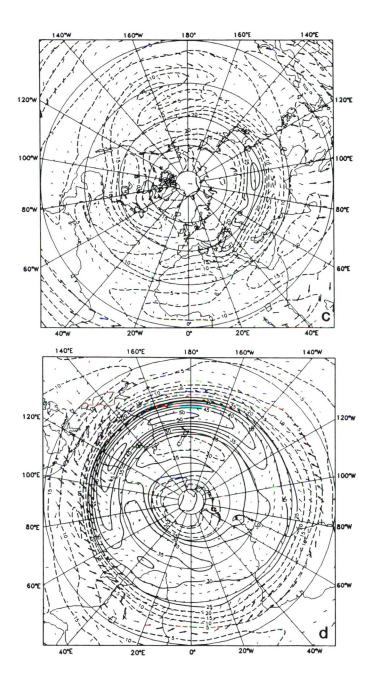

(c) Northern Hemisphere, June–August, (d) Southern Hemisphere, June–August. The periods come from 10 years of data from December 1978 through August 1989. Figures courtesy of I. James, personal communication, Dept. of Meteorology, Univ. of Reading.

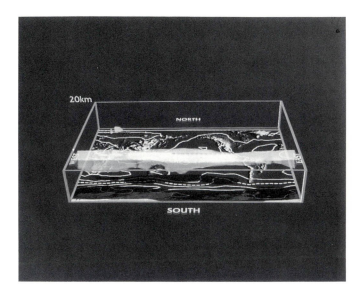

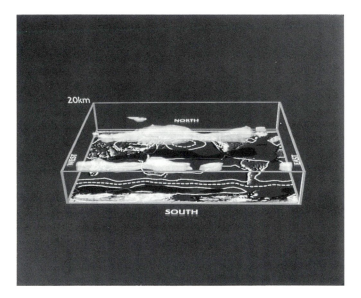

PLATE. 1 Three-dimensional depiction of the global jet streams (January 1979, above; July 1979, below). The cloud-like volumes enclose that portion of the atmosphere where the wind speed exceeds 35 m/s when averaged over the month. The vertical dimension is exaggerated; the box is 20 km deep. Colors on the *flat* box bottom depict elevation with two exceptions: white indicates elevation greater than 3050 m or permanent ice-covered surface; blue indicates water surface regardless of elevation. Contours are of sea level pressure. See text. (For color plate see front matter.)

are clearly distributed in a special way relative to the jet maxima. The poleward ageostrophic velocities occur where the jets are increasing downstream and vice versa. The link between the two is easily seen from the zonal momentum equation, which is further explored in §6.2.2.

As discussed in connection with Figure 2.6 (§2.5), significant differences can occur among different wind analyses from the same observational data. However, the general pattern, location, and relative strengths of the jets are the same in most analyses. Lau and Oort (1981) compare zonal and meridional wind analyses by U.S. NMC and GFDL; they show results at 200 hPa for Northern Hemisphere summer and winter. Differences in zonal wind maxima east of Asia exceed $10\,\mathrm{ms}^{-1}$ in both seasons, with slower winds being found in the GFDL data. The subtropical jet extends further into the Pacific in the NMC analyses. A similar trend is found off the east coast of the United States. The meridional wind is also slower over the oceans in the GFDL data (by 3 to 5 ms^{-1}, which is a sizable fraction of the time-mean value). It is difficult to compare the Lau and Oort (1981) figures (using components) with the vector magnitude plotted in Figure 5.7a because the meridional wind often reverses sign at a given location. If that problem is ignored, magnitudes calculated at selected points over the central north Pacific indicate that the ECMWF data shown here may be closer in magnitude to the NMC data. Karoly and Oort (1987) compare 200 hPa velocity components in Australian and GFDL data in the Southern Hemisphere. The zonal wind is less nearly everywhere in the GFDL analyses; in many oceanic locations the zonal wind difference exceeds $10\,\mathrm{ms}^{-1}$ in both seasons. The disagreement is particularly large across the southern Indian Ocean, which is hardly surprising given the paucity of observations there (Chapter 2). Finally, Bengtsson et al. (1982) remark that the GWE data showed the Southern Hemisphere winter circulation to be "more intense than previously assumed."

5.4   Pressure fields

Beginning in 1868, Buchan produced some of the earliest isobaric maps for the globe (e.g., Buchan, 1889). Figure 5.8 shows the global sea level pressure pattern during January and July. The equatorial trough (Figure 3.21) is weak and quite longitudinally variable. In the subtropics, the lowest pressures are over land areas during the summer in the subtropics. Some areas have persistent low pressure, the main example being the equatorial western Pacific. In the subtropics and midlatitudes the time-mean pattern is dominated by large gyres.

In the Northern Hemisphere, Figure 5.8 shows a striking reversal of the pressure pattern from winter to summer. During winter there are deep lows centered over the northern Atlantic and Pacific (the Icelandic and Aleutian lows) with high pressure over the continents (the Siberian and Canadian highs). The subtropical highs are comparatively weak during winter. The summer pattern is the reverse of this; the subtropical highs have expanded in area over the oceans and the pressure is lower

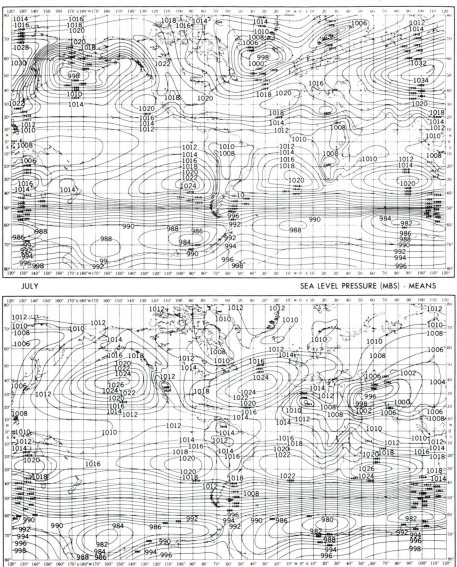

FIG. 5.8   Sea level pressure patterns in (a) January and (b) July using a 2 hPa interval. From Tanner (1990).

over the land. The Aleutian low has all but disappeared (in the mean), while the Icelandic low has shrunk and is centered over Baffin Island. The pressure pattern over the land is easily understood as a rather shallow, surface thermal effect: in winter the cold surface air is denser; in summer the warm air is less dense compared to the nearby ocean. The shallowness will be evident in the 500 hPa pattern.

The Southern Hemisphere pattern is different. The subtropical highs tend to be centered over the oceans all year. The midlatitude pattern is clearly much more zonal, with a belt of intense low-pressure cells centered near the coast of Antarctica. The pattern explains the very low surface pressures remarked upon earlier (Figure 3.21). Little seasonal change is apparent except over the continents. Le Marshall et al. (1985) produced their own ten-year climatology and found the circumpolar lows to be deeper and the subtropical highs higher than the Southern Hemisphere analysis shown either here or in Taljaard et al. (1969).

Moving to a higher level, the 500 hPa geopotential height pattern (Figure 5.9) has troughs and ridges that are quite similar to the 700 hPa temperature pattern in Figure 5.4. The similarity is hardly surprising. The thickness of the layer between two pressure surfaces is proportional to the average virtual temperature of the layer. As can be judged from Figure 3.21, the height variations are much greater at 500 hPa than at 1000 hPa. Since the wind is primarily geostrophic (outside of the tropics), the strongest height gradients correlate well with the areas of maximum wind speed (Figure 5.7). The Southern Hemisphere height pattern tends to be rather zonal with one deep (cold) low centered over Antarctica, whereas the Northern Hemisphere has a distinctly bimodal structure associated with the cold air over Canada and western Siberia. The meridional gradient is stronger in winter than in summer, especially in the Northern Hemisphere as seen in the 700 hPa temperatures. The Siberian upper-level trough is not present during summer, but the Baffin Island trough remains. Presumably, the melting of the snow cover over Siberia allows that region to warm up enough during summer to "fill" the upper-level trough; the Greenland and Baffin Island ice sheets inhibit this process (thus acting rather like Antarctica in the opposite hemisphere).

Lau and Oort (1981) and Karoly and Oort (1987) compare wintertime 500 hPa heights for the Northern and Southern Hemispheres respectively in two different analyses. In the Southern Hemisphere the GFDL and Australian WMC analyses differ in the following way: the heights are lower over Antarctica in the GFDL analyses, but generally higher over the oceans (by more than 80 m over large areas). The geostrophic winds are thereby stronger in GFDL analyses. In the Northern Hemisphere the GFDL and U.S. NMC analyses differ by 30 m over large oceanic areas; the main differences are consistent with the Asian subtropical jet differences mentioned earlier. The NMC analysis carries the jet further east. Consequently that analysis has stronger 500 hPa height gradient further east. The chart shown here (Figure 5.9a) uses the GFDL analysis.

Figure 5.10 shows the root mean square (RMS) time variances of 250 hPa heights calculated by Hoskins et al. (1989). The variances were constructed using a "high pass" frequency filter that Hoskins et al. use to mimic the $\simeq 2.5$ to 6-day band pass filter exploited by Blackmon et al. (1977). Hoskins et al. use 10 years of ECMWF data. The RMS of z'' has maxima in the middle and higher latitudes in agreement with Figure 3.22. This quantity mainly measures transient eddy activity. Regions where the eddies have large amplitude are also areas where $z''z''$ is large, except in

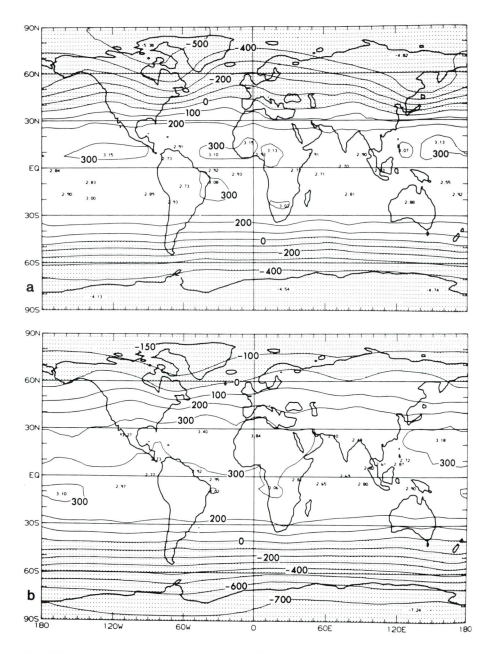

FIG. 5.9 Geopotential height contours of the 500 hPa surface; seasonal averages (a) December 1962 through February 1973 and (b) June through August from the years 1963–1973. Contour interval is 100 m and the contour labels are height values with 5,572 m subtracted. From Oort (1983).

the case where those eddies do not migrate or significantly vary in amplitude. Hence the axes of larger values approximately identify the cyclonic storm tracks (as will be seen in §5.8). At the downstream end of the Northern Hemisphere storm tracks are the locations of the climatological Aleutian and Icelandic lows. The maximum is larger over the north Atlantic than over the north Pacific during both seasons shown. The Southern Hemisphere pattern is more zonally uniform, with relative maxima in the southern Indian and Atlantic oceans; a secondary maximum is located near New Zealand. The variance changes little between winter and summer in the Southern Hemisphere in contrast to the Northern Hemisphere.

Maximum values are located just downstream from the long wave troughs, in regions where the strongest heat fluxes are found (Figure 5.4). That positioning indicates that the mature growing eddies are obtaining energy by baroclinic energy conversions (4.67). The chapters on general circulation theories explore the relation between storm tracks and long waves further (§6.5; §7.3).

Lau (1984) has calculated a similar quantity (500 hPa RMS height deviations) for two 3-month periods. Most of the features discussed above appear in Lau's analyses, with two prime exceptions in June-August data: Lau finds the maxima located in the south Pacific and the north Pacific. Karoly and Oort (1987) show 500 hPa RMS time deviations for ten Southern Hemisphere winters; the main differences from Lau's analysis are: the maximum near the Ross Sea is 45° west, the Indian Ocean maximum is 30° west, and the contours are smoother in both of their analyses. Therefore, the Karoly and Oort data look more similar to Figure 5.10. The Australian WMC analysis is more similar to Figure 5.10 than the GFDL analysis. Lau and Oort (1982) plot the same quantity (except using 300 hPa data), averaged for six Northern Hemisphere winters using GFDL and U.S. NMC data. Lau and Oort (1982) find maxima south of Greenland (similar to 5.10a), but also over the Alaskan peninsula (further north than in 5.10a) in both analyses. The NMC analysis looks more similar to Figure 5.10a than the GFDL analysis.

The Northern Hemisphere midlatitudes are often characterized by two distinct classes of circulations: transient eddies and stationary eddies. The two classes can be seen in Figure 5.11, showing the spectral decomposition of the height field at three pressure levels in the latitude band 35 N to 45 N. Three peaks are found at 500 hPa: at wavenumbers one and three (the long wave pattern of Figure 5.9) and at wavenumber six (the frontal cyclones). At 1000 hPa the frontal cyclones again show up (peak at wavenumber 6), as do the Aleutian and Icelandic lows (peak at wavenumber 2).

5.5   Cloudiness, rainfall, and humidity

Teisserenc de Bort plotted the first charts of cloudiness (January and July averages) in the middle 1880's. According to Shaw (1926, p. 292), Supan made the first global estimates of precipitation (over land areas) in 1898. Although quantifying cloud cover is immeasurably easier now using satellites, a difficult problem remains regarding

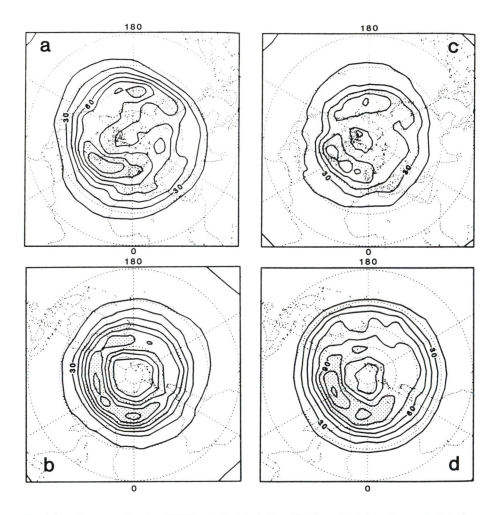

FIG. 5.10   Root mean transient 250 hPa standard deviation of "high pass" height variance calculated by Hoskins et al. (1989) from ECMWF data. Hoskins et al. use a high-pass filter intended to approximate a 2.5 to 6-day band pass filter; the field is intended to isolate the contribution by transient eddies. The seasonal averages are as follows. (a) Northern Hemisphere and (b) Southern Hemisphere during December 1978 through February 1989. (c) Northern Hemisphere and (d) Southern Hemisphere during June through August from 1979 through 1988. Contour interval is 15 m with values above 90 m shaded.

the depth of clouds (because high clouds can obscure lower clouds). Precipitation measurement remains poor; there are large variations over land due to topographic effects, and measurements over most ocean areas are infrequent, at best.

Throughout this book, most charts show time-mean fields. Time-mean fields smooth out the weather features; the smoothing is most pronounced for cloudiness. Accordingly, Figure 5.12 illustrates an instantaneous view of the cloud patterns. The image is constructed from the intensity of radiation measured by a satellite at one wavelength of infrared light. Low intensity is assigned the value white; high intensity

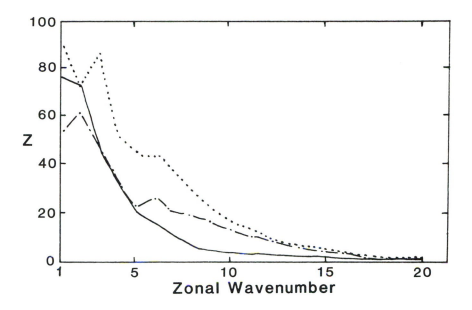

FIG. 5.11   Geopotential height amplitude as a function of zonal wavenumber at three levels: 50 hPa (solid line), 500 hPa (dotted line), and 1000 hPa (dot-dashed line). Figure based upon data in Miyakoda et al. (1972).

(warmer temperature) is assigned black. Low clouds are warmer and appear gray; high clouds are colder and appear white. This figure has been discussed earlier in regard to the cross-equatorial flow at high levels near 145 W. Another item to note is the relative absence of deep clouds in the eastern Pacific. The convection was uncharacteristically low at the moment the image was taken; even so, the figure illustrates the point that ICZ convection is often organized into localized regions of enhanced convection (such as that near 100 W). The figure also illustrates that some high cloud bands found in time-mean cloud charts (Figure 5.13) often appear as a band in an instantaneous view; an example is the south Pacific convergence zone (SPCZ). The SPCZ is located from 5 S to 40 S along 150 W in Figure 5.12. The cloud band in the subtropical north Pacific (NPCB) is located along the south side of the subtropical jet (Figure 5.7a), along the east side of a mid-ocean "tilted" trough (Figure 6.9). The image is one frame from a movie; on successive images the motion along the NPCB is obviously toward the east-northeast. Conservation of angular momentum can explain the increasing speed of the subtropical jet along this axis. The NPCB is commonly seen in winter, though it does not show up as well in Figure 5.13 as does the SPCZ.

Figures 5.13a and 5.13b are remarkable. These figures show the digitized percent of cloud amount measured by satellites averaged over two seasons during 1984 through 1987. Data from different satellites were combined. Differences in cloud detection sensitivity lead to the artificial "column" in the Indian Ocean; for example, the Meteosat satellite (viewing the center region) has a broader visible channel than other

FIG. 5.12   Infrared image from a geostationary satellite taken at 20:30 GMT on 22 January 1979. Brighter areas are high clouds, darker gray areas are lower clouds. Image is one frame in a motion picture entitled "Infrared Observational Data from the Geostationary Operational Environmental Satellites GOES-East and GOES-West," produced by the W. A. Bohan Co. (1983).

satellites. These photos may be compared with other composites, such as Plate 2, and with the figures of digitized reflectivity in Miller and Feddes (1971). The following general comments can be made. (a) The ICZ (intertropical convergence zone) is the thin band stretching across both the eastern equatorial Pacific and Atlantic. The ICZ is brighter (therefore more active) during July than during January. The ICZ is barely visible in the Indian ocean in Figure 5.13a and not apparent in Figure 5.13b. The ICZ cloud band locations correspond nicely with the convergence lines indicated in the surface winds (Figure 5.5). (b) The ICZ is quite weak compared to the regions of enhanced tropical convection, which are mainly found over land masses of the summer hemisphere. Thus, in December through February the Amazon Basin, Congo Basin, and Indonesia are areas of enhanced convection. In June through August the main areas are India and Southeast Asia, with lesser areas near the Guinea coast and Columbia. (c) The land-ocean boundary is often a demarcation between cloudy and clear sky. This is not merely a desert-ocean interface. It is seen in three other ways: first, convection enhanced over warmer land (e.g., South America in January); second, air forced to rise over mountains by prevailing winds (e.g., Madagascar in July); and third, stratus cloud due to cold ocean surface temperatures in upwelling regions (e.g., the subtropical west coasts of North and South America, and Africa). (d) Little seasonal shift occurs for the clouds over the equatorial Pacific and Atlantic, but a strong shift occurs over the Indian Ocean, and that shift is clearly related to the Asian monsoon. (e) The cloud pattern in middle latitudes of the Northern Hemisphere is most prominent in the western oceans, locations of the principal storm tracks (§5.8). The transient eddy activity in the Southern Hemisphere causes a nearly featureless time-mean cloud band. (f) The SPCZ (South Pacific Convergence Zone) is quite apparent as the diagonal cloud band to the east of Australia. Many of these features are evident in the Miller and Feddes atlas, especially (c) and (f).

An even more remarkable depiction of the time-mean clouds is shown in Plate 2. Figures 5.13a and 5.13b are constructed from visible light reflected from the earth. In contrast, the color figures in Plate 2 are constructed from infrared images (similar to Figure 5.12). Information about the cloud-top temperature has been used to assign each cloud-top emission to a temperature and therefore to an approximate pressure altitude. Colors have been assigned to different ranges of cloud-top elevation in the atmosphere: high clouds are blue, low clouds are red, with green assigned to the middle tropospheric elevations. The brightness of the color used in Plate 2 is proportional to the cloud cover percentage. Comparatively cloud-free desert regions appear black. The tropical convection along the ICZ is deep because the cloud tops reach very high elevation, as expected from the high elevation of the equatorial tropopause. The low stratus, found at the eastern edge of subtropical oceans in Figure 5.13, are clearly depicted in red. The prevalent cloud cover in the midlatitude storm belts is mainly low and middle clouds, in agreement with the application of Figure 3.26 to data in Figure 3.25. Over the arctic in summer, the cloud cover is low or middle cloud, in keeping with earlier statements about predominant stratus clouds there. In the upper figure of Plate 2 (January 1979) the SPCZ has an unusual two-banded structure; the NPCB is quite obvious, as are tropical Atlantic bands mentioned in connection with Figure 5.6.

Figure 5.13c shows percent cloudiness data compiled by Warren et al. (1991) and Warren et al. (1986, 1988) using reports by weather observers. The same data sets used to produce Figure 5.13c were used to construct Figure 3.26. Warren et al. group the data into 5 × 5 degree latitude by longitude boxes over most of the globe before constructing the contours; consequently, Figure 5.13c has less detail than Figure 5.13a. For example, the ICZ over the Pacific has broader meridional extent in Figure 5.13c. Not surprisingly, Figures 5.13a and c are quite similar. The principal differences are that the analysis by Warren et al. shows more cloudiness off the west

PLATE. 2   Mean cloud cover during January (above) and July (below) 1979. Percentage of cloud cover indicated by the intensity of the color, as indicated by the scale. The mean height of the cloud top is assigned to one of three categories, each assigned a different color. The data resolution is 2°latitude by 2.5°longitude. Data processed at the Goddard Lab. for Atmospheric Science, GSFC; plate produced by Hussey, Hall, and Haskins at the image processing lab, JPL. Plate courtesy of M. Chahine, personal communication. (For color plate see front matter.)

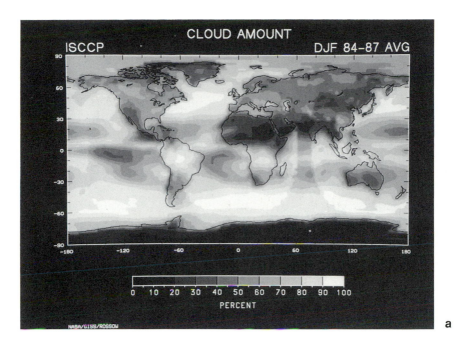

a

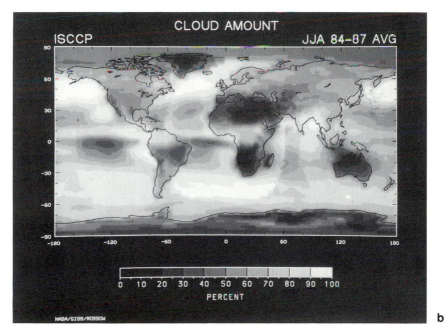

b

FIG. **5.13a and b**   Time average percent cloud amount maps for the four years of these seasons: (a) December–February and (b) June–August during 1984 through 1987. These data are compiled from 11 different satellites over the period. The various polar orbiting and geostationary satellites have different sensitivity and viewing angles, creating artificial boundaries (e.g., Indian and southern Atlantic oceans). The cloud amounts are therefore uncertain to 10%. Photographs courtesy of W. B. Rossow, NASA/GISS.

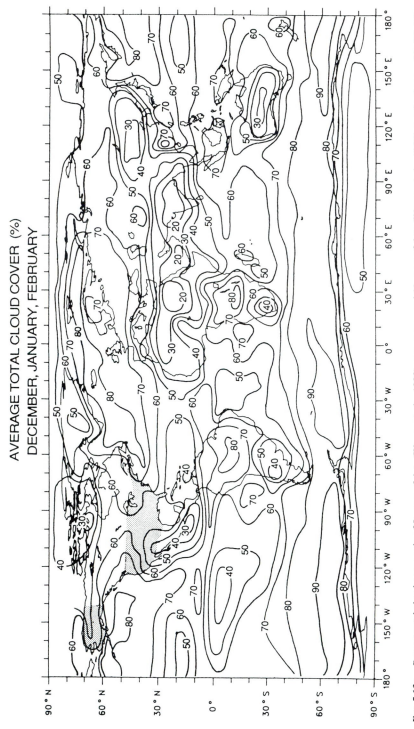

AVERAGE TOTAL CLOUD COVER (%)
DECEMBER, JANUARY, FEBRUARY

FIG. 5.13c Percent total cloud cover using 30 years of data (Warren et al., 1988) over oceans and 11 years of data (Warren et al., 1986) over land areas. Modified from Warren et al. (1991) by permission of American Meteorological Society.

188

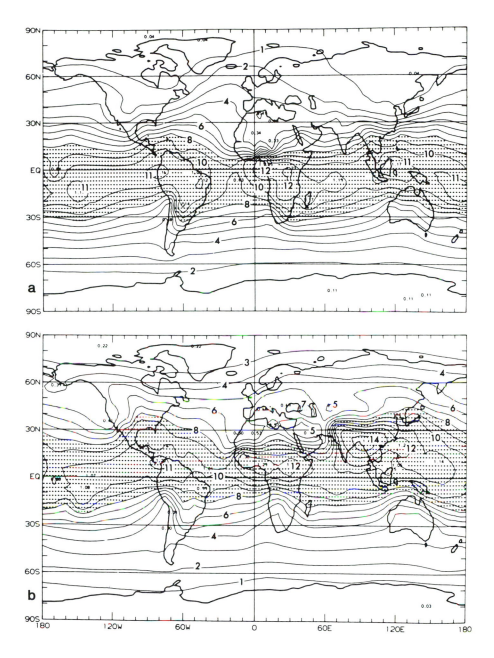

FIG. 5.14 Eleven-year seasonal averages of specific humidity at 850 hPa. (a) December through February averages starting with December 1962 and ending with February 1973. (b) June through July averages in the years 1963 through 1973. The contour interval is 1 gm/kg with values greater than 8 shaded. From Oort (1983).

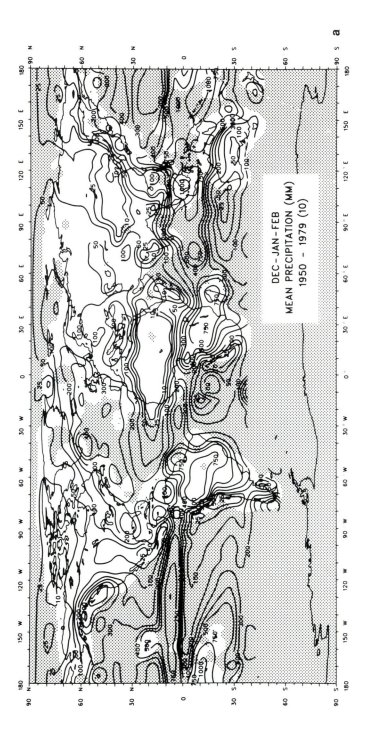

DEC-JAN-FEB
MEAN PRECIPITATION (MM)
1950 - 1979 (10)

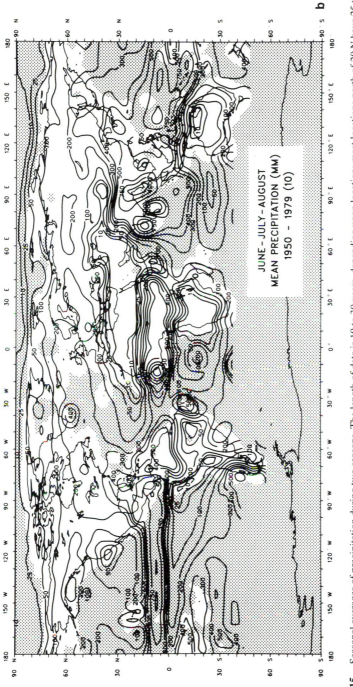

JUNE–JULY–AUGUST
MEAN PRECIPITATION (MM)
1950 – 1979 (10)

b

FIG. **5.15** Seasonal averages of precipitation during two seasons. The length of data is 10 to 30 years, depending upon location; most locations north of 20 N have 25 to 30 year records. (a) December, January, and February averages from 1950 through 1979. (b) June through August averages from 1950 through 1979. Data-sparse regions are shaded and contours reflect analyses of three previous studies as discussed in the text. The contour values are: 50, 100, 200, 300, 400, 500, 750, and 1000 mm. From Shea (1986).

coasts of South America and Australia and less cloudiness over central Australia and central America.

The cloud pattern is well correlated with the moisture content of the atmosphere (comparing Figures 5.14 and 5.13). The moisture is highest in the tropics and often largest over the land areas where the cloudiness is greatest (and low level convergence is favored, e.g., Figure 5.6a). The convergence of very moist low-level air results in heavy precipitation.

Precipitation data has fine structure like cloudiness (Figure 5.15). Shea (1986) compiled this figure from thirty years of observations. The observations were objectively analyzed, which means that observations modified a first guess. The first guess field came from Jaeger (1976), except over the Pacific (from 30 S to 60 N), where Shea used data from Dorman and Bourke (1979) and Taylor (1973). The shaded areas denote regions where the first guess predominates.

The zonal mean precipitation properties have been depicted in Figure 3.27; the zonal deviations are emphasized now. In the tropics the heaviest precipitation occurs in the areas of convergence or persistent cloudiness (with the exception of the stratus over the cold, eastern, subtropical oceans). The heaviest rainfall also follows the seasonal shift of the rising branch of the Hadley cells, especially over land areas. Maximum rainfalls occur over Borneo, Indonesia, and the eastern Pacific during both seasons shown. During Northern Hemisphere winter (Figure 5.15a), maxima are found at southern Brazil and Zaire. During Northern Hemisphere summer (Figure 5.15b), maxima are located over India, Bangladesh, and west Africa. The subtropics have a minimum in the zonal average precipitation; yet the longitudinal pattern is complex, especially over the continents. Basically, the west shores of Africa, Asia, Australia, and North and South America are still dry, but during summer the east shores of these continents are rather wet (e.g., Florida versus Baja California in Figure 5.15b). The reason for this is easily seen from the low-level flow (Figure 5.5). In summer the subtropical highs drive air from off the ocean on the east sides of the continents; this air is moist and convectively unstable. In contrast, on the west sides subsiding air is driven off the continents, suppressing rainfall. The direction of the flow around the subtropical highs is reversed at middle and higher latitudes on these continents. A noticeable feature, even on maps as smoothed as these, is that precipitation has large changes over small distances. The fine scale compounds the measurement problem.

5.6   Momentum flux fields

Figure 5.16 shows the seasonal average northward eddy momentum flux at 200 hPa calculated by Oort (1983) from 10 years of data. Away from the tropics this momentum flux is mainly due to traveling extra-tropical cyclones. Shaded areas are southward flux. Transient eddy contributions are shown in Figures 5.16a and 5.16b; stationary waves are only prominent during Northern Hemisphere winter (Figure 5.16c). In the Northern Hemisphere, the transient eddies (Figure 5.16a) have poleward momentum flux as anticipated from the tilted troughs. Such tilted troughs were shown in Figure 5.9 and will be discussed extensively in §6.2.1 (Figures 6.8 and 6.9). The maximum off the west coast of Africa (Figure 5.16a) is due to a tilted trough: *geostrophic* v' is positive (northward, from Figure 5.9a) and reinforces positive ageostrophic wind (from Figure 5.7a), and the low latitude of the subtropical jet causes u' to be large and positive, too. For the "North American" subtropical jet there are two maxima: the lesser one near Hawaii has large geostrophic v' positive (but ageostrophic v' negative), while the ageostrophic and geostrophic meridional winds reinforce each other over North America. Weaker equatorward flux occurs at high latitudes further downstream. Momentum convergence over the Atlantic is mainly from transient eddies. Along the equator weak transient eddy transport from the winter to the summer hemisphere occurs. The transient eddies carry the bulk of the transport in the Southern Hemisphere midlatitudes. Curiously, the transient eddy transport near the southern tips of Africa and South America is larger during summer; Oort gives no explantion for this result.

The stationary waves (Figure 5.16c) have very strong convergence in the northwest Pacific. The very large stationary fluxes at the Asian coast are easily anticipated from Figure 5.7a: strong northward ageostrophic winds ($v' > 0$) are co-located with the strongest jet speeds ($u' > 0$). The other maximum, occurring over southwestern Europe, has geostrophic v' and u' both negative. The only stationary eddy transport in the Southern Hemisphere (outside the Asian summer monsoon flow, not shown) appears to be along the west side of the SPCZ cloud band (Figure 5.13b).

In Chapter 2 much concern was expressed regarding the reliability of the eddy momentum flux observations. It is therefore particularly useful to compare Figure 5.16 with other analyses. Karoly and Oort (1987) compare Southern Hemisphere winter transient eddy momentum fluxes from two analyses. The Australian WMC fluxes are generally larger and differ greatly from the GFDL data (shown in Figure 5.16b). For example, some locations of the southern Indian Ocean have fluxes three times larger in the Australian analyses. Of course, the southern Indian Ocean has notoriously few observations. Even over the Australian continent the two analyses differ by as much as a factor of two. Lau and Oort (1982) plot the difference in magnitude for the vector transient eddy momentum flux. Though this difference is a slightly different quantity than the one given in Figure 5.16, it is clear that large differences occur between the GFDL and U.S. NMC fluxes. The worst disagreements are located off the east coast of the U.S. and in the central north Pacific where the NMC fluxes are half again larger than the GFDL data. The differences in the subtropical jets in these two data sets explain part of this difference (§5.3).

5.7   Heat flux fields

The vector horizontal heat fluxes at 700 hPa were described earlier in connection with Figure 5.4. The tropospheric horizontal heat fluxes tend to be strongest in the 850 to

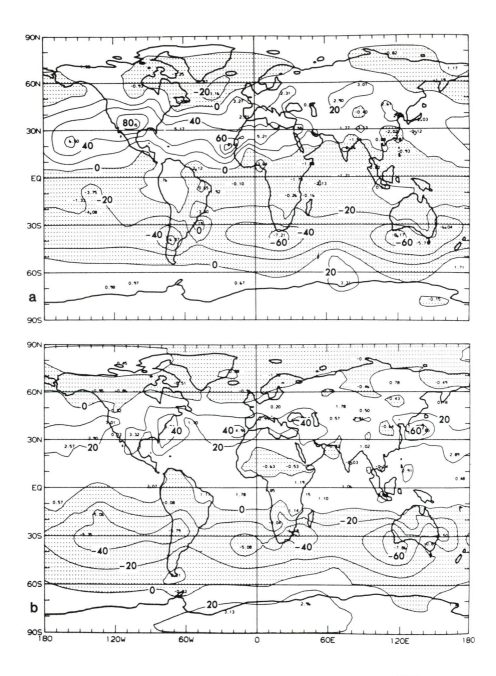

FIG. 5.16a and b Eleven-year seasonal averages of northward eddy momentum flux $(\overline{u'v'})$ at 200 hPa. (a) December through February transient eddy averages starting with December 1962 and ending with February 1973. (b) June through July transient averages in the years 1963 through 1973.

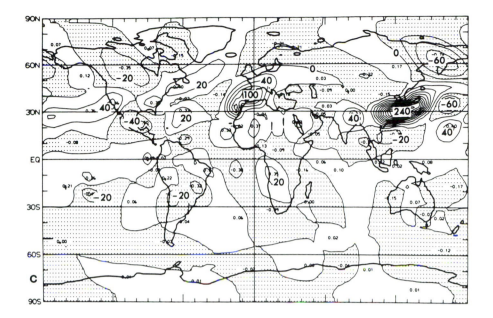

FIG. 5.16c Similar to (a) except for stationary eddies. The contour interval is $20\,\mathrm{m}^2\,\mathrm{s}^{-2}$; areas with negative values (southward direction) shaded. From Oort (1983).

700 hPa layer. Figure 5.17 shows only the meridional component of the vector heat flux at 850 hPa. This figure uses contours to depict the seasonal average meridional heat fluxes calculated by Oort (1983). This field is a major contributor to the $A_Z \rightarrow A_E$ "baroclinic" energy conversion. As anticipated above, this quantity is largest along and near the start of the storm tracks deduced from the 250 hPa height variance (Figure 5.10).

In contrast to the momentum flux fields, the horizontal heat flux fields are more consistent between analyses. The GFDL and U.S. NMC winter *vector* heat flux magnitudes agree quite well except in the north Pacific, where the NMC magnitudes are larger (Lau and Oort, 1982). In the Southern Hemisphere, the Australian WMC transient eddy heat fluxes are rather consistently less in magnitude, though the pattern is quite similar to the GFDL pattern in Figure 5.17b.

As with the momentum fluxes, the stationary waves contribute significantly during Northern Hemisphere winter. The largest poleward heat fluxes in Figure 5.17c are easily linked to the climatological Aleutian and Icelandic lows. On the east side of these lows, relatively warmer air ($T' > 0$) is advected poleward ($v' > 0$). On the west side cold air is pushed south.

It is useful to examine the Northern Hemisphere 700 hPa level vertical heat flux by transients (Figure 5.18). When this map is overlaid upon the zonal velocity field, one sees some indication that the two patterns may be related. From §4.5.2,

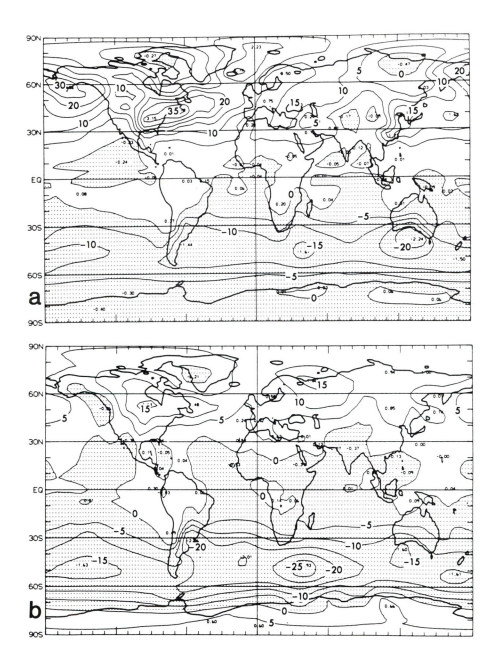

Fɪɢ. **5.17a and b** Eleven-year seasonal averages of northward eddy heat flux $(\overline{v'T'})$ at 850 hPa using the same time periods and data partitioning as in Figure 5.16. (a) Transient eddy fluxes using December, January, and February averages. (b) Transient eddy fluxes using June through August averages.

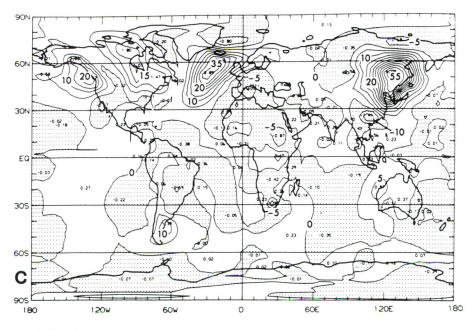

Fig. 5.17c   Similar to (a) except for standing eddies. The contour interval is 5 mK/s where regions with negative values (southward flux) are shaded. From Oort (1983).

$\omega'T'$ is proportional to the quasi-geostrophic form of the $A_E \rightarrow K_E$ "baroclinic" energy conversion. Vertical shear of the large-scale flow also enters into the quasi-geostrophic ($A_Z \rightarrow A_E$) baroclinic conversion, so it is not surprising that baroclinic eddies would co-locate large vertical heat fluxes where there are large vertical shears. The co-location is another statement of the "express bus route" energy conversion ($A_Z \rightarrow A_E \rightarrow K_E$) discussed in §4.6. The eddy vertical heat flux is one (rather crude) measure of the intensity of cyclonic storm development and therefore of cyclogenesis.

The largest values of vertical eddy heat flux seem to be downstream and slightly poleward of the 200 hPa jet maxima in the Northern Hemisphere. In the Northern Hemisphere winter, maxima are found over three areas: the east coasts of Asia and North America, and just east of the Rocky Mountains. A somewhat similar correlation exists in the Southern Hemisphere: heat flux maxima near the east coast of South America and in the southwestern Indian ocean are near and upstream from jet maxima, respectively (Figure 5.7d). The discussion of vertical heat fluxes benefits from specific examination of the energy conversions as well as the actual storm tracks.

5.8   Storm tracks

Cyclogenesis and storm tracks (frequency of occurrence) are plotted in Figure 5.19 from various sources. The Northern Hemisphere tracks start in four principal lo-

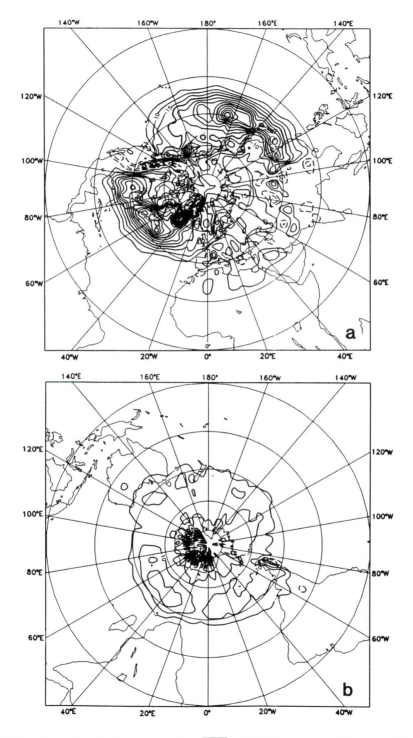

Fig. 5.18 Vertical heat flux by transient eddies ($\overline{\omega'T'}$) at 700 hPa during (a), (b) December–February during 1978–1989 and (c), (d) June–August 1979–1989. The contour interval is 5 PaK/s with the zero

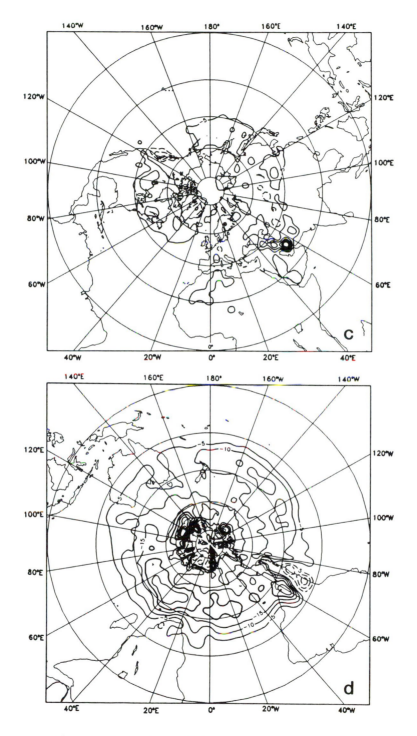

contour removed for clarity. Figures courtesy of I. James, personal communication, Dept. of Meteorology, Univ. of Reading.

cations. Two locations are near the east coasts of the continents (above western boundary currents) and two are downstream from mountain ranges (the Rockies and the Alps). Tracks are identifiable in the Southern Hemisphere and have preferred starting points, too.

Beginning with the Northern Hemisphere; Figure 5.19a shows the number of cyclones that formed in 5×5 degree boxes during a twenty-year period. Compensation is made for the convergence of meridians by multiplying counts in each box by a factor inversely proportional to the box area. Only boxes from 20 N to 85 N are used. Cyclogenesis is counted when the first "closed" isobar has been analyzed around a low pressure center. (Standard weather maps are manually examined.) This pattern matches well the vertical heat flux (Figure 5.18a). Figure 5.19b presents the number of cyclones that occurred in 5×5 degree boxes over the twenty years. The major storm tracks are across the northern oceans and across the Great Lakes. These tracks are quite broad over the western Pacific and over North America. A secondary area of storms is located over the northern Mediterranean Sea.

Individual cyclone tracks are extremely variable. Even the median track in a given season can vary greatly from year to year. Grotjahn and Lai (1989), for example, show individual storm paths across the north Pacific during 22 winters. The most common type of track, occurring perhaps half the time, is shaped like a shepherd's crook. The track, rather straight during the early stage of the cyclone's life, more often than not bends poleward, sometimes northwestward, at later stages. Often, these cyclones merge with an existing Aleutian low. As discussed in §7.2.3, the stalling and merging of transient cyclones with semi–permanent lows contributes to the maintenance of a long wave pattern.

The western boundary currents (WBCs) are regions of strong sea surface temperature gradient. In §4.5.1 (Figure 4.18) it was shown that the western boundary currents can be a source of available potential energy to developing eddies. The transient eddy development is favored by both the sensible and latent heat surface fluxes. Sensible heat fluxes (e.g., Grotjahn and Lai, 1991) reduce the static stability of the air and simple baroclinic models (§7.1) find instability to be inversely proportional to static stability. Latent heat fluxes can provide energy to the growing frontal cyclone if the heat is released in the warm air sector. Accomplishing this feat requires two steps (Grotjahn and Wang, 1989) because the latent heat fluxes are strongest in the cold air sector. Vapor is introduced into the cold air behind the first storm to pass over the WBC; the latent and sensible heating transforms this air into warm sector air for the second storm to come along, where the vapor condenses. Grotjahn and Wang (1989) found this pattern for three successive winter storms near the Kuro Shio; Grotjahn (1987) found similar behavior for the winter storms he studied over the eastern United States.

Topographic effects seem to be responsible for the frontal cyclone source regions in the lee of the Rockies and Alps. Using the fact that potential vorticity is conserved for adiabatic motions, one can show that westerly flow over a north-south oriented mountain ridge will lead to the formation of a trough on the lee side. (e.g., §7.2.1 or Holton, 1979; p. 89). This lee side trough is the seed for many midlatitude cyclones. The Alps may be a different story. The so- called "Gulf of Genoa" lows may arise when the Alps cause a pocket of warm air to be trapped over Northern Italy while a cold front attempts to move southward over Europe (e.g., Palmén and Newton, 1969; p. 349). These are but two of many mechanisms invoked to explain how topography influences midlatitude cyclogenesis; others are discussed in the review section of Hayes et al. (1987), for example.

Areas of preferred cyclogenesis are found in the Southern Hemisphere, too. Figure 5.19c portrays cyclogenesis regions by counting the frequency of "early stage" cyclones. The early stage of development is deduced by Carleton (1981) from the shapes of large-scale cloud patterns in satellite images. The numeric values given in this figure may count the same system more than once; in contrast, individual systems are counted only once in Figure 5.19a and only once per box in Figure 5.19b. The number of occurrences (Southern Hemisphere) during three months of the winter of 1958 are shown in Figure 5.19d. Different time periods, areas, normalizations, and accounting procedures are used in the diagrams of Figure 5.19. If comparable, the maximum values in Figure 5.19d would be about 1.5 those in Figure 5.19b because van Loon (1966) uses three months from one year (instead of one month each from twenty years), and his normalization area is about ten times as big as that used by Whitaker and Horn (1982). Oddly, the Southern Hemisphere cyclogenesis (Figure 5.19c) labelled by Carleton (1981) as a monthly mean is not consistent with any of the other diagrams unless "early stage" cyclones are counted more than once and/or the numeric values are for an entire winter season. Because of the inconsistency, the locations and not the magnitudes of the storm track data are emphasized here.

In the Southern Hemisphere, the areas of preferred cyclogenesis are also linked to strong sea surface temperature gradients (e.g., Barry, 1980). The oceanic surface isotherms parallel the roughly zonal Antarctic Circumpolar Current. The location of strongest surface temperature gradient roughly matches the extent of sea ice toward the end of winter (September).

The connection between the pack ice extent and the cyclone tracks seems obvious in Figures 5.19c and d; however, the mechanisms connecting the two are unclear, as demonstrated in the review by Carleton (1981; p. 33). For example, van Loon and Shea (1988) point out that the circumpolar trough (at 65 S in Figure 3.21) is deepest and close to Antarctica when the pack ice is at *both* maximum and minimum extent— but this is not necessarily a contradiction since the circumpolar trough results from mature lows, whereas the initial development stage is emphasized here. The strongest meridional sea surface temperature gradients occur to the southeast of Africa in good agreement with the $\omega'T'$ data (Figure 5.18d). The sea ice extends furthest north here as well. Using Ockham's razor, the simplest possible explanation may be that the surface air temperature gradient is greatest where the surface temperature gradient is greatest, and that happens to match the ice boundary during periods of maximum ice extent.

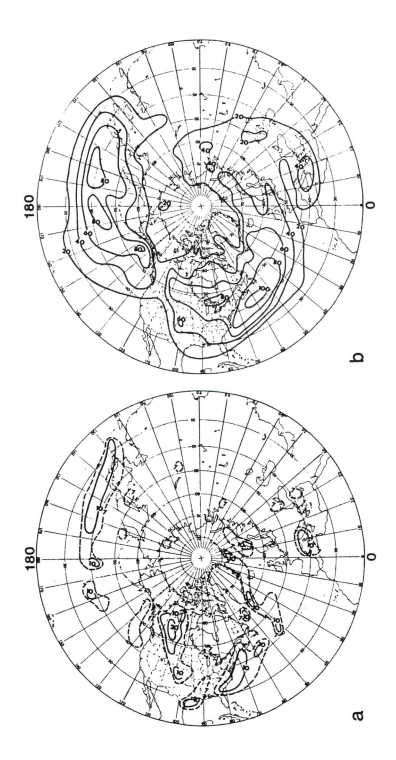

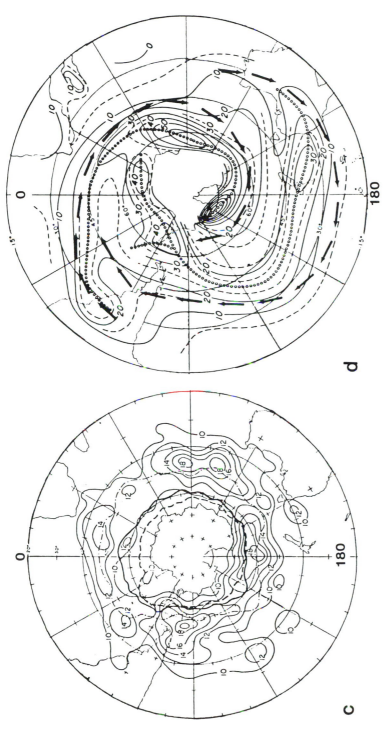

FIG. 5.19 Cyclogenesis regions and preferred storm tracks in winter. (a) Number of cyclogenesis occurrences in 5 x 5 degree areas of the Northern Hemisphere from 1958 through 1977 during the month of January. Standard sea level pressure maps were examined; cyclogenesis was noted when the first closed contour was analyzed around a low center. (b) Similar to (a) except showing the total number of cyclone occurrences in each box. Both (a) and (b) are reproduced from Whitaker and Horn (1982). (c) Contours of the number of "early stage" cyclones identified on satellite images per month (?) during five years (1973–1977) of winter months (June–September). Occurrences in 5° latitude by 10° longitude boxes are tabulated; an areal weighting correction is again used. The figure is redrawn from Carleton (1981), who referred to these values as counts of cyclogenesis; but unlike diagrams (a) and (b), the same vortex may be counted more than once. Also shown are the mean sea ice boundary during June (heavy dashed line), the September pack ice margin (heavy solid line), and the oceanic polar front position from Gordon et al. (1978; dot-dashed line). (d) Number of low centers occurring during July through September 1958 per 2.4 million km². This area is about 10 times the area used in (b), but 3 months are used in (d) instead of 20. The diagram is redrawn from van Loon (1966); circles mark the main storm tracks while arrows mark the positions of 500 hPa wind maxima.

203

Certainly this is true in the lower troposphere along the Antarctic coastline (Figures 3.11b and 5.3). Instability in simple baroclinic models (§7.1.1) is proportional to the horizontal temperature gradient, particularly the lower tropospheric gradient. What makes matters more complicated are the nonlinearity and variability of the two fluids.

The atmospheric motions strongly affect the oceanic circulations and temperature structure, as well as vice versa. For another example, van Loon (1972; p. 55) argues that the annual change of temperature *decreases* toward higher latitudes so that midlatitude gradients are *weaker* in winter. Nevertheless, the horizontal temperature gradient remains largest between 40 to 45 S along 40 E in winter (Figure 3.36 in van Loon, 1972), and that matches well the favored cyclogenesis development there. Even in the Northern Hemisphere feedbacks exist between the extratropical cyclones and the WBCs, since the WBCs are set up by great oceanic gyres that are in turn driven by subtropical atmospheric anticyclones.

Figure 5.19d counts all cyclones, revealing that many mature cyclones persist along the coast of Antarctica and that the "graveyard" for many lows is the Ross sea. The four-year climatology in Le Marshall and Kelly (1981) shows a high frequency of lows along the Antarctic coast, too. But Le Marshall and Kelly (1981) figures reveal much more spatial variability, making the tracks less easily seen than in the highly smoothed curves in Figure 5.19d. Trenberth (1991) uses band-pass 300 hPa height variance as a marker for Southern Hemisphere storm tracks. Height variance emphasizes the later stages of cyclone life-cycles and tends to miss the locations of cyclogenesis. Trenberth finds a similar pattern as in Figure 5.19d. The largest variance appears as a band across the southern Indian Ocean that spirals poleward starting at 50 S and ending in the Ross Sea. A second band of high variance starts 40 degrees equatorwards of the Ross Sea, about 40 degrees east of New Zealand.

Figure 5.19d also reveals a poleward (as well as the expected eastward) motion of the cyclonic storms. The direction of motion is characteristic of lows in both hemispheres (Figure 5.20). The high pressure systems also migrate meridionally but toward the equator. While developing lows may often merge with the climatological lows, the intervening highs between developing lows often merge with the subtropical highs. Cyclones and anticyclones start at almost the same latitudes (peak genesis frequencies at 40 and 43 N respectively). Cyclones and anticyclones migrate to 52 and 35 N, respectively (peak zonal means in Figure 5.20). These latitude positions are consistent with the interpretation of the schematic diagrams shown in Figures 4.19 and 4.20. Similar divergent paths for highs and lows may be deduced for the Southern Hemisphere (van Loon, 1972; Le Marshall and Kelly, 1981).

5.9   Energetics

Two-dimensional maps of the kinetic and available potential energy could be shown, but the kinetic energy would simply reflect the wind profiles shown before: the maxima would lie along the jets. Instead, the discussion here centers around spectra. It is

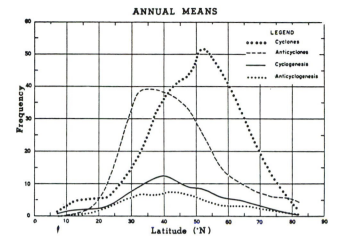

FIG. 5.20   Latitudinal distribution of cyclones (larger circles) and anticyclones (dashed line) in the Northern Hemisphere annually, using data for 20 years. Latitudes of cyclogenesis (solid line) and anticyclogenesis (smaller circles) are also indicated. From *J. Meteor.*, Klein (1958) by permission of American Meteorological Society.

common practice to examine energy-related quantities in spectral transform space. Since the earth is a sphere, it is natural to use spherical harmonics as the spectral functions.

Figure 5.21 shows kinetic energy spectra for the earth. January and July spectra of total kinetic energy are plotted in Figure 5.21a; transient and stationary components during January are shown in Figure 5.21b. Boer and Shepherd (1983) use operational global analyses from the FGGE (§2.2) archive at the U.S. NMC. One purpose in using a spectral depiction is that the type of turbulent motion present in the atmosphere can be revealed. So-called "two-dimensional" and "three-dimensional" turbulence each have simple and characteristic patterns in the kinetic energy spectrum. For example, total energy (available potential plus kinetic) for quasi-geostrophic motion is analogous to the kinetic energy equation for two dimensions (Charney, 1971b). Two-dimensional turbulence theory predicts that energy will be transferred from smaller scales to larger scales and that the kinetic energy would decrease at a rate proportional to wavenumber raised to the $-\frac{5}{3}$ power. In contrast, three-dimensional turbulence theory predicts energy transfer in the other direction and a spectrum proportional to wavenumber to the -3 power. If the energy were isotropic, it would be distributed equally among all harmonics having the same absolute wavenumber (for plane waves in Cartesian geometry); in the case of spherical harmonics, the equivalent quantity is the scale index n (Boer and Shepherd, 1983). (If m is the zonal wavenumber, then a spherical harmonic Y_n^m has $n - m$ zero crossings between the north and south poles.) There are other important aspects of turbulence theories which are beyond the scope of this text.

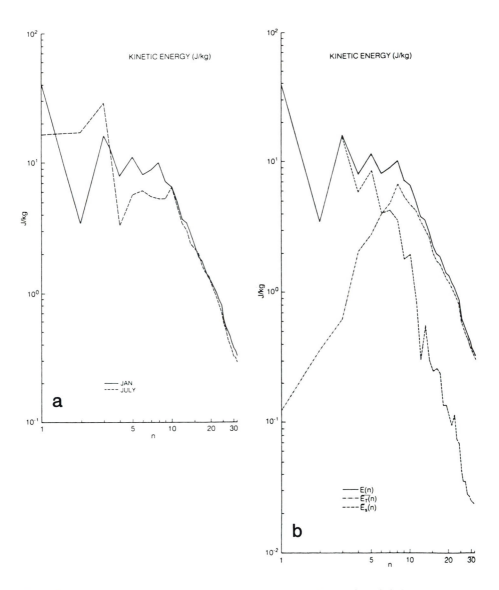

FIG. 5.21 Global kinetic energy spectra as a function of spherical harmonic scale index n. (a) January 1979 (solid line) and July 1979 (dashed line) GWE data values are plotted. (b) Total kinetic energy (solid line) is the same as the solid line in (a). Dashed line in (b) is the energy in the stationary velocity fields; dot-dashed line is transient kinetic energy. From *J. Atmos. Sci.*, Boer and Shepherd (1983) by permission of American Meteorological Society.

The kinetic energy spectra shown in Figure 5.21 are plotted using a logarithmic scale for both axes. If energy were proportional to n^b, then the spectra would be straight lines with slope b. At the higher wavenumbers ($n > 13$), the spectra do seem to follow a power law, but the slopes only approximately equal -3. In Boer and Shepherd (1983), b ranges from -1.5 (at low elevations) to -3 (near the tropopause); Baer (1972) found b to range from -3 to about -3.5 in hemispheric data for the upper troposphere. For the whole atmosphere, Chen and Wiin-Neilsen (1978) found $b \approx -2.6$ for kinetic energy and $b \approx -3$ for available potential energy. In Figure 5.21a it is interesting that the high wavenumber slope changes very little with season. The longer waves are sensitive to the season; this result is consistent with the maintenance of the longer waves primarily by land-sea contrasts and topography. (The forcing of the long waves is the subject of §7.2). Figure 5.21b further reinforces this point. The long waves are predominantly stationary waves. The shorter waves are predominantly transient waves, consistent with freely interacting turbulent eddies. Because the atmosphere is quasi-geostrophic to a fair degree (outside the tropics) one might expect some evidence of a $-\frac{5}{3}$ power law, but none is seen in these figures.

The energy can be further resolved into amounts in individual spherical harmonics (Figure 5.22). At the higher wavenumbers ($n > 13$) the contour lines in the kinetic energy are nearly horizontal, implying a nearly isotropic distribution of the energy. That conclusion can be visualized in the following way. Along the *diagonal* edge of Figure 5.22a are harmonics with zonal wavenumber m that extend from pole to pole (no zero crossings in the meridional direction) with maximum values along the equator. That makes $2n$ zero crossings along the equator. Along the *left* edge of Figure 5.22a, the harmonics are zonal means (no zero crossings in the zonal direction), and there are n zero crossings in the meridional. Since the distance from pole to pole is half the equatorial circumference, then the two modes just described have similar size and shape. Isotropy was found by others (e.g., Baer, 1972).

Spherical harmonics are useful functions to project data onto because they are eigenfunctions to LaPlace's equation on a sphere. Several important atmospheric quantities are related to each other by a LaPlacian, so the choice is obvious. However, individual weather features rarely look like a single harmonic. An exception may be the atmospheric tides, which would have amplitude at $n = m = 2$ for the most part. The tidal velocities are on the order of a few meters per second, but they do not show up in Figure 5.22b. The tides have mainly a 12-hour period that is filtered out of the data used by Shepherd (1987). Spherical harmonics with eddies centered off the equator have chains of eddies centered along *more* than one latitude. Therefore, more than one harmonic must be combined to equal a chain of eddies along a *single* latitude line off the equator. For example, a chain of 5 highs and 5 lows equally spread along a latitude line would have a kinetic energy spectrum spread along a line of constant $m = 5$. The energy would typically peak at a value of $n = 5$ if the latitude line were close to the equator. For eddies centered at higher latitudes the spread of spherical harmonic amplitudes would be more gradual and the peak centered at larger values

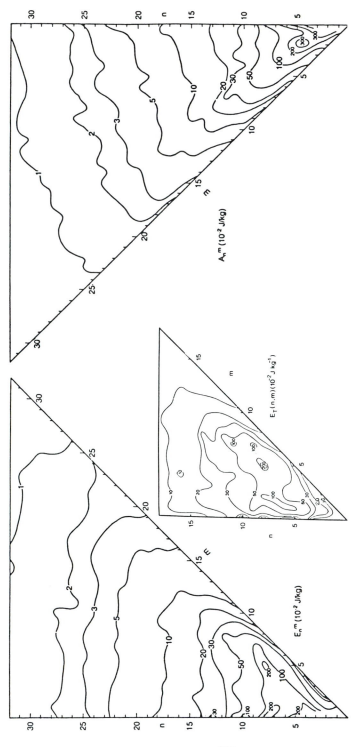
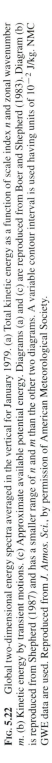

Fig. 5.22 Global two-dimensional energy spectra averaged in the vertical for January 1979. (a) Total kinetic energy as a function of scale index n and zonal wavenumber m. (b) Kinetic energy by transient motions. (c) Approximate available potential energy. Diagrams (a) and (c) are reproduced from Boer and Shepherd (1983). Diagram (b) is reproduced from Shepherd (1987) and has a smaller range of n and m than the other two diagrams. A variable contour interval is used having units of 10^{-2} J/kg. NMC GWE data are used. Reproduced from *J. Atmos. Sci.*, by permission of American Meteorological Society.

of $n > 5$. The features described in this second example are precisely those seen in Figure 5.22b.

Figure 5.22b shows transient eddy kinetic energy. There are two regions of maximum values: one near $m = 2$ to 3 and one near $m = 5$ to 7. The region near $m = 6$ is caused by the frontal cyclones for the most part. The modes are centered in the middle latitudes ($n > m$ for peak values). A peculiar aspect of Figure 5.22 is that the transient kinetic energy exceeds the total kinetic energy at $(n, m) = (11,7)$; presumably this is impossible, but the producers of the respective figures do not comment on this. The region of maximum values near $m = 2$ to 3 indicates long waves. Comparing Figures 5.22a and 5.22b, the bulk of the long wave energy is stationary; however, a significant amount of the long wave energy is transient and reveals low-frequency changes in the atmosphere, such as "blocking" patterns.

As with the kinetic energy, the available potential energy (Figure 5.22c) shows some isotropy in the shorter wavelengths ($m > 13$). Baer (1974) found similar isotropy in the related field of temperature variance. The exception is again the atmosphere's aversion for medium and long wavelengths ($15 < m < 0$) that are both confined to the tropics and centered along the equator. In the tropics the energy (both kinetic and available potential) is concentrated in the zonal mean: the maximum in Figure 5.22c near $(n, m) = (2,0)$. The aversion towards $m = n$ harmonics is not surprising since the available potential energy (APE) will also be large where the air is colder than the mean, not just where the air is warmer than the mean. So APE is large in the tropics and the polar regions, but $n = m$ harmonics are large in the tropics and very small elsewhere. In the higher latitudes, the land-ocean contrasts and the topography create a long wave distribution; as explained above, an extratropical wave train is distributed across many harmonics. In the case of long waves a diagonal "ridge line" is seen in the contours from $(n, m) = (5,2)$ to $(n, m) = (8,5)$.

The energy spectra can be further analyzed to identify how energy is flowing between the different harmonics. Boer and Shepherd (1983) show that for high wavenumbers energy is flowing toward smaller length scales. This direction of energy cascade is also consistent with the three–dimensional turbulence theory. The direction makes intuitive sense as well. Recalling the energy "box" diagrams (Figure 4.21), one sees that kinetic energy is lost by friction that may be visualized as follows. Shears in the atmosphere create small-scale turbulent eddies, the shears in these eddies create smaller eddies, etc. Friction is proportional to the wavenumber to a positive power, so the smaller eddies are more rapidly destroyed by friction than the larger eddies. The process is summarized in rhyme by Richardson. Jonathan Swift's precursor to Richardson's rhyme is given at the start of this chapter: fleas are replaced by whirls and "viscosity" is the ultimate end. For middle and long wavelengths the kinetic energy flow is toward longer waves. Shepherd (1987) has shown that the longer, mainly stationary waves draw energy from the shorter, mainly transient waves.

The horizontal distribution of vertically averaged atmospheric *eddy* energy quantities during winter are presented in Figure 5.23. Data for band pass filtered, transient' eddy available potential and eddy kinetic energies are shown in Figures 5.23a and 5.23b, respectively. The energy amounts have similar patterns; their maxima are close to the locations of maximum storm amplitude (Figure 5.10a) along the storm tracks (Figure 5.19b). Energy conversions from the *time* mean flow to the band pass transient eddies during winter are shown in Figures 5.23c and 5.23d. The dominant transfer is to eddy available potential energy, confirming the baroclinic growth of the band pass eddies. Though the time mean flow differs from the zonal mean flow, these results are also consistent with energetics (e.g., "box" diagrams) shown in §4.6. The barotropic energy conversion (Figure 5.23d) is weakly positive at the start of the storm track and negative at the end of the track. (The heavy, dashed line arrows show Holopainen's estimate of the storm tracks, which differ from tracks shown in Figure 5.19b.) Energy conversions, Figures 5.23e and 5.23f, for slowly varying (low pass) waves are calculated differently than conversions shown for the band pass eddies. In approximate terms, Figure 5.23e implies that baroclinic growth maintains the semi-permanent lows and the longer period fluctuations of the upper-level long wave troughs. In even more approximate terms, Figure 5.23f implies that slowly varying waves tend to acquire energy from the subtropical time mean jet maxima and give the energy back to the mean flow in middle latitudes, where the mean flow is weaker.

5.10   Asian summer monsoon

So far in this book, the Hadley cells have been a convenient description of the tropical meridional motion. Earlier in this chapter the connection between the Hadley cells and the subtropical highs was made clear. Discussion in §3.3 showed that the rising branch of the Hadley cells occurs in vigorous convection to overcome the moist static energy profile outside the cloud. The tropical circulations have some properties similar to the Hadley cells, but there are significant longitudinal variations. The most prominent example is the Asian summer monsoon, which has elements of a vigorous east-west circulation in addition to a Southern Hemisphere winter Hadley cell circulation.

If the association with the Hadley cell is accepted, then the summer monsoon is a thermally direct circulation driven by the warmer land areas of India and Southeast Asia in contrast with the cooler surrounding oceans. Figure 5.24 shows representative streamline patterns for July at the 200 hPa and surface levels. The general pattern is similar in five ways to the way one would expect the Hadley cell to look. (1) The lines of convergence in the surface streamlines match the cloudier regions noted in Figure 5.13 . Heavy rainfall accompanies the areas of low-level convergence. (2) Above is located a large outflow anticyclone, the Tibetan High. The shape of the Tibetan high is more accurately depicted in the stream function field shown in Figure 5.25a. (3) Large-scale cross-equatorial flow occurs over the entire Indian ocean. This flow has substantial velocity, so it is a prime region for exchange between the hemispheres. (4) Between 20 S and 25 N, the flow direction reverses between the lower and upper troposphere. That helps make actual three-dimensional trajectories

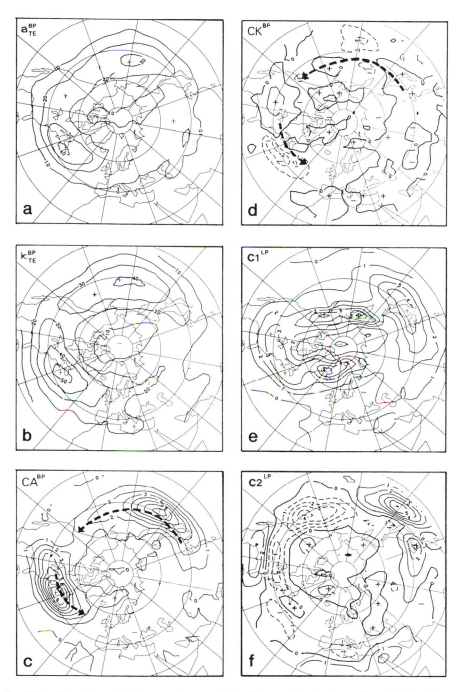

Fig. 5.23 Selected vertically integrated atmospheric eddy energetics. Band pass filtered (2.5 to 6 days) eddy (a) available potential and (b) kinetic energy. Energy conversions from the time-mean flow to the band pass eddies: (c) $(\overline{A} \rightarrow A''_E)$ and (d) $(\overline{K} \rightarrow K''_E)$. Approximate energy conversions for slowly varying (10 to 90 days) eddies; (e) and (f) are approximately related to conversion formulas used for (c) and (d), respectively. From *J. Atmos. Sci.*, Holopainen (1984) by permission of American Meteorological Society.

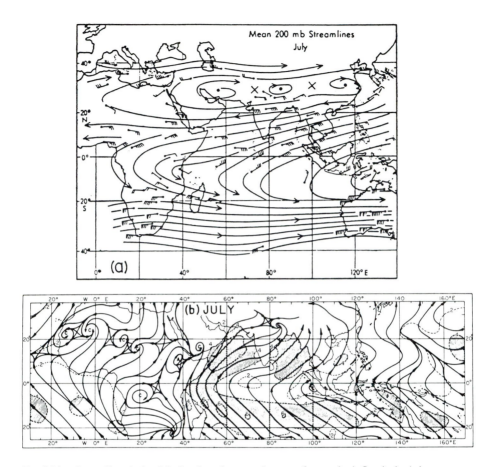

FIG. 5.24 Streamlines during July that show the general pattern of atmospheric flow in the Asian monsoon region at (a) 200 hPa and (b) surface levels. Diagram (a) is based upon data in Johnson (1970) while (b) is drawn from Ramage (1971); both are reproduced from Newton (1972, *Meteor. Mono.*) by permission of American Meteorological Society. Shaded areas are where the mean wind exceeds Beaufort force 4 (approximately 7 m/s).

quite complex; some parcels follow paths like a simple Hadley cell, but many do not. (5) The surface and upper level flows bend as they cross the equator in a manner consistent with angular momentum conservation. At low levels, southeast winds in the Southern Hemisphere become southwesterly winds in the Northern Hemisphere. These winds are so concentrated off the east coast of Africa that they form a distinct feature known as the "Somali jet." The text below elaborates upon these and other features of the summer monsoon.

Figure 5.25 shows the stream function (ψ) and velocity potential fields (χ) at 200 hPa. The χ field is calculated from the divergent part of the wind field ($\mathbf{V}_\chi$).

$$\mathbf{V}_\chi = -\nabla\chi \qquad (5.1)$$

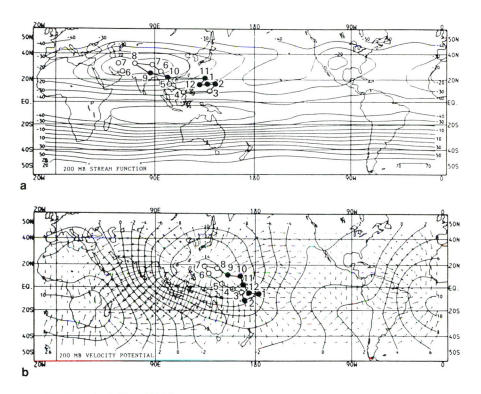

FIG. 5.25   Velocity fields at 200 hPa in the global tropical belt during July. Five years (1978–1983) of NMC analyses are averaged. (a) Stream function contours using a 10^7 m^2/s interval are plotted. Locations of the Tibetan high center during different months are marked with circles; the number accompanying each circle denotes the month, where "1" is January, etc. (b) Similar to (a) except for the velocity potential, with contour interval of 2×10^6 m^2/s. Vectors denote direction and magnitude of the divergent wind field. From Arkin et al. (1986) and Lau et al. (1988).

where

$$-\nabla^2 \chi = u_x + v_y - \frac{v}{r}\tan\phi$$

Five years of NMC data (October 1978 through September 1983) are averaged together to create the standard climatology (Arkin et al. , 1986) shown in Figure 5.25. The circles with numbers were added by Lau et al. (1988) and denote the locations of the Tibetan high center(s) (Figure 5.25a) and divergence center (Figure 5.25b) during each month of the year. (During June and July the Tibetan high has two centers.) An easterly jet is located along the southern side of the Tibetan high. The easterly jet is confined to the upper troposphere and is clearly visible along roughly 10 N in Figures 5.24a and 5.25a.

Since the gradient of χ is related to the divergent wind, the ageostrophic (divergent) flow will tend to be down the gradient of χ. This motion is indicated by the arrows in Figure 5.25b. A large "bull's-eye" is centered to the east of the Philippines; this is the divergence center. In data from the summer of 1967, Krishnamurti

(1971) found the divergence center located over southern Bangladesh. The (June through August) climatology using improved diabatic initialization of ECMWF data displayed by Hoskins et al. (1989) has some important differences. Hoskins et al. find the divergence center to be located about 20° further east and to be about one-third stronger; their minimum velocity potential is located about 40° east of the minimum over southern Africa in Figure 5.25b.

In addition to the north-south "Hadley" circulation that one expects to find, there are two east-west circulations worthy of discussion. Over the eastern Pacific the east-west motion is commonly known as the "Walker" cell (Bjerknes, 1969) after its identification in a series of reports written by G.T. Walker in the 1920s and 1930s. In contrast to Walker's original notion, the east-west circulation extends over a much wider range of latitude (from 20 S to 30 N). (The strong latent heating associated with the heavy rainfall drives the divergent winds that create the east-west circulation; since that circulation defines a long wave in the tropics, it is clear that topography, §7.2, cannot be the only mechanism maintaining the long waves.) To the west of the Tibetan high, the velocity potential indicates strong convergence over the African deserts, especially the Sahara. This upper-level convergence forces large-scale descent over the Arabian and African deserts; the consequent adiabatic warming is responsible for the high temperatures observed in this region by two means. Recalling Figure 5.2c, one sees that these regions are areas of negative net radiation, so heat must be transported into these areas. The heat source is air greatly warmed by latent heat release in regions of strong monsoon precipitation; that warmed air adiabatically descends over these deserts. The descent also inhibits the vertical motion that might be created by daytime heating of the desert surface; Blake et al. (1983) find upward motion typically in only the lowest 100 m during a summer day in Riyadh, Saudi Arabia. Additional east-west circulations have been identified in reports by several authors; some of the reports contradict each other; Hastenrath (1985; §6.9) provides an interesting review.

Figure 5.26 shows air parcel trajectories calculated from FGGE special observing period observations of the horizontal wind and a vertical wind derived using the continuity equation as a constraint in a simple variational calculus scheme proposed by O'Brien (1970). Despite these precautions, the trajectories calculated here are advisory and are not intended to be quantitative. As is clear by comparing parcels **B** and **C**, small differences in the initial location can sometimes lead to large differences in location later. The trajectories are for the period 15 to 24 June 1979. The Asian summer monsoon is active during this period. The diagram has several purposes. First, the trajectories give one an immediate impression of how far air parcels actually travel in nine days. Parcel **B** nearly completes a circuit of the equatorial anticyclone. Parcel **A** traverses the domain. Second, three-dimensional trajectories illustrate the relative magnitudes of the vertical and horizontal motions. The horizontal motion predominates except in a small percentage of the domain where the cloudiness and rainfall are greatest. (A similar behavior was seen over Brazil in Figure 5.6a.) The descent over the deserts is not picked up in this figure because those desert areas are at the back corner or outside of the domain; calculated middle and upper tropospheric trajectories over Arabia (not shown) do sink. Third, the three-dimensional trajectories show how the vertical motion often causes the speed and direction of a parcel to change dramatically. Some parcels that start in the southern Indian Ocean return to the Southern Hemisphere while others are entrained into the Northern Hemisphere westerlies. The reversal of direction of flow with height is also apparent in this figure (e.g., parcel **J**). Fourth, while the low-level flow shows clear cross-equatorial transport, cross-equatorial flow is less apparent at high levels. Parcel **H** is unusual among the 100 trajectories (most not shown) that were calculated. Fifth, the acceleration of the parcels entrained by the Somali jet and the upper-level easterly jet is evident in the wider spacing of the vertical bars in those regions. (The vertical bars connect the trajectory path and its projection at 24-hour intervals.)

The Asian monsoon is not a uniform massive zone of convection, but instead it contains spatial and temporal structure. Orographic enhancement is a famous feature and is well illustrated by the world's record rainfall amounts at Cherrapunji, India, located in the Himalayan foothills due north of the Bay of Bengal. Other weather features lead to monsoon variability: the onset vortex, monsoon depressions, monsoon breaks, the *Mei-yu*, and the "30 to 50 day wave." The onset vortex (apparently resulting from barotropic instability of horizontal shear in mid-tropospheric winds) forms at the start of the southwest monsoon over the Arabian sea and moves northeastward towards Arabia. In broad terms, monsoon depressions typically move up the Ganges river valley. Monsoon depressions most likely form via a combination of CISK (§6.7) and baroclinic (§7.1) instability mechanisms (Arakawa and Moorthi, 1982). Krishnamurti (1985) concludes that barotropic instability may be important too. Monsoon breaks occur separately over India and China. In southern China the beginning of the monsoon rainfall occurs one to two weeks after the monsoon onset over India. The onset is so distinct that it has a traditional name: the *Mei-yu* (or Plum Rain). Over southern and central China the rainfall typically decreases rapidly in July, apparently in connection with the migration of the divergence center (Figure 5.25b) to northern India, where rainfall is heavy. Various authors (e.g., Krishnamurti, 1985; Lau and Chan, 1986) have identified a *northward* propagating succession of troughs and ridges that enhance and suppress monsoon rainfall and may be linked to the monsoon breaks; the trough and ridge are separated by 10 to 20° of latitude and have a period of 30 to 50 days.

The onset of the southwest (Indian) monsoon over the Arabian sea is quite dramatic. During the monsoon experiment (MONEX; May through July, 1979), the *average* wind speed at one research vessel in the Arabian sea increased from 4 m/s to 14 m/s in couple of days. Krishnamurti (1985) diagrams average wind speed increases up to a factor of five over the Arabian sea. These winds comprise the Somali jet (the wind maximum off East Africa and south of Arabia in Figure 5.24b). The Somali jet exists at low levels and is perhaps similar to an oceanic western boundary

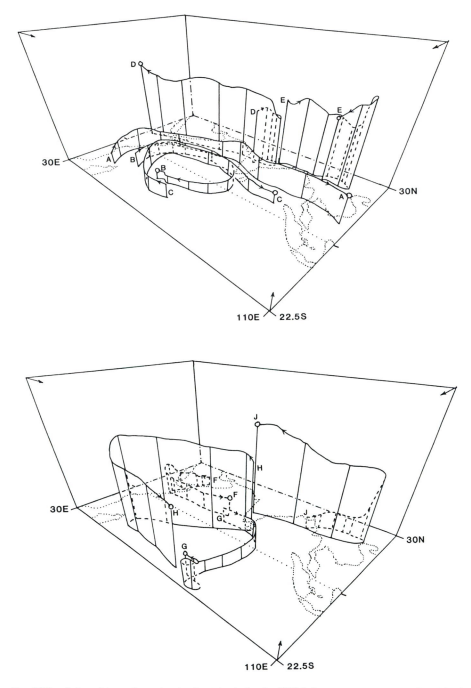

FIG. 5.26 Selected three-dimensional trajectories during the 1979 Asian summer monsoon as determined from ECMWF GWE data. Each three-dimensional path and its projection onto the earth's surface are connected by vertical lines drawn at 24-hour intervals. The trajectories cover the nine-day period from 00 GMT 15 June to 00 GMT on 24 June. The vertical axis is height, with the top of the box at 15 km elevation. Arrows show the direction of the parcels' motion; the final three-dimensional locations are marked by open circles.

216

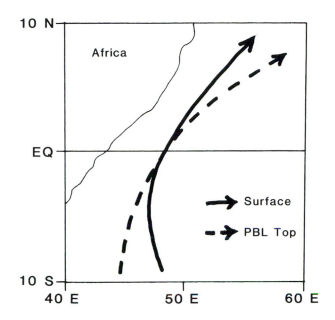

FIG. 5.27   Schematic diagram illustrating the turning of the wind in the planetary boundary layer (PBL) over the western Indian ocean during the summer monsoon. Dashed line shows the direction of the flow at PBL top; solid line shows direction of surface winds.

current. Theoretical studies have shown that the Somali jet formation requires the existence of mountains in east Africa *and* strong heating over India. The strong surface winds greatly increase the evaporation of water from the sea. It is estimated that this evaporation is as large a contributor to western India's monsoon rainfall as the cross-equatorial transport of moisture.

The cross-equatorial transport of moisture is largely within the boundary layer. Away from the equator, Ekman balance is a reasonable way to describe the vertical shear in the boundary layer. Ekman balance consists of three terms: pressure gradient, Coriolis, and friction. At the equator, the Coriolis term vanishes, and the horizontal and vertical advection terms take its place in a different type of balance. The result is summarized in Figure 5.27. Weak turning of the wind occurs near the equator. Along the Somali jet veering increases downstream. In the Southern Hemisphere, backing increases upstream.

The discussion in this section is a small sampling of the literature on the Asian monsoon. For more information, the reader might consult the review (Krishnamurti, 1985) cited above as well as books by Hastenrath, 1985; Fein and Stephens, eds., (1987), Chang and Krishnamurti, eds., (1987), and Lighthill and Pearce, eds., (1981).

6

THEORIES FOR ZONAL FIELDS

Every theory of the course of events in nature is necessarily based on some process of simplification of the phenomena and is to some extent therefore a fairy tale.

Shaw, 1926

All observers are not led by the same physical evidence to the same picture of the universe.

Whorf, 1940

Preceding chapters have drawn a detailed, comprehensive picture of the global circulations. While a few theoretical concepts were introduced in Chapter 4, the bulk of the material presented so far has been observations. On this solid observational base, theories can profitably be discussed. The theories presented here are basically simple and explain many of the gross features of the general circulation. These simple theories are well established and accepted widely. More advanced theories are beyond the scope of this text, sometimes controversial, and are covered in other monographs.

The theories are partitioned into two groups. This chapter treats theories that apply primarily to the zonal average description of the atmosphere. In chapter 7, theoretical explanations of zonally varying structures are the focus. Of course, the subjects of these chapters overlap. For example, the discussion of Jeffreys' work (§6.2.2) demonstrates that the eddies are necessary to maintain the zonal mean field; since the emphasis is still upon the maintenance of the zonal average flow, that historical discussion is included in this chapter.

6.1 Simple radiative-convective models

The observations shown in Chapters 3 and 5 began with radiation and temperature fields. Accordingly, the theoretical discussion begins with two simple models that explain the gross features of the vertical profile of temperature. The first model uses

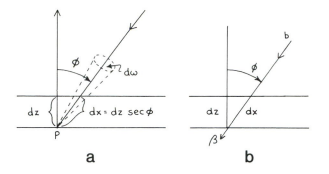

FIG. 6.1   Geometry and notation used by the radiative models in the text.

radiative equilibrium, which develops super adiabatic lapse rates in the lower atmosphere. The super adiabatic lapse rates are removed by simple convective adjustments in the second model. Though crude, these models can be used to illuminate (a) why the atmospheric lapse rates vary with altitude, (b) why the height of the tropopause varies with latitude, and (c) why the lapse rates vary with latitude. Explaining the last of these three variations requires a further embellishment: incorporating heat transports by large-scale circulations. To keep the discussion as simple as possible, important and complex effects due to clouds are ignored. Hence, the discussion is not intended to be complete. Several books (e.g., Paltridge and Platt, 1976) treat radiation and radiative convective models with greater care and in greater depth than the scope of this text allows.

The original derivation of a radiative balance model for the atmosphere is attributed to Emden (1913). The standard development can be found in both editions of Goody and Yung (1989). The discussion here is streamlined and based upon comprehensible works by Houghton (1977) and Paltridge and Platt (1976).

Absorption of radiation along a vertical path may be described by Lambert's law (or Bouguet's law). Lambert's law states that as a beam of radiation travels through a slab of atmosphere it will be absorbed at a rate proportional to (a) the thickness of the slab dz, (b) the angle of the beam relative to the slab ϕ, (c) the density of the absorber ρ, and (d) the intensity of the beam of radiation R. The path length through the absorber is thus $dX = \sec \phi dz$, as shown in Figure 6.1a. Thus

$$dR = -Rk\rho dX \qquad (6.1)$$

dR is the loss per unit horizontal cross-sectional area and k is the absorption coefficient. Radiation may be diffuse or parallel beam. Parallel-beam radiation describes the solar radiation reaching the earth to a good approximation. The earth and its atmosphere emit radiation in all directions, so one must treat terrestrial radiation as diffuse. The distinction between parallel-beam and diffuse radiation will be made clear after a brief discussion of the solar radiation.

Since the sun is very far away (compared to its radius), solar radiation reaching the earth can be treated as parallel-beam radiation to high accuracy. In that case, R is the solar *irradiance*. The solar radiation $R(z)$ is directed downward (z *decreasing*) so that $R(z)$ must increase with height. Integrating (6.1) in the vertical from height z to ∞ gives the irradiance arriving at level z.

$$R(z) = R_0 \exp\left(-\sec\phi \int_z^\infty k\rho dz\right) \qquad (6.2)$$

where R_0 is the intensity of incident radiation at the top of the atmosphere. The integral

$$\tau = \int_z^\infty k\rho dz \qquad (6.3)$$

is commonly known as the *optical depth*. (Some authors use the optical *path*, which is ususally defined as $\tau \sec\phi$.) The optical depth stretches the vertical coordinate as a function of the amount of absorber.

The atmosphere emits as well as absorbs radiation. The emission, e, is not parallel-beam radiation but is diffuse. The emission is a function of solid angle and is termed *radiance*. Irradiance, E, is calculated by integrating the diffuse radiance over all solid angles. One may approximate the atmosphere by plane parallel layers, each indexed by a subscript. When radiance is isotropic, then the irradiance in one direction (e.g., downward) from layer i is half of the total, E_i. For all directions and using radiance e_i, then

$$\int_0^{2\pi}\int_0^\pi e_i \sin\phi d\phi d\lambda \equiv \int_\pi e_i d\omega = 2\pi e_i = E_i \qquad (6.4)$$

where ω is an element of solid angle and ϕ is the angle of incidence relative to the unit horizontal area (Figure 6.1) and λ is the azimuth angle. The next to last step in (6.4) assumes that e_i is isotropic. From the Stefan-Boltzmann law, irradiance is proportional to the temperature (T_i) of the emitter ($E_i = \sigma T_i^4$).

The net radiative transfer dI through the slab is the difference between the radiation absorbed by the slab and the radiation emitted by the slab. I and b are radiances, thus

$$\beta = b\left(1 - k\rho dx\right) I k\rho dx$$

using Figure 6.1b. Since

$$\beta - b \equiv dI = (Ik\rho dz - bk\rho dz)\sec\phi$$

then

$$\cos\phi \frac{dI}{d\tau} = I - b \qquad (6.5)$$

Kirchhoff's law states that the absorptivity equals the emissivity, so k is used for the emission and the absorption. The factor of $\cos\phi$ is introduced because all intensities from all directions are now expressed in amounts incident upon a unit horizontal area. The b indicates emission from a source outside the slab, and equals the Planck emission B if scattering is ignored.

The net energy flux through a unit horizontal area is denoted F. Integrating over all solid angles,

$$F \equiv \int_\pi I \cos\phi \, d\omega \tag{6.6}$$

$$I^\omega \equiv \frac{1}{4\pi} \int_\pi I \, d\omega \tag{6.7}$$

where I^ω is the "mean intensity" (Goody and Yung, 1989) through a unit horizontal area, and is integrated over all angles in a sphere centered upon point P in Figure 6.1.

Multiplying (6.5) by $d\omega$ and integrating over all solid angles yields

$$\int_\pi \cos\phi \frac{dI}{d\tau} d\omega = \int_\pi \frac{d}{d\tau}(I\cos\phi)\,d\omega = \frac{d}{d\tau}\left(\int_\pi I\cos\phi\,d\omega\right)$$
$$= \frac{dF}{d\tau} = 4\pi I^\omega - 4B \tag{6.8}$$

where (6.6) and (6.7) are used in the last step. The integral can be pulled inside the derivative above because τ as defined by (6.3) is not a function of solid angle. B is irradiance; $\pi b = B$ for emission from a plane surface.

The radiation field can be divided into upward (U) and downward (D) components (using the sign convention of ϕ in Figure 6.1). Multiplying (6.5) by $\cos\phi\,d\omega$ and integrating over all solid angles yields

$$\int_0^\pi \frac{dI}{d\tau}\cos^2\phi\,d\omega \equiv \int_0^{\pi/2} \frac{dU}{d\tau}\cos^2\phi\,d\omega + \int_0^{\pi/2} \frac{dD}{d\tau}\cos^2\phi\,d\omega$$
$$= F - \int_0^{2\pi}\int_0^\pi B\cos\phi\sin\phi\,d\phi\,d\lambda = F \tag{6.9}$$

The B term disappears because it is a source term from one direction; consequently its contribution entering from one hemisphere (from above, say) cancels the amount leaving the opposite hemisphere (from below). If the upward and downward radiation are isotropic but not necessarily equal, then the left-hand side of (6.9) becomes

$$-\left(\frac{2\pi}{3}\right)\frac{dI^\omega}{d\tau}\cos^3\phi\Big|_0^\pi = \left(\frac{4\pi}{3}\right)\frac{dI^\omega}{d\tau} = F \tag{6.10}$$

Taking $d/d\tau$ of (6.8) and combining the result with (6.10) yields the radiative transfer equation sought.

$$\frac{d^2 F}{d\tau^2} = 3F - 4\frac{dB}{d\tau} \tag{6.11}$$

If the atmosphere is in radiative equilibrium, then the net radiation through each level is the same. Consequently F would be a constant and (6.11) reduces simply to

$$3F = 4\frac{dB}{d\tau} \tag{6.12}$$

Radiative equilibrium has several unrealistic properties. The first odd feature is the discontinuity in temperature at the top and bottom. Substituting the definitions of U and D made in (6.9) into (6.6) yields

$$F = \pi(U - D) \tag{6.13}$$

and

$$I^\omega = \frac{1}{2}(U + D) \tag{6.14}$$

where U and D are isotropic in their respective hemispheres. If U is eliminated from (6.13) and (6.14), then the result may be substituted into (6.8) to eliminate I^ω. The result is

$$-\frac{1}{2}F + \frac{1}{4}\frac{dF}{d\tau} + B = \pi D$$

For radiative equilibrium, $dF/d\tau = 0$, and the last equation reduces to

$$-\frac{1}{2}F + B(\mathrm{T}_g) = \pi D \equiv B(\tau_1)$$

at the ground. The mixed subscripts used for the argument of B indicate emission downward from the base of the atmosphere (where $\tau = \tau_1$) and upward from the surface of the earth (with temperature T_g). Rearranging obtains

$$B(\mathrm{T}_g) - B(\tau_1) = \frac{1}{2}F \tag{6.15}$$

Equation (6.15) is a statement of energy balance at the surface of the earth. Solar radiation is absorbed by the earth's surface. (The atmosphere is assumed to be transparent to solar radiation in this model.) Averaged over time, the solar radiation absorbed is proportional to F, and that same amount must be emitted for energy balance. (Surface sensible and latent heat fluxes are ignored for this first model.) The radiative balance at the surface in (6.15) requires that the emission back down from the atmosphere be less than the emission upwards by an amount equal to the amount of solar radiation absorbed. Otherwise, the ground would heat up. Since F is positive, then the ground must be hotter than the air. Therefore, a discontinuity is created between the surface temperature and the surface air temperature.

The atmosphere is emitting radiation downward, which is also absorbed by the ground. That requires the surface temperature to be greater ($B(T_g)$ greater) than it would be if no atmosphere were present. The situation is sometimes called the "greenhouse effect."

At the top of the atmosphere (where $\tau \to 0$) the temperature asymptotically approaches a value commonly called the "skin temperature" of the planet. To compute that, one must eliminate U and I^ω from (6.13), (6.14), and (6.8). That obtains

$$B(\tau = 0) = \frac{1}{2} F \tag{6.16}$$

since the downward directed flux from outside the atmosphere has been assumed to be zero. Eqn. (6.16) states that the net emission outwards at the top of the atmosphere just equals the amount of radiation absorbed below. In this case the atmosphere was transparent to solar radiation, so the amount leaving the top equals the amount absorbed by the ground ($F/2$).

Of course, every level in the atmosphere has the same F under radiative equilibrium. Since every level must be emitting more radiation upward than comes downward, the temperature of the atmosphere must decrease with height. The radiative equilibrium temperature profile can be found by returning to (6.12). Making an indefinite integral of (6.12) gives

$$\int_0^\tau dB = \frac{3}{4} F\tau$$

since F is independent of τ. The general solution is

$$B(\tau) = B(0) + \frac{3}{4}\tau F \tag{6.17}$$

The radiative temperature at height given by τ is easily obtained by substituting (6.16) into (6.17) and using the Stefan-Boltzmann law. To find the temperature of the ground, (6.15) is used to eliminate $B(\tau)$. The temperature of the air at the level τ is therefore given by

$$\sigma T^4 \equiv B(\tau) = \frac{F}{2}\left(1 + \frac{3\tau}{2}\right) \tag{6.18}$$

The temperature of the earth's surface can be determined from

$$B(T_g) = \frac{F}{2}\left(2 + \frac{3\tau}{2}\right) \tag{6.19}$$

If the profile of τ is known (including the value at the bottom of the atmosphere, τ_1) then the temperature profile can be found. Goody and Yung (1989) use a simple exponential profile that is proportional to the distribution of *water* vapor (not dry air) density in the atmosphere.

$$\tau(z) = \tau_1 \exp\left(-\frac{z}{H}\right) \tag{6.20}$$

where $H = 2\,\text{km}$ is an appropriate scale height for water vapor according to Goody and Yung (1989). Substituting (6.20) into (6.18) gives

$$T(z) = \left[\frac{F}{2\sigma}\left\{1 + \frac{3}{2}\tau_1 \exp\left(-\frac{z}{H}\right)\right\}\right]^{\frac{1}{4}} \tag{6.21}$$

Before temperature profiles, skin temperatures, and surface temperatures may be estimated, one needs to know τ_1 and F. For radiative equilibrium, the solar energy absorbed by the earth must balance that lost to space by infrared emission. The solar flux received at the earth's average orbital distance from the sun is $R_0 = 1380\,\text{W m}^{-2}$. The average earth albedo (Figure 3.7) is $\alpha = 0.31$. If F_a is the solar energy absorbed, then for energy balance the skin temperature is

$$T_s = \left(\frac{F_a}{2\sigma}\right)^{\frac{1}{4}} = \left(\frac{R_0[1-\alpha]}{8\sigma}\right)^{\frac{1}{4}} = 214\text{K} \tag{6.22}$$

where the factor of four in the denominator arises because the solar flux hits the cross-sectional disk of the earth (πr_e^2 where r_e is the earth's radius) but the emission is assumed uniformly spread over the whole earth's surface ($4\pi r_e^2$).

Radiative equilibrium temperature profiles are easily calculated from (6.18) and (6.19) once the optical depth is known. The optical depth will respond to different absorbers, which have different vertical profiles. Equation (6.20) uses an exponential variation of water vapor. In contrast, naturally occuring ozone has highest concentrations in the middle stratosphere. Figure 6.2a approximates the ozone profile with a simple function

$$\rho_{O_3} = Az^2 \exp(az) \tag{6.23}$$

The corresponding optical depth, due only to ozone, is

$$\tau_{O_3} = \frac{kA}{a}\left(2\frac{z}{a} - z^2 - \frac{2}{a^2}\right)\exp(az) \tag{6.24}$$

Radiative equilibrium profiles for the ozone and water vapor absorptions, separately and together, are illustrated in Figure 6.2.

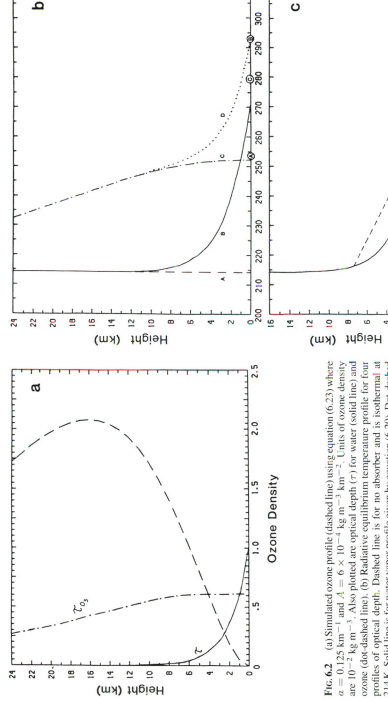

FIG. 6.2   (a) Simulated ozone profile (dashed line) using equation (6.23) where $a = 0.125$ km^{-1} and $A = 6 \times 10^{-4}$ kg m^{-3} km^{-2}. Units of ozone density are 10^{-2} kg m^{-3}. Also plotted are optical depth (τ) for water (solid line) and ozone (dot-dashed line). (b) Radiative equilibrium temperature profile for four profiles of optical depth. Dashed line is for no absorber and is isothermal at 214 K. Solid line is for water vapor profile given by equation (6.20). Dot-dashed line is for ozone profile using $k = 1.0$ and the profile shown in (a). Dotted line is for both water vapor and ozone absorption combined. Letters indicate ground temperatures correponding to each profile of the same letter label. (c) Radiative-convective model solution for water vapor (dashed line) assuming 6.5 K/km lapse rate. Solid line same as in (b).

The solutions asymptotically approach $214\,\mathrm{K}$ as $\tau \to 0$ (Figure 6.2b, dashed line). For $\tau \to 0$ the atmosphere is nearly transparent and is isothermal at the skin temperature; the ground temperature is $40.5\,\mathrm{K}$ larger.

For water vapor, the equilibrium temperature profile follows the skin temperature in the high atmosphere because there is no vapor there. At low levels the temperature changes rapidly with height (Figure 6.2b, solid line). The temperature discontinuity at the bottom is $23.6\,\mathrm{K}$. The rate of change of temperature with height increases as z decreases and increases for larger values of τ_1. Near the surface, the lapse rate rapidly increases with τ; that diminishes the temperature discontinuity at the same time. Goody and Yung (1989) make a similar calculation and find the discontinuity between surface and air temperatures to be $11.3\,\mathrm{K}$ for $\tau_1 = 4$, but $16.7\,\mathrm{K}$ for $\tau_1 = 2$. The lapse rates near the surface become so large that they exceed the dry adiabatic lapse rate when τ_1 is greater than about 0.5. Obviously, such convectively unstable lapse rates are another unrealistic feature of a radiative equilibrium model. Superadiabatic lapse rates are partly a reflection of the sharp exponential increase of the water vapor (2 km scale height).

The ozone profile chosen has a smoother variation with height (8 km scale height). The ozone profile does not develop superadiabatic lapse rates (Figure 6.2b, dot-dashed line). Because the ozone concentration diminishes rapidly in the troposphere, the optical depth does not change much either. Consequently, the equilibrium temperature profile does not change much in the troposphere. The surface temperature discontinuity is $27.8\,\mathrm{K}$.

When the ozone and water vapor absorptions are combined (dotted line), the result exhibits properties of both absorbers. In these calculations solar absorbtion is absent. Ozone is an effective absorber of solar radiation in the middle atmosphere. Incorporation of that effect can lead to increasing temperature with height in the lower stratosphere, a feature not present in Figure 6.2b.

As mentioned above, the two main unrealistic features of the radiative equilibrium model are (1) the development of superadiabatic lapse rates near the ground and (2) the temperature discontinuity at the ground. The second model introduces convective adjustment to eliminate both of these problems. Such a model is called a radiative-convective model. The common approach has been to specify a lapse rate from the surface up to a height z_t, the model's "tropopause." Using average conditions over the earth, then T is given by (6.21) in the model's "stratosphere" above z_t and by

$$T(z) = T(z_t) + \Gamma(z_t - z) \tag{6.25}$$

in the troposphere. The tropospheric lapse rate may be specified to be several values. A constant lapse rate $\Gamma = 6.5\,\mathrm{K\,km^{-1}}$ is used in Figure 6.2c (the dashed lines). Near the surface, this lapse rate would be similar to the moist adiabatic lapse rate for typical middle latitude conditions. The actual moist adiabatic lapse rate varies with the moisture content of the air; it can be as little as 3 K km^{-1} near the surface in the tropics and approaches the dry adiabatic rate (9.8 K km^{-1}) in the upper atmosphere.

The radiation flux is no longer constant in the model troposphere, so that equation (6.11) must be used instead of (6.12). In addition, the convective layer extends higher than the level at which the radiative temperature profile (6.21) has lapse rate equal to Γ (dot-dashed line in Figure 6.2c). This occurs because the flux at the top of the convective layer must match the radiative equilibrium flux F in the stratosphere. If the dot-dashed profile in Figure 6.2c is used, the temperature in the troposphere is everywhere less than the radiative equilibrium profile, hence the flux out of the troposphere would be too small. Even so, the surface temperatures can be greatly reduced by the convection. The resulting temperature profile is similar to observed annual average temperature. Of course part of the agreement is built in by using a value of Γ that is similar to the observed lapse rate. Needless to say, the height of the tropopause is not specified, but is derived and is comparable to observed heights (Figure 3.11).

A pair of papers (Manabe and Strickler, 1964; Manabe and Wetherald, 1967) introduce several further modifications to the radiative-convective model. Their model differs from that just shown in the following ways: (a) a specific latitude (35 N), time of year (April), and local albedo (0.102) are used; (b) observed profiles of H_2O, O_3, and CO_2 are employed; and (c) the atmosphere is allowed to absorb solar radiation. As is seen in Figure 6.3, the model arrives at the equilibrium profile by iteration. Isothermal profiles are assumed at the start and the lapse rates are adjusted so as never to exceed $\Gamma = 6.5$ K km^{-1}. The radiative and radiative-convective models have one obvious difference from before. Temperature now increases with height in the stratosphere; the effect is created primarily by the O_3 absorption of the solar radiation (Figure 3.2 or 3.3a) mentioned earlier in this section.

Further refinements have been made to these types of models. For example, the specification of the tropospheric lapse rate can be avoided by use of a convection model. Using such a model makes sense since the dry adiabatic lapse rate is steeper than 6.5 K/km. Lindzen et al. (1982), for example, make this refinement and predict the Γ profile with moist and dry convection models.

Clouds have a complex effect upon the temperature profile. Figure 6.4 shows results from Liou and Ou (1983), whose model includes a convective model, water vapor, ozone, and clouds. They find, as in previous studies with clouds (e.g., Manabe and Strickler, 1964), that middle clouds reduce the temperatures throughout the troposphere and lower stratosphere. Low clouds reduce the temperatures even more. In contrast, high clouds (located just below or even above the tropopause) can increase the temperatures throughout the troposphere. The net effect varies with cloud altitude because of two competing effects. First, clouds reflect solar radiation (a cooling effect). Second, clouds can trap infrared terrestrial radiation (a heating effect). Because the cloud is opaque, it absorbs nearly all the radiation emitted from the earth's surface; the trapping occurs because the typical cloud emits less radiation to space than does

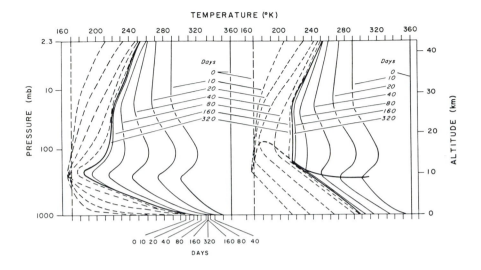

FIG. **6.3**   Temperature profiles from Manabe and Strickler (1964) asymptotically approached by the purely radiative (left chart) and the radiative-convective (right chart) models. The heavy solid lines are the final profiles. Intermediate states are shown as well. The models used here differ from the model shown in the previous figure in several ways, such as incorporating local effects and observed profiles of radiatively active gases. From Liou (1980).

the earth's surface because the cloud is cooler. The first effect, the reflection effect, dominates for low and middle clouds. For high clouds, their temperature is so low that the second effect, the trapping effect, dominates. Clouds occur at many elevations in the atmosphere. On a global average, the cooling effect from reflection dominates, according to satellite observations (e.g., Rossow and Schiffer, 1991). However, in the tropics the heating and cooling are very large but nearly canceling. The global average cloud cooling is largely due to middle latitude clouds.

Convection is a vertical transport of heat. Another refinement is to incorporate the *horizontal* transports of heat in the atmosphere. These transports are expected from the net radiation curves described in §3.1. Liou and Ou (1983) compare solutions from two types of models. The temperature profiles obtained by their one-dimensional model (using climatological cloud distributions that vary with latitude) are plotted as the dashed lines in Figure 6.5. To improve the simulation, horizontal heat transports between three latitudinal belts are incorporated by Liou and Ou (1983) in their "two-dimensional" model. (The observed transport of sensible heat may be estimated by multiplying the field plotted in Figure 4.31 by C_p and atmospheric density.) The heat transport is poleward for the three belts. Below about 4 km the transport into the midlatitude belt from the tropics exceeds the transport out, towards the polar belt. Higher up, the transport out to the polar belt slightly exceeds the input from the tropical belt. One can anticipate the following changes: the net transport out of the tropics will lower the tropospheric temperatures in that belt; net convergence in the

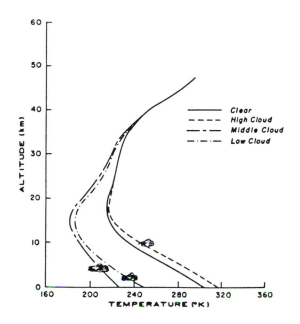

FIG. 6.4 Radiative-convective profiles for clear sky, and for clouds present at one of three levels. From *J. Atmos. Sci.*, Liou and Ou (1983) by permission of American Meteorological Society.

polar belt and in midlatitudes (below 4 km) will raise the temperatures in those belts. These changes appear in Figure 6.5 (dotted lines).

Several other refinements have been made to radiative-convective models. More recent uses of the models focus on the study of "greenhouse" gases (e.g., Ramanathan et al. , 1987). Changes in the concentrations of these trace gases are inserted and new temperature profiles calculated. A very different application is by Held (1982), who attempted to relate the height of the tropopause to the baroclinic instability of the flow (§7.1). He used a simple radiative-convective model with fixed tropospheric lapse rate. A marginally baroclinically unstable flow will have a certain ratio of horizontal to vertical temperature gradients. The degree of instability varies with Coriolis parameter as well as vertical shear. By exploiting these connections he obtained a tropopause that decreased with increasing latitude. Figure 3.5 shows such a decrease.

A class of simple models exists for investigating surface temperature. Energy balance climate models (EBMs) simulate surface temperature for various, mainly climate change, scenarios (e.g., North, 1975). An EBM balances input and output of radiation at the surface. The solar flux is multiplied by an albedo that is dependent in some *empirical* way on temperature (T) and latitude ϕ. Terrestrial emission depends upon T^4 from the Stefan-Boltzmann law, and also upon an empirical coefficient that depends upon T and ϕ, too. The models have time-dependent temperature except when input and output match. The matches are equilibrium temperatures: three, five, or more matches are found in the EBM, but usually just two are stable states. One

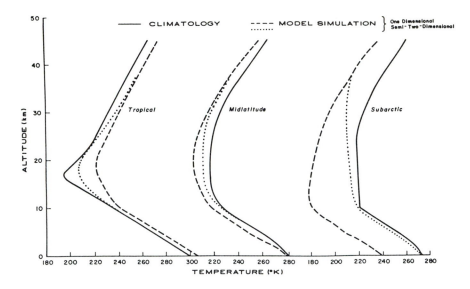

FIG. 6.5   Radiative-convective profiles for three latitude regions, where the regions are distinguished by the annual average solar zenith angle chosen and by the albedo used; both increase with latitude. The principal difference between models is that the "semi-two-dimensional" model (dotted lines) includes observed horizontal heat transports from Oort and Rasmussen (1971) between latitude bands. From *J. Atmos. Sci.*, Liou and Ou (1983) by permission of American Meteorological Society.

stable state matches the observed temperature distribution if the empirical factors are properly chosen. A second stable state corresponds to an ice-covered earth.

Before leaving this section, it is instructive to introduce some properties of the stratosphere. Here, local radiative balance was first proposed to explain the stratospheric temperatures, with convection invoked in the troposphere. With some embellishment from horizontal heat fluxes, calculated temperature profiles can reflect the observed latitudinal differences. Those heat fluxes are necessary not just because the incoming radiation (Figure 3.6) has a greater latitudinal variation than the outgoing radiation. Local radiative balance may not be compatible with a balanced circulation. For example, Shine (1987) shows that equilibrium temperature fields in other seasons do not have solutions able to satisfy gradient wind balance. According to McIntyre (1987), the summer hemisphere radiative equilibrium temperature field is balanced only by a very strong *surface* circumpolar vortex. Nonetheless, radiative equilibrium gives a good approximation to the stratospheric temperatures. One consequence is that stratospheric temperatures respond to solar radiation more directly than do tropospheric temperatures. The troposphere is quite sensitive to earth surface temperatures, which take more time to warm up (or cool down) as summer (or winter) arrives. In addition, the stratosphere contains few clouds, whose seasonal changes partially oppose the seasonal changes in radiation (Figure 3.7). However, the much-publicized "ozone hole" above the South Pole illustrates that the radiatively active constituents can vary greatly from place to place and over time in one locale. Even so, the seasonal changes of tropospheric weather lag the solar cycle by about a month, but little lag is apparent in stratospheric data (e.g., Geller and Wu, 1987).

6.2 Zonal mean motions from eddies

6.2.1 Historical background: Part II

Some of the earliest theories about the structure of the general circulation were described in Chapter 1. One motivation for discussing those early papers was to introduce the terms "Hadley" and "Ferrel" cells, which were useful when describing the observed circulations and transports. The earliest papers were primarily empirical with little theoretical basis. Most of the papers to be discussed now are more mathematically rigorous and so have been left until this point in the text. A convenient starting point is to begin where the zonal average view of the general circulation was first challenged by the suggestion that eddies are important.

The first paper to suggest that eddies are a required component of the general circulation is probably one by Dove (1837). Dove was briefly discussed in Chapter 1 because his work was observational. In Dove's time the "general circulation" meant "zonal average circulation" to most people. Toward the end of the 1800s, strong objections were being raised about equating the general circulation with the zonal average circulation.

One objection has to do with the lack of strong westerlies. Ferrel's later theory has an upper flow extending from equator to pole (Figure 6.6, or 1.6). This circulation accomplished the high latitude heat transport, but conservation of angular momentum could cause the upper-level westerlies in middle and high latitudes to be very large, as illustrated in §6.2.3. Such wind speeds were not observed. To alleviate this objection, Lorenz (1967) points out that one might assume the upper-level poleward motion to be so small that *molecular* friction could keep the westerly acceleration in check, but apparently this idea was not accepted. (A model based on this idea is introduced in §6.2.3.) Possibly people thought that the poleward flow would not be fast enough to transport enough heat. However, Helmholtz (1888, 1889) had a means to alleviate this problem. He postulated that equatorward moving (Ferrel cell) air would meet poleward moving (Hadley cell) air at high altitudes in midlatitudes, forming a sharp discontinuity in atmospheric properties. At this discontinuity, there would form vortices that would mix the air vertically (and horizontally), thus diffusing the momentum. One might think that Helmholtz's vortices refer to large-scale cyclonic storms, because he states that the waves would be "not only small" but "many kilometers" in length. However, Helmholtz emphasizes density differences across the "strata" as a driving mechanism, suggesting that he was writing about gravity waves, not frontal cyclones. Lorenz (1967) also infers that Helmholtz was describing the phenomenon known as "Kelvin-Helmholtz instability," which is a small-scale, gravity wave, shear instability. But even so, Helmholtz introduced the concept of turbulent

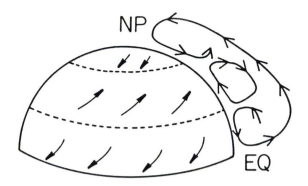

FIG. 6.6   Ferrel's later (1859) description of the general circulation; similar to Figure 1.6. The refinement over Figure 1.5 is the high altitude branch of the Hadley cell extending all the way to the pole.

viscosity, which is a required process in current theories of the general circulation. A coefficient of turbulent viscosity is much larger than the molecular viscosity, as mentioned in §4.1. Hence, this mechanism could allow larger values of poleward drift speed without creating unrealistically strong westerlies. For turbulent viscosity coefficients of 10 to $100 \, \text{gm} \, \text{cm}^{-1} \, \text{sec}^{-1}$, Lorenz calculates an upper-level poleward current of 1 to $10 \, \text{cm/s}$, which brackets observed values in the Hadley cell (Figure 3.16) and provides Ferrel (or Thomson) with a feasible mechanism to overcome the first objection.

A second objection to the Ferrel (Thomson) zonally symmetric theory was the lack of evidence for a poleward current at high altitude in midlatitudes. The mean meridional velocity is hard to measure (discussed in regard to Figure 3.16) because it is so small on the zonal average. Bigelow (1900, 1902) used cloud observations to formulate a scheme for the general circulation that is not so different from Dove's (65 years before). He did not stop at observational evidence (as did Dove) but inferred that warm, poleward-moving and cold, equatorward flows accomplish the necessary heat transport. He linked these motions to the cyclones and anticyclones and stated that the vertical mixing of momentum in the cyclones and anticyclones reduced the westerlies aloft. While he provided for energy balance, a few items are missing. Lorenz (1967) points out that Bigelow left off horizontal tilt of the cyclone and anticyclone axes, and he turned out to be incorrect—in hindsight of later observations from the last 40 years—in ignoring variations in the flow above 3 km. Defant (1921) carried the point further, treating the heat transport by the eddies as the result of large-scale turbulence.

To wrap up the short summaries of early studies, four other studies are worth noting. The first two are discussed in Lorenz (1967). First, Exner (1925) is credited with originating the idea that a poleward or equatorward motion over many latitudes over all longitudes could not be supported by a uniform pressure gradient. The reason is that an eastward or westward pressure gradient would extend around the entire globe, which is impossible (since that is a closed circuit). Exner maintained that zonally asymmetric east-west pressure gradients would develop that could keep the westerlies from growing too strong. Second, Jacob and Vilhelm Bjerknes each wrote a series of papers that bear upon the general circulation. V. Bjerknes (1937) was aware of Helmholtz's work, and it spurred him on to the correct identification that the disturbances forming upon the unstable discontinuity (described by Helmholtz) were actually the traveling midlatitude cyclones. The observed verification that such cyclones can occur as large-scale waves upon the polar front is a well-known discovery of J. Bjerknes. An important point here is the separation of phenomena: that there is a zonally-averaged circulation that is unstable to the growth of eddies. J. Bjerknes felt that the zonal average meridional circulation (at least outside the tropics) would be a statistical residual and nearly zero, and that the eddies were responsible then for the heat and momentum transports.

Monin (1986) discusses Kochin (1936) and its extensions by Blinova beginning in 1943 (see Blinova, 1976). Kochin tried to balance Coriolis and viscous forces. Heat and momentum transports by synoptic systems were parameterized in their models. Temperature, surface pressure, and viscosity are specified from observations, and equations for momentum, density, and geopotential height are solved. Monin indicates that zonal and monsoonal-type circulations were obtained by them from the input data and after including land and sea contrasts.

The short descriptions of historical interest stop at this point for two reasons. First, more detailed analyses of later papers are made in this and the succeeding chapter. Second, the pace and amount of theoretical development grows very rapidly after World War II.

6.2.2 Zonal momentum: Maintenance against friction

In 1926 a noteworthy paper by Jeffreys appeared. He began by noting that over a long period of time, there must be conservation of planetary angular momentum. Friction would be constantly destroying momentum, so there must be transfer of momentum to maintain the observed flows. He showed that the zonal average flow could not accomplish the transport. He next showed that transport of momentum in a frictional boundary layer would not transport enough momentum. Finally, it followed that an *eddy* horizontal momentum flux was the major mechanism for maintaining the angular momentum balance. Eddies could transport the momentum even if the winds are geostrophic. In short, he demonstrated that the eddies are *fundamental* to the general circulation. His arguments are sketched below using observed magnitudes of observations.

The earth is nearly spherical and, to be strictly correct, curvature terms should be included. However, a Cartesian coordinate system is acceptable for illustrative purposes. The region is assumed cyclic and only linear momentum and friction are considered. From the linear momentum equation, Jeffreys proceeded to determine what the rate of dissipation would be. He chose $u_0 = 4 \text{ m s}$ for the horizontal average

value of surface zonal velocity. Using (4.9) in Cartesian geometry, the frictional loss
by an air column interacting with the surface of the earth may be approximated by

$$\int_{\text{vol}} F_x d\text{vol} = \int_{X_0} \int_{Y_0} \mu \rho_0 u_0 \left[u_0^2 + v_0^2 \right]^{\frac{1}{2}} d\text{Area} \tag{6.26}$$

Approximate values may be substituted into (6.26) to estimate the rate of extraction
from a column of unit horizontal area. The meridional velocity is assumed to be 1 m/s.

$$\int_{\text{vol}} F_x d\text{vol} \sim (4 \times 10^{-3})(1 \text{ kg m}^{-3})(4.123 \text{ m}^2\text{s}^{-2})(10^{-4} \text{ m}^2)$$

$$\simeq 0.66 \times 10^{-5} \text{ kg m s}^{-2} \tag{6.27}$$

The total amount of (linear) momentum in the column is

$$M_L = \iiint \rho u \, dz \, d\text{Area} = \int_0^{P_0} \iint \frac{u}{g} \, dP \, d\text{Area}$$

$$= P_0 \left(\frac{\overbrace{u}}{g} \right) \iint d\text{Area}$$

where the last step follows because Jeffreys let $u = \overbrace{u}$, a vertical average value
that he also took to be 4 m/s. This value is too small (Figure 6.7). Nonetheless he
obtained

$$M_L = \frac{(4 \text{ m/s})(10^7 \text{N m}^{-2})(10^{-4}\text{m}^2)}{(10\text{ms}^{-2})}$$

$$= 4 \times 10^2 \text{N s} \tag{6.28}$$

As a fluid spins up or down, the rate of loss changes. But, for the purpose of this first
estimate, a constant rate of (6.27) is used and is assumed to be spread proportionally
throughout the depth of the atmosphere. The constant rate provides a minimum time
for spin down. The time to extract all this momentum at the rate (6.27) is 6×10^5
seconds or about 7 days. By Jeffreys's estimate the atmosphere in middle latitudes
loses all its momentum in about a week.

As noted, this derivation has some questionable simplifications. Two that are
easily fixed are: (1) to estimate more accurately the linear momentum (Figure 6.7),
and (2) to make the rate of extraction a function of the instantaneous velocity. From
Figure 6.7, one expects M_L to be 2 to 4 times as large as (6.28). However, given
the rather crude estimate of the frictional loss, it hardly makes any difference to the
discussion: most of the momentum is still lost on a time scale of a week or two.

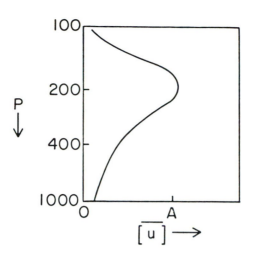

FIG. 6.7   Jeffreys (1926) used a value of vertical and zonal average of $U = 4$ m/s, but much stronger winds are typically found at high altitudes, so the true U value is closer to 10 to 15 m/s in middle latitudes. A schematic diagram of zonal and time average zonal wind $[\bar{u}]$ is plotted here. The value of A varies with latitude; in middle latitudes A is between 20 and 50 m/s.

Incorporating speed-dependent friction is more interesting. The frictional loss in (6.27) is proportional to the velocity squared. Therefore, spin down of the flow could be governed by

$$u_t = -au^2 \qquad (6.29)$$

where a is a constant controlling the rate of damping and b^{-1} defines the initial amplitude at time $t = 0$. The solution to (6.29) is simply $u^* = u(t) = (b+at)^{-1}$. The velocity u^* asymptotically approaches zero, so one must discuss the decay in terms of a fractional change in the velocity. For comparison purposes, the time over which 80% of the original momentum is lost is used. For an asymptotic decay given by the u^* formula, the time is $(0.8\, b)/(0.2\, a)$. Using values for b and a that correspond to the velocity in (6.28) and rate of decay in (6.27) at the initial time, then

$$b = u(0)^{-1} = 0.25\text{s m}^{-1} \qquad (6.30)$$

and

$$a = \left(\mu = 4 \times 10^{-3}\right) \times \left(\text{scale height} \simeq 8.4\,\text{km}\right)^{-1}$$
$$\simeq 5 \times 10^{-7}\,\text{m}^{-1} \qquad (6.31)$$

where the measure of the density-weighted depth of the atmospheric column is the scale height. From (6.30) and (6.31), the time to lose 80% of the momentum is about 2×10^6 seconds or 23 days when the friction varies as au^2. In contrast, the constant rate given by (6.27) loses 80% of the initial momentum (6.28) in about $5\frac{1}{2}$ days. The amount of momentum lost by these two rates of frictional loss are equivalent at about 1.2 days.

Returning to Jeffreys's analysis, something must be replenishing the momentum at a rate of M_L per week. The transport into the domain from the meridional wall flux term in the angular momentum tendency equation (4.4) must balance this rate of loss.

$$R \int_t \int_z \int \rho uv \, dx \, dz \, dt \simeq \iiint \rho_0 R \mu u_0 \left(u_0^2 + v_0^2\right)^{\frac{1}{2}} dx \, dy \, dt \qquad (6.32)$$

where the flux through the meridional wall is on the left-hand side of (6.32). The flux in (4.4) has a second part: the $f[\overbrace{v}]$ part of the flux through the meridional wall is assumed to be negligible over time scales of a year since such a net shift of mass would violate mass conservation. The sign in (6.32) is for northward flux across the lower latitude wall. The transport across the higher latitude is zero if it is located at the pole; alternatively it can be incorporated into the left-hand side of (6.32) too. Following the discussion in §4.1, the mountain torque usually has the same sign as the friction, so mountain torque can be assumed to be incorporated into (and to increase) the friction term on the right-hand side of (6.32).

Jeffreys imagined a model with geostrophic winds above an Ekman boundary layer. Geostrophy precludes a zonal average meridional motion (away from the possible zonal pressure imbalances across mountains). Hence the only meridional transport must occur in the boundary layer of such a model. Therefore, the vertical integral in (6.32) is zero outside that boundary layer. A steady state may be assumed, which eliminates the need for the time integrals. For simplicity, Jeffreys let the velocities be constant within and outside of the boundary layer. Equation (6.32) becomes

$$R \rho u_0 v_0 \int_z \int dx \, dz \simeq \rho_0 R \mu u_0 \left[u_0^2 + v_0^2\right]^{\frac{1}{2}} \iint dx \, dy \qquad (6.33)$$

Again, $u = u_0 \simeq 4$ m/s is used throughout the depth of the atmosphere, and $v = 0$ except in the boundary layer, where $v = u/4 = 1$ m/s is used. For a region extending from 30 N to the North Pole with a 1 km deep boundary layer, the numeric values are

$$(1 \text{ m/s}) \cos(30)(1 \text{ km}) \underbrace{\simeq}_{?} (12 \times 10^{-3} \text{ m/s})(0.5)(6.6 \times 10^3 \text{ km}) \qquad (6.34)$$

where the common $R\rho u_0$ is factored out and where value 0.5 is the average cosine of latitude for the domain and 6.6×10^3 is the distance from $30°$ to the pole. A question mark is inserted above since the right-hand side of (6.34) is about 50 times greater than the left-hand side. So the frictional dissipation exceeds the boundary layer transport

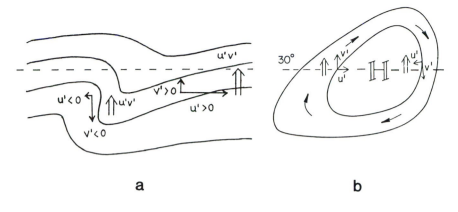

FIG. 6.8 Schematic diagrams showing how geostrophic winds (thin arrows) around non-circular eddies can cause poleward $[u'v']$ flux (double-shafted arrows). Figure 4.9 is similar. Examples shown are Northern Hemisphere: (a) upper level tilted trough and (b) surface subtropical high.

by roughly a 50 to 1 ratio! No reasonable refinement in the numerical estimates can reduce this ratio below about 20.

Balance can be achieved without having to abandon geostrophy if there are tilted troughs, as in Figure 6.8a. From the Appendix, the flux across the latitude boundary wall has a contribution from $[u'v']$ that is not zero for a tilted trough. As mentioned in §4.2.2, a circular low (or high) has eddy momentum fluxes on the east side that cancel those on the west side. The importance of the tilts is illustrated by this simple example from the Northern Hemisphere. When the trough (or high) is "stretched" so that it has a southwest to northeast axis tilt, then the locations of the maximum and minimum values of $u'v'$ around the low are rotated in a counterclockwise manner about the low center compared to the locations for a circular low. This rotation allows positive values of $u'v'$ to occur on both the east and west sides of the tilted low along the central latitude of the low. The largest negative values of $u'v'$ are rotated to the south and north sides of the tilted low. The resultant zonal mean eddy fluxes would create convergence of eddy momentum to the north of the central latitude. Consequently, a southwest to northeast tilted trough in the Northern Hemisphere subtropics would create zonal mean eddy momentum convergence at the latitude of the subtropcial jets.

For a tilted trough like Figure 6.8a, one can reasonably let $u' = v' = u_0$ on the east side of the trough and $u' = v' = -u_0$ on the west side. If this is done just in the boundary layer, then the left side of (6.34) is four times and the right side is $\sqrt{2}$ times as large. It is reasonable to assume that the tilted troughs extend through the depth of the troposphere ($\sim 10\,\text{km}$). If the decrease of density with height is ignored and velocity is allowed to increase with height, then the presence of tilted troughs as in Figure 6.8a can bring (6.34) into balance. Since density (ρ) decreases exponentially with height, then the tilted troughs must be deeper or sharper. An obvious solution is to have the tilted troughs become stronger with height in the troposphere. Obviously, the stronger flux is consistent with other figures such as Figure 4.11. What makes the calculation so important is that Jeffreys deduced *without upper air observations* that the atmosphere must have deep tilted waves in the subtropics and middle latitudes in order to maintain the general circulation!

The results of the analysis by Jeffreys were carried by Starr (1948) to what must have been a very gratifying conclusion. The required $u'v'$ covariance meant that the shape as well as the type of mechanism was specified to fulfill the angular momentum budget. The simplest patterns that can transport the momentum over a wide range of latitudes are ones with long waves. Long waves have since become well known to synopticians. Figure 6.9 gives an example of the long wave pattern. The tilted troughs in the subtropics are more easily seen in the stream function, rather than in the height field. As mentioned in §5.10, the subtropical tilted troughs are typically found over the oceans. The southwest to northeast tilt places the trough much further east at higher latitudes as can be seen by comparing Figures 6.9 and 5.9. Starr points out that even the shape of subtropical *surface highs* contribute to the momentum flux (e.g., Figure 6.8b). Observations reported by Starr and White (1951) showed quite clearly that the angular momentum transport was primarily accomplished by the eddies. White (1951) found that the geostrophically calculated heat transports were close to those required for balance. Benton and Estoque (1954) showed that the eddy transport of water vapor seemed sufficient to satisfy balance as well.

Recent theoretical models of linear instability on a sphere (e.g., Baines, 1976, or Fredericksen, 1978) also contain poleward momentum fluxes produced by the most rapidly growing solutions. The nonlinear extensions of these simple theoretical models show this transport even better since the poleward transport is enhanced at high levels compared with linear models (e.g., Simmons and Hoskins, 1978).

6.2.3   *Subtropical jets: Roles of Hadley cells and eddies*

6.2.3.1   *Transient eddies versus time-mean cells*

By 1950, the evidence was clear that transient eddies were important in maintaining the heat and momentum transports in middle latitudes. But while the eddies transport enough momentum to maintain the middle latitude westerlies, it would not be correct to conclude that the mean meridional cells play no role in *organizing* the westerlies into jets. Examination of the December 1949 issue of *Journal of Meteorology* reveals prominent scientists of the time clashing over the issue of the extent to which eddies and meridional cells govern the jets.

The different roles of the eddies and meridional cells are easily illustrated by examining the tendency equation for time-mean zonal momentum. The time-mean pattern has zonal variations. Hence there must be at least one other term in the momentum equation that can balance the $\partial \overline{u}/\partial x$ term. Namias and Clapp (1949) suggest a balance between the ageostrophic flux of planetary vorticity and the advection of

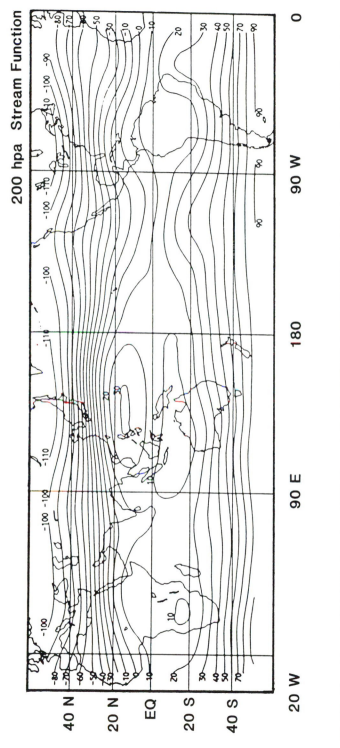

200 hpa Stream Function

FIG. 6.9 Five-year mean of December, January, and February stream function at 200 hPa showing mid-ocean tilted troughs in the manner illustrated by Figure 6.8a. Data from October 1978 through August 1983 period. Contour interval is 10^7 m^2 s^{-1}. The midlatitude troughs are located near the east coasts of the continents, but in the tropics the trough axes bend well to the west. From Arkin et al. (1986).

zonal wind.

$$u\frac{\partial \overline{u}}{\partial x} = f(v - v_g) = f\overline{v_a} \tag{6.35}$$

where v_g is the geostrophic meridional wind. In fact, (6.35) is a reasonable approximation, as Figure 5.7 demonstrates. Figure 5.7, showing zonal wind at 200 hPa, illustrates
the strong correlation between $\overline{u}$ variation (the contours) and $\overline{v}_a$ (the arrows). It is
hard to estimate $\overline{v}_a$ accurately from observations; but Blackmon et al., 1977) were
able to verify (6.35). The discussions by Namias and Clapp and by Blackmon et al.
infer vertical motions from these ageostrophic motions; Blackburn (1985) criticizes
the *vertical* circulations inferred in this manner, but these details are left for §6.5
after the forcing of the meridional cells has been examined in §6.3.

The ageostrophic, meridional motion in the Namias and Clapp (1949) model is
still accepted as valid and offers an interesting interpretation of the localization of the
subtropical jets. In the confluence regions (where the jet flow is accelerating), there
is poleward $\overline{v}_a$, consistent with poleward extension of the Hadley cell. Downstream,
in the diffluent region (where the jet decelerates), equatorward $\overline{v}_a$ is consistent with
the presence of a Ferrel cell. The circulation in the diffluent region is similar to
the schematic diagram for decaying frontal cyclones (Figure 4.22). Eddies in fact
do drive the Ferrel circulations, as the Kuo-Eliassen equation demonstrates in §6.3.
Thus, mechanisms are able to drive both circulations, with the eddies forming in the
jet entrance region and becoming prominent in the diffluent zone. From baroclinic
instability theory, the most rapid growth tends to occur where the strongest vertical
shears are, and a minimum vertical shear is needed before the instability is initiated.
All the pieces in this puzzle seem to fit nicely together.

While (6.35) may describe how the zonal variations are maintained, it will not
describe the formation of the *zonal average* jets so easily. Taking a time and zonal
average of the zonal equation of momentum, one finds another balance.

$$\frac{\partial}{\partial y}\left([u'v'] + [u''v'']\right) = \left(f - \frac{\partial [\overline{u}]}{\partial y}\right)[\overline{v}] \tag{6.36}$$

(e.g., Wallace, 1978a). The balance changes for two reasons. First, the left-hand side
of (6.35) must vanish on a zonal average because it is a perfect differential. Locally,
it is observed to be two or three times as large as the eddy momentum convergence
retained in (6.36). Second, the ageostrophic motions are also locally large, but they
nearly sum to zero on a zonal average in midlatitudes. Since $[v_g] = 0$ by definition,
then $[v]$ in (6.36) is actually the ageostrophic velocity. (To simplify the discussion,
friction has been neglected in both (6.35) and (6.36).)

6.2.3.2 *Hadley cell motions and the jet*

To conclude this sub-section, a simple model is devised that can explain some gross
features of the $[u]$ field. The description of the jet entrance region in §6.2.3.1 relies

upon poleward ageostrophic motion. From angular momentum conservation, pole-ward moving air experiences westerly acceleration relative to the ground. The model employs angular momentum conservation, vertically varying friction, and a Hadley cell meridional motion. The model is only intended to show the westerly accelera-tion. The model is too simplistic for the Hadley cell because vertical motions are not incorporated directly. However, a conceptual picture of the vertical motions could be that all parcels move up or down in thin layers at the north and south boundaries and a *fraction* of the interior air parcels at each latitude sink or rise so as to maintain mass continuity. Sections 6.3 and 6.5 formulate better descriptions of the Hadley cell.

From the definition of angular velocity (4.1), one can write the zonal mean angular momentum as

$$[M] = [\rho R(\Omega R + u)] \tag{6.37}$$

Here, ρ is the air parcel density, Ω is the angular velocity of the earth, $R = r\cos\phi$ and r is the earth's radius. The conservation law for $[M]$ is given by (4.2). Ignoring mountain torques, then (4.2) becomes

$$\frac{d[M]}{dt} + R[F_x] = 0 \tag{6.38}$$

Friction is modeled using a Rayleigh formulation.

$$F_x = -\alpha u \rho \tag{6.39}$$

This form of friction means that the local derivative and frictional terms in the zonal momentum equation are

$$\frac{\partial u}{\partial t} = \alpha u$$

which has exponentially decaying solutions for negative Rayleigh coefficient, α.

The constant rate (6.27) and the inverse-time rate (6.29) have equal loss of momentum at about 1.2 days; α ($\alpha_b = -0.155274$ days^{-1}) is chosen to match at that time too. This value of α might be appropriate for the atmospheric boundary layer, but frictional decay in the free atmosphere at the "top" of our "Hadley" cell would be much less. Reasoning by analogy with scaling arguments for Ekman layers in theoretical models (e.g., Grotjahn et al. 1992), $\alpha_f = -4.16 \times 10^{-3}$ days^{-1} is used here.

Substituting (6.37) and (6.39) into (6.38) yields

$$\left(\rho[u] + 2\rho R\Omega\right)\frac{dR}{dt} + \rho R\frac{d[u]}{dt} = \alpha\rho R[u] \tag{6.40}$$

Following the motion, the total derivative can be transformed into a latitudinal derivative; time (t) is equivalent to a latitude location (ϕ) that differs from the starting point ($t = 0$ and $\phi = \phi_0$).

$$\int_0^t [v]d\tau = r \int_{\phi_0}^{\phi} d\phi \tag{6.41}$$

$[v]$ is set to a constant in space and time, leaving

$$dt = \frac{r}{[v]}d\phi \tag{6.42}$$

After substituting (6.42) into (6.40), the factor $\rho[v]\cos\phi$ is divided out to obtain

$$\frac{d[u]}{d\phi} - [u]\underbrace{\left(\tan\phi + \frac{\alpha r}{[v]}\right)}_{G} = \underbrace{\left(2\Omega r \sin\phi\right)}_{H} \tag{6.43}$$

Equation (6.43) is a linear first order ordinary differential equation. The general solution (e.g., Beyer, 1984; p. 315) of (6.43) is found by integration from an initial state with zonal velocity u_0. The general solution is

$$u(\phi) = \exp\left(-\int_{\phi_0}^{\phi} G(\phi)d\phi\right) \times \left\{u_0(\phi_0) + \int_{\phi_0}^{\phi}\left(H(\phi)\exp\left(\int_{\phi_0}^{\phi} G(\phi)d\phi\right)\right)d\phi\right\}$$

Evaluating the integrals of G yields

$$u(\phi) = \exp\left(\frac{r\alpha(\phi - \phi_0)}{[v]} - \ln(\cos\phi) + \ln(\cos\phi_0)\right)$$
$$\times \left\{u_0(\phi_0) + \int_{\phi_0}^{\phi} 2\Omega\sin\phi\, r\exp\left(-\frac{r\alpha(\phi - \phi_0)}{[v]}\right.\right. \tag{6.44}$$
$$\left.\left. + \ln(\cos\phi) - \ln(\cos\phi_0)\right)d\phi\right\}$$

The integrals in (6.44) can be evaluated numerically using a sum of trapezoids and small increments of latitude. It is useful to consider the time-dependent form of (6.44) by substituting (6.42). The result is

$$u(\tau) = \exp\left(\alpha\tau - \ln\left(\cos\left(\frac{[v]\tau}{r} + \phi_0\right)\right) + \ln(\cos\phi_0)\right)$$
$$\times \left\{u_0(0) + \int_0^{\tau} 2\Omega\sin\left(\frac{[v]t}{r} + \phi_0\right)[v]\exp\left(-\alpha\tau\right.\right. \tag{6.45}$$
$$\left.\left. + \ln\left(\cos\left(\frac{[v]t}{r} + \phi_0\right)\right) - \ln(\cos\phi_0)\right)dt\right\}$$

Several cases are now examined.

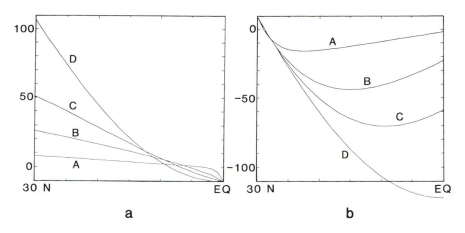

FIG. 6.10   Example solutions for various magnitudes of the meridional velocity. (a) Solutions for poleward motion; $[v]$ equals 5 mm/s (curve A), 2 cm/s (curve B), 5 cm/s (curve C), and no viscosity (curve D). The viscous solutions use $\alpha = -4.82 \times 10^{-8}$ s^{-1}. (b) Solutions for equatorward motion; $[v]$ equals 50 cm/s (curve A), 2 m/s (curve B), 5 m/s (curve C), and no viscosity (curve D). The viscous solutions use $\alpha = -1.8 \times 10^{-6}$ s^{-1}.

The first case has no change of latitude. Since $[v] = 0$ and $\phi = \phi_0$, one must use the form given in (6.45). The integral in (6.45) vanishes in this case, leaving simple exponential decay of the velocity, $[u(\tau)] = [u(0)] \exp(\alpha\tau)$.

Another special case excludes friction. Either (6.44) or (6.45) may be used, but the velocity profile is independent of time at a given latitude, so it is instructive to use (6.44). The velocities change rapidly with latitude to compensate for the changing planetary rotational velocity. This solution is plotted in Figures 6.10a (for poleward motion) and 6.10b (for equatorward motion).

A more general, third case is poleward motion in the upper troposphere (using α_f and $[u] = -10$ m/s at the equator). Several choices of $[v]$ are plotted in Figure 6.10a. For slower meridional motion, the Rayleigh friction has more time to reduce the wind speed. For the value of α_f used here, a northward motion of 3 cm/s at 30 N results in $[u]$ ($\simeq 35$ m/s) that is close to the observed value. But 3 cm/s is about an order of magnitude smaller than the annual and vertical average observed $[v]$ in the upper branch of the Hadley cell (Figure 3.16).

Equatorward flow is occurring in the surface boundary layer. Figure 6.10b ($\alpha = \alpha_b$) shows solutions that commence at 30 N with $[u] = 10$ m/s. Reasonable zonal velocities occur for much larger meridional velocities than in the third case because friction is now much stronger. Friction was crudely specified by Jeffreys, and that in turn specified the α_b used here; these admittedly crude estimates of friction still allow $[v]$ velocities that are comparable with observed values without creating unrealistically large (negative) values of $[u]$. In short, if the free atmospheric friction is increased in the simple model here, then the observed $[u]$ and $[v]$ profiles in the tropics and subtropics can be reasonably matched, along with the required heat transports.

The small $[v]$ speeds in the upper and lower troposphere could transport sufficient heat, as can be easily calculated. From Figure 3.9, the observed annual sensible heat transport in the subtropics is about $F = 2.5 \times 10^{15}$ W. (Newton (1972) calculates sensible heat transport using the DSE (dry static energy) $= C_p T + g\Phi$ part of the moist static energy.) From Figure 3.20a, a reasonable average value for the dry static energy (using upper tropospheric values at 20 N) is DSE $\simeq 3.4 \times 10^5$ J/kg. The *upper tropospheric* heat flux is therefore

$$F \sim \text{DSE} \, [v] \, \Delta P \cos\phi \, 2\pi r \, g^{-1}$$

where the vertical integral over pressure has been approximated using vertical average values for the integrand. To make a comparison, it is assumed that the transport occurs in an atmospheric column that is 800 hPa thick, encircling the earth at 20 N. The meridional heat flux for $[v] = 3$ cm/s is only 4.8×10^{14} W. The heat flux in the boundary layer is about four times larger ($[v] = 1$ m/s, $\Delta P = 100$ hPa). The net transport in this simple model happens to be extremely close to the observed values. The discrepancy between this model value and observations is small enough to suggest that the scheme described here is a feasible model of the subtropical zonal mean horizontal winds. While there is plenty of room to fudge the parameters to make the transports and velocities match, such fudging is somewhat circular. By fudging further, one is merely adjusting α_f and α_b to give $[\,\widehat{v}\,]$ and $[u]$ that are close to observations. Better ways to illustrate the mechanisms that drive the mean meridional circulations are contained in the models of §6.3 and §6.6.

6.2.4   *The stratospheric "Hadley" circulation*

A common theme threading throughout this book is that zonal mean circulations are intrinsically linked with eddies. Here are some examples encountered so far. In §6.2.1 the eddies were proposed as the driving mechanism for the Ferrel cell, a relationship that is expanded upon in §6.3. The linkage is built into the analysis by Kuo (1956) and Eliassen (1952), to be discussed in §6.3. The rationale for striking a balance between eddy fluxes and mean meridional circulations was further formalized by Charney and Drazin (1961), Andrews and McIntyre (1976), and others into the so-called "non-interaction theorem." In §6.2.2, Jeffreys's calculations proved eddies to be necessary to maintain the zonal mean westerlies. In §6.2.3 the transient eddies and the time-mean meridional cells were shown to be interconnected in explaining the longitudinal variation as well as the zonal mean existence of the subtropical jets. Further details about this matter are included in §6.5, as well. It is precisely because the eddies have several links to zonal mean circulations that it is hard to separate our discussion into neatly bordered theoretical analyses. Consequently, this section (§6.2) identifies several main linkages; subsequent sections in Chapters 6 and 7 will each treat parts of the interaction in greater detail.

In this subsection the links are adumbrated between eddies, diabatic heating and cooling, and mean meridional circulations in the stratosphere. To understand how the investigations of these links came about, it is useful to begin with some "mysteries" in the observations.

The first mystery concerned the distribution of ozone. Dobson set up a network of observations of ozone in the late 1920s (e.g., Dobson, 1930). He found the largest amounts in a column of air to be located in higher latitudes, reaching maximum value in the spring and minimum value in the fall. The original assumption (see review in Mahlman et al. , 1984) was that diffusive separation of gases would occur throughout the stratosphere. To be consistent with that assumption, ozone would be greatest in the tropics and subtropics, areas having the greatest solar radiation (Figure 3.7a). (Later, the assumption was further strained when it was found that radioactive debris from bomb tests did not stay in the stratosphere as long as anticipated.)

Observations of water vapor in the stratosphere were also puzzling. The lowest mixing ratios (around 4 ppmv) occurred along the base of the stratosphere. It is diffi-cult to imagine how such dry air can be formed without it's being previously located at a much lower temperature or at a lower pressure. Inspection of Figure 3.11 reveals the coldest temperatures to be at the tropical tropopause and at the winter pole. The sim-plest explanation for such low mixing ratios (especially in the summer hemisphere) is that the air had to have passed through the tropical tropopause. Brewer (1949) pro-posed a circulation in which midlatitude, lower stratospheric air was freeze dried (so to speak) by passing through this equatorial "cold trap." Dobson (1956) proposed that the same circulation could explain the unexpected ozone concentrations in the same regions. The circulation they proposed looks similar to a *stratospheric* "Hadley" cell (Figure 6.11a). They proposed one cell in the lower stratosphere of each hemisphere having rising in the tropics and sinking in midlatitudes and polar regions. Direct mea-surement of the mean meridional circulations is difficult as discussed in §3.3. Even so, the observations (e.g., Vincent, 1968) clearly showed two cells; the second cell has rising in polar regions and sinking in the middle latitudes (Figure 6.11b).

In an early attempt to model the stratospheric circulations, Murgatroyd and Singleton (1961) use a zonal mean form of the thermodynamic equation to deduce the circulation that would result for a specified temperature pattern [T] and diabatic heating rate [Q]. The model consists of two equations and significantly, it does not include eddy fluxes. The thermodynamic equation is

$$[v]\frac{\partial[T]}{\partial y} + [w]\left(\frac{\partial[T]}{\partial z} + \frac{\kappa[T]}{H}\right) = \frac{[Q]}{C_p}$$

while the continuity equation is

$$\frac{\partial[v]}{\partial y} + \frac{1}{\rho_0}\frac{\partial}{\partial z}\left(\rho_0[w]\right) = 0$$

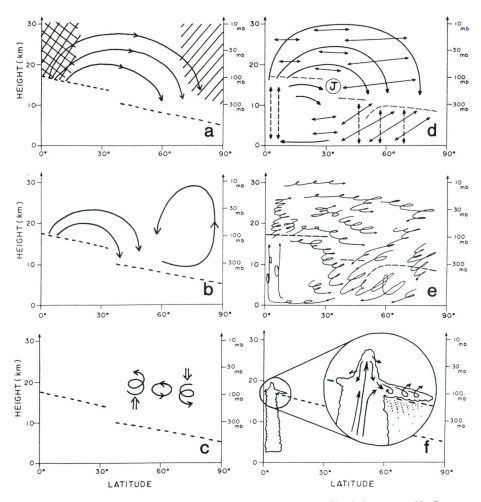

FIG. 6.11 Stratosphere zonal mean meridional circulation features. (a) Circulation proposed by Brewer (1949) and Dobson (1956). Cross-hatching indicates area of diabatic heating from ozone absorption. Hatching indicates area of stronger radiational cooling. (b) Observed mean meridional circulations. (c) Thin arrows show individual parcel trajectories; double-shafted arrows show Stokes drift from nonlinear parcel motions along planetary waves. (d) Schematic motions in general circulation model (thick arrows), horizontal mixing along θ surfaces (thin arrows), and vertical mixing (dashed line arrows). (e) Parcel trajectories showing looping paths similar to (c) in the model. Diagrams (d) and (e) from Kida (1983b). (f) Schematic view of precipitation (dots) from cirrus anvil clouds in deep tropical convection leading to cooling and drying of air just above tropopause.

The resulting diabatic circulation is quite similar to what Brewer and Dobson had proposed. The cell is a direct cell in that rising motion occurs where there is net diabatic heating; sinking occurs where there is net cooling. Figure 6.11a illustrates the sense of the diabatic circulation during an equinox season.

To recap, Murgatroyd and Singleton find a direct, meridional circulation that seems consistent with the observations of ozone and water vapor in the lower stratosphere, but observations show a second (indirect) cell as in Figure 6.11b. Murgatroyd and Singleton did not find the second cell, but they also did not include the eddy fluxes, which turn out to be large. The resolution of this conundrum lies in the near cancellation between the circulations by the eddies and by the indirect cell.

The resolution lies in the so-called "non-interaction theorem." Simply put, the theorem states that for adiabatic, small amplitude, propagating waves in a zonal mean flow, when there is no net effect by the eddies upon the mean flow, the waves develop meridional circulations that cancel the flux convergences by the waves. The statement of the theorem is almost self evident: since the eddies cannot grow, there cannot be any net change, so the eddies must develop a circulation to cancel their fluxes. The cancellation appears to be more robust than the stated restrictions might imply. Cancellation is seen in general circulation model (GCM) studies; in the lower stratosphere meridional cell and eddy flux convergences nearly cancel for trace constituents (Hunt and Manabe, 1968) and heat fluxes (Smagorinsky et al. , 1965). The way in which the cancellation occurs is through "Stokes drift."

The opening paragraph of this subsection (§6.2.4) lists several ways in which eddies are linked to zonal mean flows. In addition to forcing zonal mean flows, the eddies can create zonal mean *drift of trajectories* when the eddies have asymmetric structure. Air parcel trajectories in the eddies may follow looping paths in the meridional plane that do not quite return to the same point at the end of each loop. The first scientific discussion of the phenomenon was by Stokes (1847); he examined the net drift of floating debris that followed looping paths with the passage of water waves. This phenomenon is now commonly referred to as "Stokes drift." The debris has a net motion because the horizontal displacement at the wave crest is greater than the opposite displacement at the wave trough. The difference in displacement follows from the decreasing wave amplitude with depth.

Stokes drift can occur for long waves, which have significant amplitude penetrating into the stratosphere (Figure 6.11c). Wallace (1978b) describes the situation in detail. The general idea is as follows. Most of the stratospheric air parcels are above the steering level (the level at which parcel speed equals the wave propagation speed). Because parcels tend to sink in the troughs and rise in the ridges as they pass through the long wave, the parcels trace out a counterclockwise path in the meridional plane (when viewed looking westward). The long waves that penetrate deeply into the stratosphere have largest amplitude near 60° latitude (e.g., Kida, 1983a; Geller and Wu, 1987). As a consequence, parcels that are on the poleward side of the long wave sink further than they rise during each loop. Parcels on the equatorial side of the long wave rise further than they sink during each loop. The looping paths are schematically drawn in Figure 6.11c. The net Stokes drift is shown by the double-shafted arrows.

Dunkerton (1978) shows how the Stokes drift can exactly cancel the mean meridional motion in the absence of diabatic heating. Mathematical details of the analysis are left to §7.4.1. The cancellation is clearly evident by comparing the higher latitude net flows in Figures 6.11b and c. The cancellation leaves only the diabatic terms to drive a net circulation. Consequently, the "diabatic" circulation calculated by Murgatroyd and Singleton (1961) is still an appropriate description even if one includes wave dynamics. There are some subtleties. One subtlety is that the mean circulation in Figure 6.11a is the *Lagrangian* mean. A Lagrangian mean is an average following sets of moving air parcels (e.g., Andrews, 1987). The indirect cell seen in the observations results from the *Eulerian* mean constructed by averaging data along latitude lines instead of flow lines.

Another subtlety is that the isopleths of the ozone and water vapor do not have as great a meridional slope as the "diabatic" circulation may imply (Mahlman et al., 1984). The slopes are closer to the isentropes, as would be expected from meridional mixing along isentropes by the eddy motions. GCM simulations by Kida (1983a,b) illustrate the isentropic mixing. *Higher* mixing ratios occur in the middle stratosphere than along the stratosphere's base. The isentropes are nearly horizontal in the stratosphere, the high static stability resists the broad uplift suggested by Figure 6.11a. So, one anticipates that air parcels crossing the tropical tropopause will be mixed across latitudes just above the tropopause. Kida (1983b) performed an experiment to test this transport mechanism using a GCM. His GCM covers a hemisphere, uses a specified land-sea thermal contrast, no topography, and makes assumptions about how air parcels are vertically mixed. Most parcels rapidly mix poleward and downward above the tropopause and most parcels cross the tropopause in middle and high latitudes. Some fraction of parcels remain behind to be mixed into higher elevations of the stratosphere. The parcels exit the stratosphere much faster than they would if the "diabatic" circulation operated alone. All the parcels released at the tropical tropopause enter the stratosphere; about half return to the troposphere (in middle latitudes) within one year of simulated time. The net circulation and schematic trajectory paths are shown in Figures 6.11d and 6.11e, respectively. The eddy motions are quite apparent in the looping paths followed by the parcels.

Another detail to work out is that even the tropical troposphere temperatures are not low enough. The minimum temperatures in Figure 3.11 are about 195 K. The minimum temperatures needed for 4 ppmv are even colder. Holton (1984) has an excellent review of this subject. Newell and Gould-Stewart (1981) propose that the temperatures are low enough where there is very active, very deep convection. The "cold trap" over Indonesia in January and the Indian Monsoon region in July can lower the mixing ratio to 2 ppmv. Even these temperatures may not be sufficient when horizontal mixing around the globe is considered. Danielsen (1982) proposes that a combination of cloud physics and radiative cooling may finally be enough. Very strong convective updrafts would cause air parcels to overshoot their level of neutral buoyancy and penetrate the stratosphere. The parcels would be much colder than the lower stratosphere and would start to sink while mixing with the stratospheric air. Anvil clouds are large and long-lived in the tropics. The top of the anvil cloud is a better long-wave emitter to space than clear air, so it cools radiatively at a faster rate than clear air. The strong cooling at the anvil top leads to further convective overturning.

Meanwhile, ice particles continue to precipitate out, leading to the necessary drying. The process is illustrated schematically in Figure 6.11f.

6.3   Kuo-Eliassen equation

6.3.1   *Derivation*

The "Kuo-Eliassen" equation is so named to recognize Kuo's (1956) extension of a paper by Eliassen (1952). The equation shows the interlinkage between various types of forcing phenomena and the resultant mean meridional circulations. These forcings can be from eddy processes (such as eddy momentum convergence) and from diabatic processes (such as friction). The strength of the forcing also depends upon the properties of the mean flow, a result already seen in all of the energy and momentum balances described in Chapter 4. As pointed out in §6.2.4, by including all eddy fluxes, the resultant mean meridional circulation is an *Eulerian* mean, such as would be obtained by averaging along latitude lines as was done with observations in §3.3.

A Cartesian coordinates form of the equation is derived here; the spherical coordinates form can be derived by analogy (Kuo, 1956). Gill (1982, p. 368) derives a related model of buoyancy-driven flow, but the Kuo-Eliassen equation includes non-diabatic processes (eddy fluxes) that are crucial for understanding the mean meridional cells. The Kuo-Eliassen formulation is a standard model of introductory dynamics (e.g., Holton, 1979; §10.4).

One begins by defining a stream function ψ as

$$[v] = \frac{\partial \psi}{\partial p} \qquad \text{and} \qquad [\omega] = -\frac{\partial \psi}{\partial y} \tag{6.46}$$

so that a simple continuity equation is satisfied.

$$\frac{\partial [v]}{\partial y} + \frac{\partial [\omega]}{\partial p} = 0$$

The first governing equation used is the zonal momentum equation written in flux form

$$\frac{\partial u}{\partial t} + \frac{\partial u^2}{\partial x} + \frac{\partial uv}{\partial y} + \frac{\partial u\omega}{\partial p} - fv + g\frac{\partial Z}{\partial x} + F_x = 0$$

where the variables have their usual meanings. The zonal average is applied next. Topographic (e.g., "mountain torque") effects are excluded, so that all zonal integrals are complete circles around latitude belts. Consequently, each quadratic term (such as $[uv]$) can be expanded into a zonal mean circulation ($[u][v]$) and eddy flux ($[u'v']$) part. Finally, a pressure derivative is applied to the whole equation so that the time derivative term can later be eliminated by means of a thermal wind relationship. The result of these steps is

$$
\frac{\partial}{\partial t}\left(\frac{\partial[u]}{\partial p}\right) + \underbrace{\frac{\partial^2[u][v]}{\partial p \partial y} + \frac{\partial^2[u][\omega]}{\partial p^2}}_{(D)} + \underbrace{\frac{\partial}{\partial p}\left(\frac{\partial[u'v']}{\partial y} + \frac{\partial[u'\omega']}{\partial p}\right)}_{(A)}
$$

$$
\underbrace{- f\frac{\partial[v]}{\partial p}}_{(B)} + \underbrace{\frac{\partial[F_x]}{\partial p}}_{(C)} = 0
$$

(6.47)

- *Term (A)* in (6.47) is the divergence of the flux of relative westerly angular momentum due to eddies. Convergence here can intensify the zonal westerly jet $[u]$.
- *Term (B)* can develop $[u]$ motion from meridional circulations that are deflected by the Coriolis effect.
- *Term (C)* is friction, the first diabatic process so far.
- *Term (D)* is momentum divergence due to the mean meridional cells. This term will contribute to the linear operator upon ψ and to coefficients C and B in (6.50).

The second governing equation is the thermodynamic equation.

$$
\frac{\partial\theta}{\partial t} + \frac{\partial u\theta}{\partial x} + \frac{\partial v\theta}{\partial y} + \frac{\partial\omega\theta}{\partial p} = \frac{[\theta]}{[T]}Q
$$

Q is the heating rate. As was done for the zonal momentum equation, a zonal average is applied and the quadratic terms are each split into two parts: one for zonal mean circulations, one for eddy fluxes. Derivatives with respect to x again vanish because the zonal average integrates around complete latitude circles. A meridional derivative is taken so that the time derivative can eventually be eliminated by application of a thermal wind equation. The result is

$$
\frac{\partial}{\partial t}\frac{\partial[\theta]}{\partial y} + \underbrace{\frac{\partial^2[v][\theta]}{\partial y^2} + \frac{\partial^2[\omega][\theta]}{\partial y \partial p}}_{(C)} + \underbrace{\frac{\partial}{\partial y}\left(\frac{\partial[\theta'v']}{\partial y} + \frac{\partial[\theta'\omega']}{\partial p}\right)}_{(A)}
$$

$$
= \underbrace{\frac{[\theta]}{[T]}\frac{\partial[Q]}{\partial y}}_{(B)}
$$

(6.48)

- *Term (A)* in (6.48) is the divergence of heat flux due to eddies. This and term (B) are forcing functions of the zonal mean state. The analogous terms are friction and eddy momentum transport in (6.47).

- *Term (B)* is the input of *differential* heating. This is another diabatic term; many processes are included in Q.

- *Term (C)* is heat flux divergence by the mean meridional circulation. This term also contributes to the linear operator upon ψ in the Kuo-Eliassen equation, namely the coefficients A and B in (6.50).

Steady-state solutions are sought, so the time derivatives in (6.47) and (6.48) must be eliminated. As anticipated above, a thermal wind relation will be used for that purpose. A quasi-geostrophic thermal wind equation is derived as follows. The geostrophic wind (u_g) may be defined as

$$u_g = \frac{-g}{f} \frac{\partial Z}{\partial y} \Rightarrow \frac{\partial u_g}{\partial p} = \frac{-g}{f} \frac{\partial}{\partial y} \left(\frac{\partial Z}{\partial p} \right)$$

where f is a variable Coriolis parameter. Using the hydrostatic, then ideal gas relations obtains

$$\frac{\partial u_g}{\partial p} = \frac{g}{f} \frac{\partial}{\partial y} \left(\frac{1}{g\rho} \right) = \frac{R}{f} \frac{\partial}{\partial y} \left(\frac{T}{p} \right)$$

The chain rule is applied and the second term is identically zero because p is one of the coordinates. That is, the y derivative is evaluated upon a surface over which p is held constant.

$$\frac{\partial u_g}{\partial p} = \frac{R}{fp} \frac{\partial T}{\partial y} - \frac{RT}{fp^2} \frac{\partial p}{\partial y} = \frac{R}{fp} \frac{\partial T}{\partial y}$$

Introducing the Poisson equation obtains

$$\frac{\partial u_g}{\partial p} = \frac{R}{fp} \frac{\partial}{\partial y} \left(\frac{\theta p^\kappa}{P_{00}^\kappa} \right) = \frac{Rp^{\kappa-1}}{f P_{00}^\kappa} \frac{\partial \theta}{\partial y} + \frac{R\theta\kappa p^{\kappa-2}}{f P_{00}^\kappa} \frac{\partial p}{\partial y}$$

The last term again vanishes since $\partial p/\partial y = 0$. Applying a zonal average gives the thermal wind relation sought.

$$\frac{\partial [u_g]}{\partial p} = \frac{Rp^{\kappa-1}}{f P_{00}^\kappa} \frac{\partial [\theta]}{\partial y} \equiv \Upsilon \frac{\partial [\theta]}{\partial y} \tag{6.49}$$

Υ is defined by (6.49) and is a function of pressure and latitude.

One multiplies (6.48) by Υ and subtracts (6.47) from the result. Taking the forcing terms over to the right-hand side (RHS), leaves $[v]$ and $[\omega]$ terms on the left-hand side.

$$\Upsilon \frac{\partial^2 [\theta][v]}{\partial y^2} - \frac{\partial^2 [u][\omega]}{\partial p^2} + \Upsilon \frac{\partial^2 [\theta][\omega]}{\partial y \partial p} - \frac{\partial^2 [u][v]}{\partial y \partial p} + f \frac{\partial [v]}{\partial p} = RHS$$

where

$$RHS = -\Upsilon \left(\frac{\partial}{\partial y} \left(\frac{\partial [\theta' v']}{\partial y} + \frac{\partial [\theta' \omega']}{\partial p} \right) \right) + \frac{\Upsilon [\theta]}{[T]} \frac{\partial [Q]}{\partial y} + \frac{\partial [F_x]}{\partial p}$$
$$+ \frac{\partial}{\partial p} \left(\frac{\partial [u' v']}{\partial y} + \frac{\partial [u' \omega']}{\partial p} \right)$$

Some cancellations result from the thermal wind equation (6.49) and continuity equation as follows. Differentiating the left-hand side by parts finds

$$\Upsilon \frac{\partial}{\partial y} \left(\underbrace{[\theta] \frac{\partial [v]}{\partial y}}_{(T1)} + \underbrace{[v] \frac{\partial [\theta]}{\partial y}}_{(T2)} \right) - \frac{\partial}{\partial p} \left(\underbrace{[u] \frac{\partial [\omega]}{\partial p}}_{(T3)} + \underbrace{[\omega] \frac{\partial [u]}{\partial p}}_{(T4)} \right) + f \frac{\partial [v]}{\partial p}$$

$$+ \Upsilon \frac{\partial}{\partial y} \left(\underbrace{[\theta] \frac{\partial [\omega]}{\partial p}}_{(T5)} + \underbrace{[\omega] \frac{\partial [\theta]}{\partial p}}_{(T6)} \right) - \frac{\partial}{\partial p} \left(\underbrace{[u] \frac{\partial [v]}{\partial y}}_{(T7)} + \underbrace{[v] \frac{\partial [u]}{\partial y}}_{(T8)} \right) = RHS$$

Terms T1 and T5 cancel from the continuity equation, as do terms T3 and T7. To simplify the left-hand side, one applies the chain rule to terms T2 and T8.

$$T2 = \Upsilon \frac{\partial}{\partial y} \left([v] \frac{\partial [\theta]}{\partial y} \right) = \Upsilon \frac{\partial [v]}{\partial y} \frac{\partial [\theta]}{\partial y} - [v] \frac{\partial \Upsilon}{\partial y} \frac{\partial [\theta]}{\partial y} + [v] \frac{\partial}{\partial y} \left(\Upsilon \frac{\partial [\theta]}{\partial y} \right)$$
$$T8 = -\frac{\partial}{\partial p} \left([v] \frac{\partial [u]}{\partial y} \right) = -\frac{\partial [v]}{\partial p} \frac{\partial [u]}{\partial y} - [v] \frac{\partial^2 [u]}{\partial y \partial p}$$

The last term in these two equations cancels using (6.49). Similarly, terms T6 and T4 can be expanded using the chain rule.

$$T6 = \Upsilon \frac{\partial}{\partial y} \left([\omega] \frac{\partial [\theta]}{\partial p} \right) = \Upsilon \frac{\partial [\omega]}{\partial y} \frac{\partial [\theta]}{\partial p} - [\omega] \frac{\partial \Upsilon}{\partial p} \frac{\partial [\theta]}{\partial y} + [\omega] \frac{\partial}{\partial p} \left(\Upsilon \frac{\partial [\theta]}{\partial y} \right)$$
$$T4 = -\frac{\partial}{\partial p} \left([\omega] \frac{\partial [u]}{\partial p} \right) = -\frac{\partial [\omega]}{\partial p} \frac{\partial [u]}{\partial p} - [\omega] \frac{\partial^2 [u]}{\partial p^2}$$

The last term in both equations again cancels from relation (6.49). Substituting the stream function definition (6.46) and collecting like terms obtains the Kuo-Eliassen equation.

$$A\frac{\partial^2 \psi}{\partial y^2} + 2B\frac{\partial^2 \psi}{\partial y \partial p} + C\frac{\partial^2 \psi}{\partial p^2} + D\frac{\partial \psi}{\partial y} + E\frac{\partial \psi}{\partial p} = \Upsilon\frac{\partial H}{\partial y} + \frac{\partial \chi}{\partial p} \qquad (6.50)$$

where A, B, C, D, E, H, and χ are defined below. All eddy heat flux and diabatic heating processes are included in H, where

$$H = -\left(\frac{\partial [\theta' v']}{\partial y} + \frac{\partial [\theta' \omega']}{\partial p}\right) + \frac{[Q][\theta]}{[T]}$$

All eddy momentum convergence and diabatic friction processes are included in χ, where

$$\chi = [F_x] + \frac{\partial [u'v']}{\partial y} + \frac{\partial [u'\omega']}{\partial p}$$

The coefficient A is proportional to the static stability of the zonal mean atmosphere.

$$A = -\Upsilon\frac{\partial [\theta]}{\partial p}$$

The coefficient B is proportional to the baroclinic stability of a ring of air around a latitude circle, since it depends upon the zonal mean vertical shear from (6.49).

$$B = \frac{\partial [u]}{\partial p} = \Upsilon\frac{\partial [\theta]}{\partial y}$$

The coefficient C is a measure of the absolute vorticity of the zonal mean zonal flow.

$$C = f - \frac{\partial [u]}{\partial y}$$

C is also proportional to inertial instability of the zonal mean flow (e.g., Holton, 1979; p. 216). For the mean flow to be inertially stable, C must be positive throughout the Northern Hemisphere. If C were negative, then the inertial instability would create eddy motions that would mix the fluid in the meridional direction, and by so doing would reduce the horizontal shear until C again became posititve.

The coefficients of the first-order terms in (6.50) arise from the variable coefficient (Υ) used in the thermal wind relation (6.49).

$$D = \frac{\partial \Upsilon}{\partial p}\frac{\partial [\theta]}{\partial y}$$

and

$$E = -\frac{\partial \Upsilon}{\partial y} \frac{\partial [\theta]}{\partial y}$$

Υ is inversely proportional to the Coriolis parameter and to pressure raised to a negative power. Hence, E will be negative and D positive in the Northern Hemisphere.

6.3.2 *Discussion*

The solution of (6.50) is the stream function ψ, from which the zonal mean meridional circulation can be deduced. It is important to note that (6.50) relates the type of circulation to (a) zonal mean atmospheric properties (as given by coefficients A, B, C, D, and E), and (b) forcing of the zonal mean atmosphere by terms representing diabatic processes and zonally asymmetric eddy fluxes. An analogous equation is derived by Kuo (1956) except using spherical geometry and including a different form of the thermal wind balance. Pfeffer (1981) diagnosed the meridional circulation derived using Kuo's mathematical formulation and observations of zonal mean fields and eddy heat fluxes in Oort and Rasmussen (1971), and of diabatic heating in Newell et al. (1970).

The Kuo-Eliassen equation is a two-dimensional, linear, second-order partial differential equation with non-constant coefficients. This second-order equation can be classified into one of three canonical forms depending upon the sign of its discriminant. The terms multiplied by D and E are first-order terms; they do not affect the canonical form of a second-order equation (Young, 1972; p. 62). For $B^2 - AC > 0$, the equation is *hyperbolic*. For $B^2 - AC = 0$, the equation is *parabolic*. For $B^2 - AC < 0$, the equation is *elliptic*.

The discriminant determines both the nature of the solution and the method of obtaining it. The number of boundary conditions required depends upon the canonical form. For hyperbolic equations, the solutions are generally wave-like, or oscillatory in character; the wave equation is an example. Hyperbolic equations tend to be initial value-type problems. Elliptic equations are typically boundary value problems—all four boundaries must be specified and even then (if the boundary conditions are Neuman or mixed) the solution may not be unique. An elliptic equation example is LaPlace's Equation. A classic problem in meteorology is encountering an equation where the discriminant changes sign in the domain; because of the different boundary requirements, such a problem is not well-posed. An example is the solution of the nonlinear balance equation, which in midlatitudes involves finding the horizontal stream function from the geopotential field. Observations often make this equation mixed hyperbolic and elliptic, and strictly speaking, unsolvable.

In practice, the Kuo-Eliassen equation (6.50) is elliptic. Three scenarios (or their combination) could cause (6.50) to become hyperbolic. (a) Negative static stability (which makes $A < 0$) could develop in some region, but presumably convection would eliminate any negative static stability on the time and length scales considered here. (b) Strong horizontal shear could occur such that $\partial[u]/\partial y > f$ making $C < 0$. This is possible in low latitudes. Typically, the Coriolis parameter is about five times the magnitude of the horizontal shear on the large scale in middle latitudes. However, the condition of $C < 0$ should violate the necessary condition for inertial instability. Hence, waves should be initiated that preclude $C < 0$. (c) Strong baroclinicity might be combined with weak static stability (so that $B^2 > AC$), but it is more likely that the instability would initiate cyclonic development, which ultimately limits the vertical shears.

Actually, if (6.50) were hyperbolic, then two assumptions made during the derivation would be violated. If C is multiplied by v, then one obtains two terms in the u momentum equation. Quasi-geostrophy requires that the Coriolis term be much larger than the horizontal advection, hence C cannot be negative. Similarly, if $\partial[\theta]/\partial p$ is positive, then hydrostatic balance is also violated; if it is very small, then geostrophy is again violated. In short, one can easily justify assuming that (6.50) is elliptic. The story can be different near the equator, however. Stevens (1983; p. 886) discusses the ellipticity of a similar equation in spherical geometry, finding that no horizontal shear can be present at the equator (if friction is neglected), otherwise the flow is unstable and non-elliptic.

If (6.50) is elliptic, then the left-hand side is proportional to the opposite sign of the dependent variable (ψ). If one represents the linear differential operator upon ψ by the symbol L, then

$$-\psi \sim L(\psi) = RHS \tag{6.51}$$

RHS is the right-hand side of (6.50). Since only the qualitative features of the mean meridional circulations are of concern, it is sufficient to directly deduce the field ψ implied by each forcing term in RHS. A proportionality like (6.51) might break down if ψ has a linearly varying component in the meridional or vertical; for a global domain, it clearly cannot. To show how a relationship like (6.51) is valid, imagine decomposing the field ψ into a Fourier Series (for Cartesian geometry). The Fourier series illustrates a limitation of (6.51). The second-order derivatives in (6.50) emphasize the small-scale variations more than the larger-scale variations, whereas (6.51) treats all scales equally.

Typically (6.50) is solved by assuming $\psi = 0$ on all boundaries (no normal flow). This works fine if the boundaries are at logical places—the poles, the equator, and the bottom and top of the atmosphere, for example.

The following discussion parallels that found in Wallace (1978a), with some updating based upon more recent data and theories. Figure 6.12 shows the relationship between stream function ψ and the circulation pattern anticipated by (6.46). Note that $\omega = dp/dt$ is proportional to minus the vertical velocity; $\omega < 0$ implies upward motion.

RHS has four contributors: eddy momentum convergence, eddy heat flux convergence, differential diabatic heating, and friction. They are considered one at a

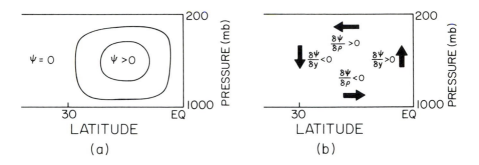

FIG. 6.12   Meridional cross-sections of (a) ψ and (b) the implied meridional circulation. A "Hadley" cell has $\psi > 0$, as indicated in this example.

time, then in combination. Schematic diagrams are presented first, followed by numerical solutions based on observations. The interpretations that follow are not so much explanations as statements of how the mean meridional cells and eddy circulations are consistent. The Ferrel and Hadley cells are needed to balance eddy heat and momentum fluxes in addition to the spatially-varying diabatic processes.

6.3.2.1   *Eddy momentum flux divergence*

The annual mean observed distribution of eddy horizontal momentum flux looks similar to the schematic presentation in Figure 6.13a. The observed seasonal distributions were shown in Figures 4.10e and f. The distribution of flux implies convergence in midlatitudes and divergence in the subtropics, as shown in Figure 6.13b. The flux magnitude increases with height in the troposphere, so the derivative with respect to pressure looks like Figure 6.13c. From this, the stream function field shown in Figure 6.13d is inferred. The eddy momentum flux divergence is driving Hadley and Ferrel cell circulations. Kuo (1956) compiled some observed values of $\partial \chi / \partial p$ where the eddy momentum divergence drawn here is the main contributor. Kuo also finds evidence for a Hadley and Ferrel cell. Ignoring the vertical eddy momentum fluxes seems reasonable based upon general circulation model output (Stone and Yao,1987).

The mechanism might be visualized as follows. To conserve angular momentum, air given westerly acceleration by the eddy fluxes will drift equatorward and appear like the upper branch of the Ferrel cell. In the subtropics and tropics there is divergence of zonal momentum, an easterly acceleration of the upper-level mean westerlies. The deceleration induces a poleward drift consistent with the upper branch of a Hadley circulation. (The vertical momentum flux convergence is neglected here.) In Chapter 4 the eddy momentum convergence could accelerate the mean flow. But acceleration of the mean flow is not allowed here because steady-state solutions have been sought. Accordingly, air parcels must drift to a different latitude in order to conserve angular momentum in the face of the eddy fluxes.

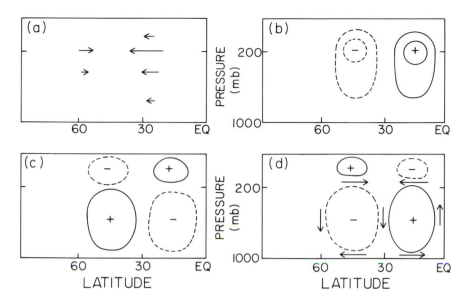

FIG. 6.13 Schematic meridional cross-sections of the eddy momentum flux contribution to the Kuo-Eliassen equation (6.50). (a) Eddy momentum flux is $[u'v']$. (b) Eddy momentum divergence is $\partial[u'v']/\partial y$. (c) Pressure derivative of the eddy momentum divergence contribution to the RHS of (6.50). (d) The resultant stream function and motion forced by the eddy momentum fluxes.

6.3.2.2   *Eddy heat flux convergence*

Attention is restricted to the annual mean horizontal heat transport, which is diagrammed in Figure 6.14a. The seasonal observed heat fluxes, shown in Figure 4.26, have a maximum in the lower troposphere and lower stratosphere in middle latitudes. The middle troposphere has a relative minimum. As a first guess, the meridional second derivative of this heat flux would have a pattern something like that in Figure 6.14c, and the stream function follows in Figure 6.14d. The eddy heat flux pattern creates a Ferrel cell circulation and more weakly implies a Hadley and possibly a Polar cell. While vertical momentum fluxes can be neglected, it is less clear that the vertical fluxes of heat can be neglected.

Vertical eddy heat fluxes also create a Ferrel cell. Figure 5.18 shows $\omega'T'$ correlations at 700 hPa. Above and below this level the correlation is generally lower. If one mentally constructs a zonal average of the maps in Figure 5.18, the correlation has its maximum amplitude (but negative sign) in the middle troposphere in midlatitudes. The second derivative with respect to pressure will produce a principal, positive maximum in the midlatitude, middle troposphere. Thus, ψ will have a negative sign and a maximum amplitude in the same region, implying the existence of a Ferrel cell.

One might visualize these heat fluxes as acting in a similar way as the momentum convergence (but now preserving the zonal mean temperature gradient). The heat fluxes act to destroy the meridional gradient of the zonal mean temperature. The rising

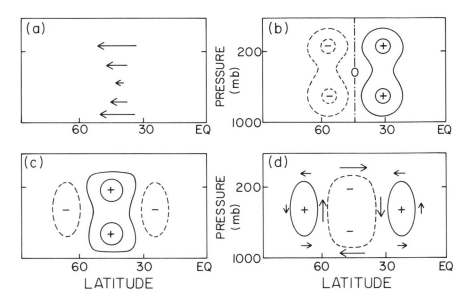

FIG. 6.14   Similar to Figure 6.13, except for eddy heat fluxes. (a) $[v'T']$. (b) Divergence of eddy heat flux. (c) Eddy heat flux contribution to (6.50). (d) The implied circulation.

and sinking motions in the Ferrel and Hadley cells counteract changes brought by the eddy heat fluxes. In middle latitudes the gradient is reduced by eddy transports of heat from the subtropics into the polar regions. The heating at high latitudes is negated by rising motion in the Ferrel cell. The rising air adiabatically cools because the zonal mean state is statically stable. In the subtropics, the sinking air is adiabatically heated and that counteracts the eddy fluxes there.

These interpretations of the circulation response to eddy heat and momentum fluxes are also linked to the thermal wind relation (6.49). If the eddy fluxes were allowed to change the mean flow, then each side of (6.49) would change as well. However, the momentum fluxes would change the left-hand side of (6.49) in the *opposite* way as the heat fluxes change the right-hand side. Obviously, such opposing changes would violate thermal wind balance. The Kuo-Eliassen equation maintains the balance by making sure that neither side of (6.49) changes.

Pfeffer (1981) calculates the mean meridional circulations needed to balance the eddy heat and momentum fluxes. His results are derived from Kuo's (1956) formulation in spherical geometry, and they are shown in Figure 6.15a. The eddy fluxes generate Ferrel and Hadley circulations. The diagnosed circulations vary with season in a realistic way, but their amplitudes are between a third and a quarter of the circulations estimated from observations of winds. Figure 6.15 shows solutions for Northern Hemisphere data where positive latitudes are winter season and negative latitudes are summer season. Pfeffer's annual average solution has the Hadley cell almost twice as strong as the Ferrel cell. Since the eddy momentum convergence is

STREAM   FUNCTION   WINTER-SUMMER

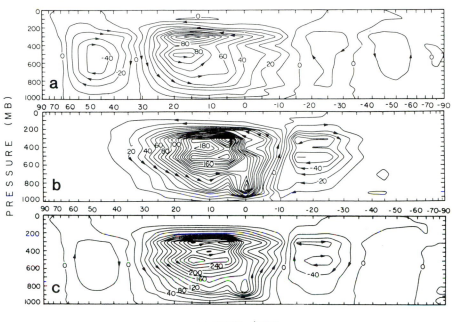

LATITUDE   (DEG)

FIG. 6.15   Circulations calculated from spherical coordinates form of the Kuo-Eliassen equation. The left hand side of each chart shows winter conditions; the right sides show summer. (a) Meridional circulation from eddy fluxes of heat and momentum. (b) Meridional circulation from diabatic heating field in Newell et al. (1970). (c) Circulation combining solutions in (a) and (b). From *J. Atmos. Sci.*, Pfeffer (1981) by permission of American Meteorological Society.

most strong in the latitudes of the Ferrel cell, one concludes that the assumed solution form (6.50) has a latitudinal bias. The bias implies that stream functions shown in Figures 6.13d and 6.14d would be more accurate if multiplied by a function that decreases with increasing latitude.

The eddies create both Hadley and Ferrel cells withoutexplicitly requiring diabatic heating. Of course, the diabatic processes set up the currents upon which the eddies can draw their energy. On the other hand, one would not expect the observed radiational heating to set up a Ferrel cell. The existence of the Ferrel cell in the data and its presence here due to eddy processes forges a link between eddies and Ferrel cell circulations. The link is strengthened further when the diabatic processes are examined.

6.3.2.3   *Diabatic heating*

The heating comes from three main contributors: radiational heating and cooling, latent heat release, and sensible heat transfer. Heating by friction is ignored. Figure 6.16 shows the diabatic heating rate estimated from six recent years of ECMWF data. Hoskins et al. (1989) determine the heating rate as a residual in the thermodynamic equation. (a) Radiational cooling is large near the tropopause level and greatest in the high latitudes. The atmospheric radiation is emitted mainly from the tops of the clouds. The direct absorption of solar radiation is presumably fairly evenly distributed throughout the tropical troposphere (except for the cloud-shaded areas of the ICZ). Radiational cooling appears to be present in the middle troposphere subtropics. In the tropical stratosphere, one sees radiational heating, as anticipated in §6.2.4. (b) Latent heat release occurs mainly in the equatorial troposphere (e.g., Figures 3.27 or 3.28). However, a secondary maximum lies in the troposphere at middle latitudes, and it is associated with the cyclonic storms. (c) Sensible heating occurs in the boundary layer. Sensible heat input from the ground is probably most important in the subtropics. Oceans can transport heat (Figure 3.9), and large fluxes of heat into the air occur over western boundary currents in middle latitudes. All these features are summarized in Figure 6.17a. The implied meridional circulations are also shown in that figure. The diabatic heating (mainly latent heating from ICZ rainfall) suggests a strong Hadley circulation. There are also Ferrel and Polar cells; but they are largely induced by the *eddies*. In midlatitudes, eddies are responsible for the clouds and precipitation that cause the heating patterns that create the Ferrel and Polar cells in Figure 6.17.

Pfeffer (1981) calculates the meridional circulations due only to diabatic heating in the Kuo-Eliassen equation. His resultant circulation is given in Figure 6.15b. His results are for winter and summer seasons, but it is still feasible to compare them with the anticipated circulations given in Figure 6.17. The most striking difference is that Pfeffer does not find a Ferrel cell. The reason for the discrepancy appears to lie in his use of a significantly different midlatitude diabatic heating field. His heating field is based on data from Newell et al. (1970) and is similar to Figure 4.28. Comparing Figures 4.28 and 6.16 reveals significant diagreement in the midlatitude middle troposphere; the latter figure has positive heating there, while Newell et al. find strong cooling. Hoskins et al. (1989) show maps of the heating rate; maximum values in the lower troposphere lie along the oceanic storm tracks. Those locations are consistent with latent heating creating the positive values in Figure 6.16, and with Figure 4.28 since Newell et al. data were restricted to radiosonde observations that do not sample the oceanic storm tracks.

6.3.2.4 *Frictional retardation*

Slowing down westerly motion means $F_x < 0$; slowing down easterlies is positive. The principal contributors shown schematically in Figure 6.18a are surface drag, which is largest in the boundary layer, "cumulus" momentum transport, and turbulence created by wind shear. (a) Boundary layer drag is positive in the tropics and negative in the middle latitudes due to the prevailing surface winds. A key point employed in Figure 6.18 is that the magnitude of the surface friction increases with pressure. (b) The discussion in Wallace (1978a) describes cumulus *friction*. At that time some scientists believed that the updrafts in a cloud would transport momentum *down* the gradient of wind; the cloud would distribute momentum in such a way as to reduce

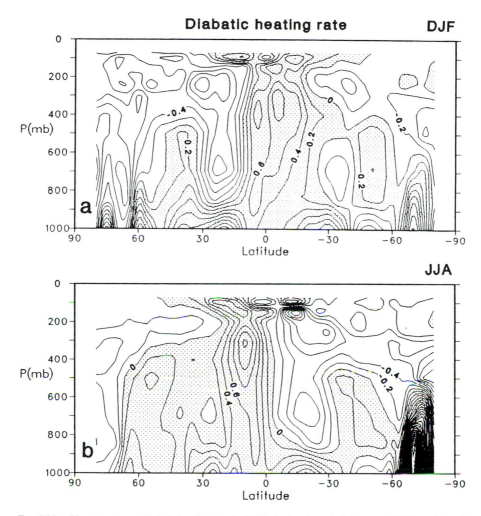

FIG. 6.16   Six-year average diabatic heating rate for (a) December through February and (b) June through August. Diabatic heating calculated as a residual in the thermodynamic equation. From Hoskins et al. (1989).

vertical shears and therefore act analogously to "friction." Where wind speed increases with height, cumulus friction might also be visualized as the top of the convective cloud moving slower than its environment and thus obstructing that high-level flow. More recently, the opposite has proven true. Instead of a down gradient flux, the observed transport in tropical convective clouds is *up* the gradient of the vertical shear (perpendicular to squall lines; e.g., LeMone et al. , 1984). Momentum flux up the gradient is also seen in laboratory and theoretical models of Benard convection (Krishnamurti and Zhu, 1991). Simulations by a tropical general circulation model were markedly improved by incorporating up-gradient fluxes into the convective

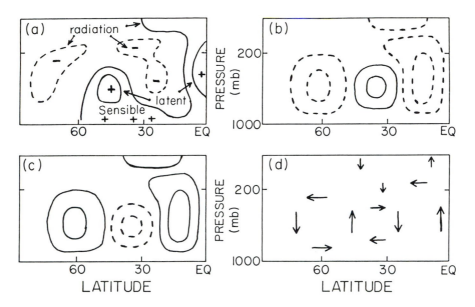

FIG. 6.17   Similar to Figure 6.13, except for contributions to the Kuo-Eliassen equation from diabatic heating. (a) Contributors to $[Q]$ in H. Sensible heating occurs near the surface; latent heating is mainly in the mid to upper troposphere; radiation cools except in the tropical stratosphere. (b) Meridional derivative of $[Q]$; (c) implied stream function and (d) circulation.

parameterization (Krishnamurti et al. , 1989). Their parameterization scheme makes the momentum flux proportional to the velocity difference between the layer reached by the cloud and the 700 hPa level where moist static energy (Figure 3.20) is a minimum. Cumulus convection tends to become deeper as the equator is approached. In the tropics, but away from the equator, the high-level wind is westerly, while easterlies prevail at cloud base (Figure 3.15). At the equator the vertical shear is not as great on a zonal and annual average. All these elements are combined in the highly qualitative "negative friction" distribution ascribed to cumulus clouds in Figure 6.18a. (c) Turbulence will be associated with the strongest vertical and horizontal shears, and thus with the jet streams. Turbulence acts to reduce that westerly motion.

Schematic contours aid the visualization of $\partial[F_x]/\partial p$ in Figure 6.18. Since the magnitude of the friction increases with pressure in the boundary layer, the magnitude of ψ increases with pressure as well, leading to the surface horizontal motion shown in Figure 6.18d. The cumulus momentum transport tries to maintain the meridional circulations associated with the Hadley cell, so they encourage the Hadley cell found in Figure 6.18. By reducing the vertical shear, the jet stream turbulence implies equatorward motion at tropopause level and poleward motion in the middle troposphere; the pattern is vaguely similar to a Ferrel cell. The resultant motions may look like a Ferrel, but one notable difference is that the "Ferrel" cell in Figure 6.18d is centered

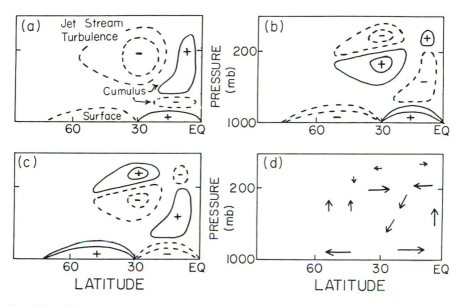

FIG. 6.18   Similar to Figure 6.13, except for contributions to the friction term in (6.50). (a) $[F_x]$ contributors. (b) $\partial[F_x]/\partial p$ which increases in magnitude with increasing pressure in the boundary layer. (c) Implied stream function. (d) Implied meridional circulation.

at the subtropical jet stream latitude (30°), whereas the Ferrel cell is normally placed further poleward.

6.3.2.5 *Combined influences*

The arguments presented above may be combined into the following picture. The derivation of (6.50) was based upon (6.49), that is, upon thermal wind balance (which assumes geostrophic and hydrostatic balance). Thus the mean meridional cells act to maintain this balance for any forcing function distribution. For example, the eddies are acting to build the upper-level westerlies with their maximum westerly momentum convergence, a process that increases the westerly vertical shear and, via a thermal wind argument, should *increase* the meridional temperature gradient. But the eddies also transport heat, which acts to *reduce* the meridional temperature gradient. Clearly, these eddy fluxes are acting to take the atmosphere out of thermal wind balance. The balance is maintained by the presence of a Ferrel cell circulation, which does two things. First, the equatorward flow at high levels decelerates the westerly flow high up (by introducing easterly momentum), while the poleward moving lower branch gives westerly acceleration to the low-level flow. The acceleration and deceleration follow from conservation of angular momentum. Second, rising and sinking motions cause air parcels to change temperature at a different rate than the local lapse rate because the atmosphere is statically stable on the large scale. The sinking branch warms the air in the subtropics due to adiabatic descent and cools the air adiabatically in its rising branch. This acts to build the meridional temperature gradient. The rising air in the Ferrel cell is moist, and when it becomes saturated, there is latent heat release, which partly compensates for this cooling due to rising motion. That means that the Ferrel cell needs to be even more vigorous. That is how the latent heating contribution to the $\partial H/\partial y$ term in (6.50) also forces a Ferrel cell circulation. Thus the Ferrel cell acts to reduce the westerly vertical shear and increase the meridional temperature gradient. Both effects counteract eddy fluxes and eddy diabatic processes in this model.

Similar arguments explain the Hadley cell in the tropics. Once again, the meridional cell exists to maintain thermal wind balance. First, the divergence of the eddy heat flux in the subtropics (more is transported from middle latitudes than from the tropics) is acting to cool the subtropics—a process that *increases* the latitudinal temperature gradient there. At the same time, the eddies are decreasing the vertical shear since there is a strong divergence of eddy momentum flux in the tropics—a process that *reduces* the westerly vertical shear. Once again the eddy fluxes act to destroy the thermal wind balance. Using the same reasoning as before, we find that the mean meridional Hadley cell maintains the thermal wind balance. (The cell's upper-level, poleward branch gives westerly acceleration; its downward branch adiabatically heats the subtropical air; its equatorward, low-level branch gives easterly acceleration, etc.) The Hadley cell, being thermally direct, counters the eddy fluxes by reducing the meridional temperature gradient and increasing the westerly vertical shear. Of course, the Hadley cell does not require eddies to exist. The heating differential with latitude sets it up.

This model is a convenient, simple description of the general circulation. It is not entirely a theory, in the sense that it does not mandate a particular circulation; $[u]$ and $[\theta]$ are specified. However, this model does show how the observed eddy fluxes produce a mean meridional cell structure that is similar to that observed. It does not, for example, explain why the eddy forcing occurs where it does or how it does—but it was never intended to do so. More sophisticated models, that allow the zonal wind to adjust, are described in §6.6.

6.4 Some possible feedbacks

Before the Kuo-Eliassen equation was derived, it was introduced as an equation that demonstrates the connections between complex diabatic and eddy features and the mean meridional circulation. Carrying that theme a step further, we find that changing one parameter may cause complicated and unanticipated changes in the atmosphere. Some such feedbacks are presented in a paper by Stone (1973).

Stone (1972, 1973) investigates some atmospheric feedbacks with a simple model constructed from the time and zonal average potential temperature conservation equation:

$$\frac{\partial [\overline{v\theta}]}{\partial y} + \frac{\partial [\overline{\omega\theta}]}{\partial z} = \frac{[\overline{\theta_r}] - [\overline{\theta}]}{\tau}$$

where τ is a characteristic radiative relaxation time for the atmosphere and $[\overline{\theta_r}]$ is the radiative equilibrium potential temperature profile. The radiative equilibrium model he uses is very similar to (6.21) and (6.22), derived in §6.1. Solar constant, albedo, and optical depth can be varied. The horizontal and vertical transports of heat are parameterized in terms of other quantities, primarily temperature gradients. Specifically, he uses the meridional and vertical gradients of $[\overline{\theta}]$; the former is considered a measure of baroclinic instability, the latter a measure of static stability. Rotation rate and other parameters can be varied in the parameterizations of the eddy heat fluxes.

Stone's model is a highly simplified view of the atmosphere. While the model successfully simulates some features of the atmospheres of Earth and Mars, it is not surprising that the model has some significant inaccuracies. A prime problem is that his predicted values for tropospheric static stability are about three times too big. The static stability is a key parameter in this type of analysis (since it affects baroclinic wave development), and the discrepancy casts some doubt upon the details of his analyses. Stone remarks that the discrepancy is mainly caused by his exclusion of latent heat transport processes. Radiation is the only diabatic process; no latent heating is incorporated. Since moisture content of the atmosphere increases nonlinearly with temperature, and the amount and distribution of cloud cover could vary a lot with the moisture content, the lack of latent heat parameterization seems to be a critical drawback. Despite these caveats, his results are interesting examples of complex interactions that may hold in the real atmosphere.

The feedbacks Stone examines are present in the Kuo-Eliassen equation, though he does not specifically consider that equation. The overall conclusion he draws is that the eddy *heat* fluxes have a strong negative feedback and that feedback strongly inhibits climate change by changes in the parameters he examines. Here are some specific cases. (a) He finds a strong negative feedback between changes in the solar "constant" and the eddy heat fluxes. When the incoming radiation increases, the implied cyclonic storm activity takes care of the greater heat transport required by the differential solar heating. Thus the "equilibrium" (i.e., time-mean) conditions of the atmosphere are little changed. In terms of the Kuo-Eliassen equation, greater differential solar heating leads to a canceling change in the eddy heat fluxes so that the mean meridional cells are only moderately changed: the two terms contributing to $\partial H/\partial y$ in (6.50) are locally canceling in midlatitudes. (b) The zonal and time average static stability strongly resists change. A 25 percent change in the external parameters of heating or rotation rate change the static stability only by a few percent. (c) An increased solar constant leads to a slight increase in static stability, increased baroclinic instability [term B in (6.50)], and stronger zonal velocity, especially at high levels. (d) If the short-wave radiation absorbing ability of the atmosphere increases, then the static stability decreases, the baroclinic instability again increases, and zonal winds become stronger, especially at high levels. (e) If the rotation rate increases, the static stability increases, the baroclinic instability decreases, and the zonal wind becomes weaker, but vertical velocities in the eddies increase.

Implicit in the model devised by Stone (1973) is the expression of the eddy heat fluxes in terms of zonal mean flow properties. Such a scheme appears in earlier studies (e.g., Green, 1970) and in numerous later studies. One reason for all the attention is that zonally averaged computer models of the atmosphere are easier and faster to solve than fully three-dimensional models. The advantages allow a variety of parameter studies to be performed more easily (e.g., Genthon et al. , 1990). Another reason for the attention is that a fundamental issue in wave dynamics is at the heart of Stone's formulation. The size of the eddy *heat* fluxes are proportional to the meridional gradient of θ in his formulation. Since the early studies of baroclinic instability, it has been widely held that a minimum vertical shear (meridional θ gradient) is needed to initiate the instability. Since the unstable waves are the source of the eddy fluxes, it is a small step to infer the following scenario. Diabatic differential solar heating builds a meridional θ gradient until a critical gradient is reached. After that, baroclinic instability is unleashed and the eddy heat fluxes then reduce the gradient to a marginally unstable state. Stone (1978) refers to this scenario as "baroclinic adjustment."

During the intervening time, various researchers have commented upon the baroclinic adjustment hypothesis. Using a laboratory model (§7.1.2), Pfeffer et al. (1980) found that the time-mean radial temperature gradient decreased as the eddy heat flux increased. In their experiments the temperature difference imposed across the annulus was fixed but by varying the rotation rate, eddies could be encouraged or discouraged from developing. Stone et al. (1982) and Lorenz (1979) point out that the eddy heat fluxes and zonal flow vertical shear are *not* positively correlated on short time scales. That is, the strongest zonal flows are not immediately followed by the strongest eddy fluxes. Instead there is a mixture of eddy heat flux intensities and zonal flow intensities that are positively correlated only on longer time scales, say more than a month. A simple example of the longer time scale is that the zonal wind shear and eddy fluxes are both stronger in winter than in summer. On the much shorter synoptic time scale, their result means the following. Eddies having diverse structure occur over time. Some eddies are very efficient at transporting heat, others are less so. The point is that these "random" eddies do not have their initial structure linked to the particular state of the zonal mean flow. Instead, the heat fluxes and vertical shear are *negatively* correlated; the mean flow responds to the eddy heat fluxes and not vice versa. When an eddy comes along that has stronger than normal heat fluxes, then that eddy reduces the mean flow shear more strongly than normal. Hence large eddy heat fluxes are linked to small vertical shear; weaker eddy heat fluxes are linked to larger vertical shear.

The occurrence of a mixture of waves raises another issue, namely the possibility of nonlinear interactions between waves. Such nonlinear interaction may determine the types of eddies more strongly than the mean flow properties, according to Vallis (1988). Some early calculations allowing nonlinearity (e.g., Vallis, 1988) led researchers to question the applicability of the baroclinic adjustment concept since excessively unstable vertical shears appeared to form. However, Cehelsky and Tung

(1991) concluded that the supercritical shears were for *saturated* waves and not for the waves that perform the baroclinic adjustment. After considering a wide range of radiative forcing and allowing wave-wave interactions, Cehelsky and Tung (1991) conclude that a baroclinic adjustment process also occurs between a mean flow and nonlinear large-scale waves.

As pointed out by Stone et al. (1982), the eddy fluxes (and the diabatic heating) could modify the *vertical* gradient of θ, i.e., the static stability. Static stability changes affect the baroclinic instability of the flow, so the critical shear needed for instability may change without any changes in the vertical shear. Gutowski (1985) shows that the static stability changes are probably occurring. Baroclinic adjustment predicts the meridional temperature gradient to be a minimum near the ground because the eddy horizontal heat fluxes are largest near there. Careful inspection of the observations shows the opposite to be true; the effect is visible in Figure 3.11. Gutowski (1985) points out that the vertical eddy heat fluxes transfer heat upwards, increasing the static stability in the lowest atmospheric layers. Higher static stability weakens the baroclinic instability (e.g., Grotjahn and Lai, 1991). Unrealistic surface temperature drops occur in his simple model, but the basic premise appears to be correct. This scheme has been incorporated into a model having only zonal mean fields explicitly; the resulting circulation was compared to the zonal average state from a fully three-dimensional version of the same model (Genthon et al., 1990). The comparison shows that the eddy horizontal heat fluxes were successfully mimicked, but the vertical fluxes were much too small.

The changes to vertical shear and static stability are unified by expressing the baroclinic adjustment scenario in terms of waves that reduce the zonal mean state's potential vorticity gradient $[Q_y]$.

$$
\begin{aligned}
[Q_y] &= \beta - \frac{f^2}{\rho} \frac{\partial}{\partial z} \left(\frac{\rho}{N^2} \frac{\partial [u]}{\partial z} \right) - \frac{\partial^2 [u]}{\partial y^2} \\
&= \beta + \frac{f^2}{N^2} \frac{\partial [u]}{\partial z} \left(\frac{1}{H} - \frac{\partial}{\partial z} \left(\ln \frac{\partial [u]}{\partial z} \right) + \frac{\partial \ln N^2}{\partial z} \right) - \frac{\partial^2 [u]}{\partial y^2}
\end{aligned}
$$

where β is the meridional gradient of the Coriolis parameter, H is the scale height, and $N^2 = gd\ln[\theta]/dz$ is the Brunt-Väisälä frequency. Gutowski et al. (1989) adopt the second form above to make the separation between static stability and vertical shear changes more clear. Heat fluxes from a developing eddy reduce the vertical shear, $1/N^2$, and $\partial \ln N^2/\partial z$ in the lower atmosphere. They conclude that the potential vorticity gradient is changed about equally from static stability and vertical shear modifications. They carry the analysis a bit further by incorporating surface heat flux and surface friction. When the eddies transport heat vertically, the surface air temperatures decrease, but that increases the air to ground temperature difference, leading to larger surface heat fluxes that partially offset the static stability modification.

Surface friction, by slowing down the horizontal winds, reduces the horizontal heat fluxes, thus reducing the vertical shear modification.

While sensible heat transport leads to a smaller meridional θ gradient, the opposite is the case for momentum transport. Leach (1984) finds that smaller than average $[u]$ precedes larger eddy momentum convergence $(-\partial[u'v']/\partial y)$, which builds larger than average $[u]$ approximately two days later. The situation is complicated by competing baroclinic and barotropic processes. Large jet stream winds probably also have large vertical shears, so both heat and momentum eddy fluxes would be responding to the mean flow.

Baroclinic adjustment is just one item in Stone's (1973) original analysis. Changes in rotation rate have been examined with much more sophisticated, three-dimensional general circulation models. Hunt (1979) finds that as the rotation rate increases, the Hadley cell shrinks in meridional extent, the latitudinal temperature gradient increases, and the subtropical jet stream diminishes and moves closer to the equator. The main jet follows the boundary of the Hadley cell and seems linked to a maximum critical speed that would occur from conservation of angular momentum. Poleward of that location, eddies predominate and are less efficient (in Hunt's model) at transporting heat, leading to the higher meridional temperature gradient. Del Genio and Suozzo (1987) find similar results with their GCM intended for any general planetary atmosphere. In addition, Del Genio and Suozzo find that the eddy momentum flux reverses direction. For rotation rates similar to the earth's, the flux is largely up the gradient (especially on the equatorward side of the subtropical jet); for slower rotation rates (e.g., 16 days per revolution) the eddy momentum flux is equatorward, indicative of barotropic *growth*. The changes in rotation rate are proportionally large in these studies, hence rotation rate has a very large impact on the general circulations found.

Williams (1978, 1979) varies the rotation rate in barotropic and two-layer baroclinic models. He is most interested in how well concepts from two-dimensional turbulence and energy cascade in barotropic systems apply to planetary atmospheres. A key concept is a dynamical transition wavenumber, k_β, defined by

$$k_\beta^2 = \frac{\beta}{2\,\widehat{U}}$$

where $\widehat{U}$ is a root mean square average wind over the domain. Waves smaller than k_β tend to merge, forming eddies with larger scale, but their vorticity develops progressively smaller scale structure. According to the theory (Rhines, 1975) the cascade to larger scales is much slower for $k < k_\beta$ so that energy tends to accumulate at the transition wavenumber length scale. While that is happening, some kinetic energy has reached the zonal mean current which begins to distort the eddy shapes. The differing Rossby wave phase speeds (because β varies with latitude) also begin to distort the eddies. The eddies begin to line up into zonally oriented chains of highs and lows. If these chains are dynamically stable and there is little interaction with the planetary surface (as is true for Jupiter, but not the Earth), then the result is a series of zonally oriented bands as seen in Figure 2.9. The same effect can be seen on Earth when lows migrate poleward (and highs equatorward) as they move eastward (Figure 5.20). The theoretical model simulations to be shown in §7.3 reveal the same effect. Since the terrestrial β is about five times the Jovian β, one might expect this theory to predict strong meridional cells in addition to the Hadley cell; Williams (1978) proposes that strong surface drag limits this possibility for the Earth.

Stone (1973) also considered changes of the solar constant. Solar heating can be altered in other ways, either by changing the distribution, concentration, and type of atmospheric absorbing gases, or by altering the albedo of the earth. One way in which these changes have been examined is to study climate change scenarios. Some simulations of ice age conditions find a weaker Hadley cell, while other studies find a stronger Hadley cell. An example is Rind (1986), who studies five possible climates: two ice age scenarios, two warmer-than-present scenarios, and present climate. Many features (absorbing gases, continental positions, sea surface temperatures, ice extent, etc.) are quite different between the five simulations, so it is very hard to isolate the feedbacks in Rind's results. Nonetheless, Rind finds that the Hadley cell, jet stream, and mean precipitation patterns are little different in the experiments. On the basis of these results, one might conclude that strong feedbacks are in place that limit the sensitivity of the Hadley cell and the subtropical jet to climatic changes.

6.5   Reconciliation of momentum flux and jet stream positions

The largest contribution in the zonal mean momentum budget is the eddy transport (Figure 4.7). The eddy flux is generally up the gradient of $[u]/R$ (e.g., Figure 4.12). The latitude of the maximum eddy momentum convergence does not quite line up with the latitude of the time and zonal average jet; the jet is located equatorward. The discrepancy is now explained. The Kuo-Eliassen equation provides a sufficient basis to reconcile the momentum fluxes and the zonal average jet position. Additional information about eddy life cycles and the maintenance of the long waves turn out to be crucial and are touched upon as needed; their complete discussion is left for Chapter 7.

Before presenting the explanation it is useful to consider briefly the "typical" life cycle of a developing cyclone. The cyclone life cycle details are covered in §7.4. Generally, cyclones most often develop in regions where the horizontal temperature gradient, and thus the vertical shear in their environment, is large. When the cyclone is initially formed it has small amplitude compared to its environment. It may be growing rapidly and efficiently transporting heat, but in dimensional terms it does not have a major impact until it reaches some appreciable amplitude. A low usually moves poleward as well as eastward while developing. This has several effects. First, the motion and growth cause the time average eddy fluxes to be spread out along the storm track, so the maximum fluxes are not necessarily associated with the most rapid growth rate. This conclusion follows because fluxes are proportional to amplitude squared, but the fastest growth rate occurs when the eddy amplitude is small (e.g., Grotjahn and Lai, 1991). Second, as the low grows to large amplitude, it significantly distorts the flow. This distortion can cause a reinforcement of the flow on the equatorward side of the low, or in other terms, the low appears to separate from the *surface* frontal zone. The low migrates into the cold-air side of the front not only to carry out its poleward transport of heat, but because of the favorable cyclonic shear on the poleward side of the jet stream. It is at this late stage that the eddy momentum convergence becomes especially strong, operating as one of the main causes of the storm's decay.

Many of these features are seen in Figure 6.19. The area encompassing North America plus the North Atlantic illustrates the features. The maximum heat flux and momentum convergence occur off the coast of New England. The maximum of the band-pass variance of the height fields (Figure 6.19c) is located nearby, and this field looks very much like the heat fluxes. The height variance responds to the amplitudes of the typical traveling cyclones isolated by the band-pass filter. Yet most storms arriving off New England with large amplitude originate over the land or coastal waters of the southeastern United States. When these figures are compared with the time mean jet patterns (Figure 6.19d), some interesting relationships appear. The maximum eddy momentum convergence is downstream and poleward of the maximum time mean zonal wind. The maximum eddy heat flux is clearly poleward of the time mean jet stream, as is the maximum variance of band-pass height values.

How can the eddy momentum convergence be a maximum downstream of the jet maximum instead of upstream, as one might expect? The answer to that question will appear while a more general question is addressed: why do the zonal mean jets appear where they do? The answer to this question has several parts: (a) angular momentum conservation creates zonal winds from the meridional cell circulations; (b) eddies set up secondary (Ferrel cell) circulations; (c) large-scale orography and localized heating can concentrate the winds in certain preferred regions; and (d) time averaging dilutes flows that vary strongly (over space and time) where the eddies have large amplitude.

6.5.1   *Angular momentum conservation*

Two or three decades ago a commonplace view held that the zonal mean subtropcial jet lay at the boundary between the Hadley cell and the Ferrel cell. This viewpoint is illustrated, for example, in Palmén and Newton (1969; their Figure 4.2). The basic idea is that convergence occurs at high levels where the Ferrel and Hadley cells meet. The convergence builds the high altitude thermal gradient and that leads to a jet stream at that boundary. Earlier in this chapter, a simple model employed conservation of angular momentum to create largest westerlies at the poleward edge of a Hadley cell. Either classical argument creates a jet between the Ferrel and Hadley cells. This view is not wrong; it is just not complete.

6.5.2 *Eddy-induced circulations*

As recently as a decade ago, it was commonly believed that an eddy-induced Ferrel cell played a key role in slowing down the jet in the so-called "jet exit" region. The article by Blackmon et al. (1977) illustrates this view. The eddies induce a mean meridional circulation that acts (by advecting planetary easterly momentum once again) to reduce the westerly acceleration brought about by the convergence of eddy momentum. Therefore, the greatest mean zonal velocities would be maintained equatorward or (possibly) poleward of the maximum eddy momentum convergence since the latter is near the center of the Ferrel cell. This description is not wrong either, but it is incomplete on two grounds. Pfeffer (1981) found the Ferrel cell induced by the eddy fluxes to be too weak to do the job. Other studies, like Valdes and Hoskins (1989), find that the eddy heat and momentum fluxes tend to cancel each other out: the building of the vertical shear is negated by the weakening of the temperature gradient.

As discussed in §6.3, diabatic effects also create meridional circulations necessary to maintain a steady flow. One can use the diabatic heating, eddy fluxes, and implied meridional circulations to calculate where these contributors create the greatest tendency in the zonal momentum equation. Pfeffer (1981) makes such a calculation and finds these contributors to be maximum at the same level, but about 8° poleward of the jet location. It is plausible that the diabatic contribution to the Ferrel cell calculated by Pfeffer is much too weak; it may be that an improved estimate of midlatitude diabatic heating is sufficient to line up these contributors with the location of the jet. (See §6.3.2.3.)

The mean meridional circulations contain divergences, hence the motions are ageostrophic and thus cross-isobaric. Blackmon et al. (1977) calculate the magnitude of the cross-isobaric flow at 250 hPa (Figure 6.20). The maximum ageostrophic velocities are roughly ten times larger than their zonal average values (Figure 3.16). The jet stream axis is shown with a heavy arrow in Figure 6.20. (The downstream end of each heavy arrow did not bend equatorward in the original source of this figure; the bend reflects more recent and presumably more accurate data; the bend is also more consistent with recent models and concepts.) The ageostrophic motion (as in Figure 5.7) is somewhat orthogonal to the jet with a poleward direction in the region where the jet speed is increasing and equatorward where the jet speed is decreasing. This description is very much like the scheme proposed by Namias and Clapp (1949) and described in equation (6.35). Blackmon et al. (1977) take the description a bit further by incorporating eddy fluxes. Blackburn (1985) has criticized the procedure by which Blackmon et al. calculate their ageostrophic velocities. While the magnitude may be in error, the sign and the magnitudes of the velocities shown in Figure 6.20 should be suffciently accurate for the general discussion here.

The description is summarized schematically in Figure 6.21. "Jet entrance" regions are indicated by transects labelled "a" in Figure 6.19d and are where the jet velocity increases downwind. "Jet exit" regions are indicated by the "b" transects and

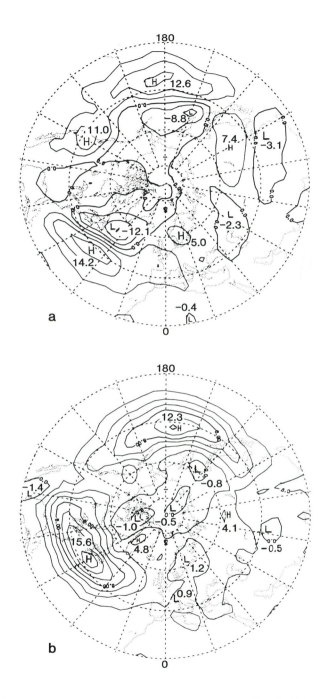

FIG. 6.19 Band-passed eddy statistics from nine winters: (a) 500 hPa $[u'v']$; (b) $[v'T']$ at 850 hPa; (c) geopotential height variance at 300 hPa. (d) Time average wind speed at 500 hPa. The letters "a" indicate the locations of "jet entrance" regions. The letters "b" are for "jet exit" regions. (a) and (b) are only partly

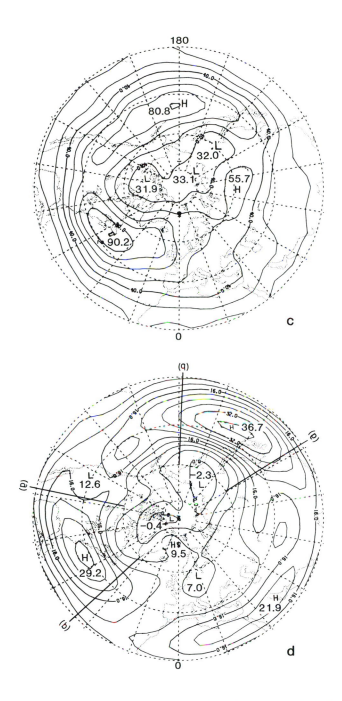

representative of upper tropospheric patterns; these two panels should be compared with Figures 7.23 and 7.22, respectively. From *J. Atmos. Sci.*, Blackmon et al. (1977) by permission of American Meteorological Society.

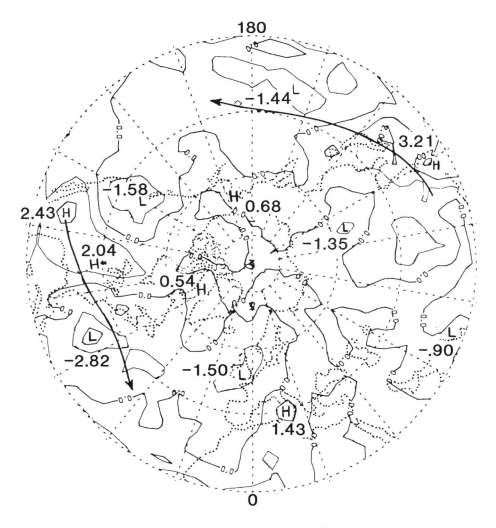

FIG. 6.20   Contours of cross-isobaric (ageostrophic) flow at 250 hPa calculated over nine winters. Contour interval is 1 m/s. The darker arrows indicate the axes of the subtropical jets. Redrawn from *J. Atmos. Sci.*, Blackmon et al. (1977) by permission of American Meteorological Society.

are where the jet velocity decreases downstream. The first thing to note about the zonal mean subtropical jet is the dominance of the Asian jet. The time-mean wind speeds are so large (Figure 5.7) that the jet maximum near North America approximately equals the zonal mean!

6.5.3   *Orographic and thermal localizing of winds*

Because the time-mean Asian jet dominates the zonal average, the perturbation stream function pattern to the zonal mean flow has a very strong north-south dipole structure

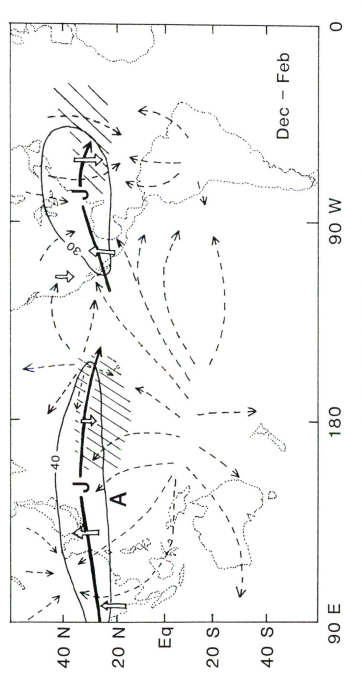

Fig. 6.21 Schematic illustration linking the subtropical jet maxima ("J") with divergent motions (dashed arrows; Figure 6.20 (open-shaft arrows), and eddy momentum flux divergence (hatching). Northern Hemisphere winter conditions. Axis of the time-mean subtropical jet indicated by the heavy arrow. "A" denotes high in stream function field (Figure 6.9). Selected isotachs are shown: 40 m/s contour near Asia, 30 m/s contour near North America.

along the Asian coast with high heights to the south and relative minimum to the north. A similar, but weaker pattern is found over eastern North America (Figure 7.12b). The perturbation pattern also has a relative maximum to the east of each of these relative minima. The pertubation pattern is nearly zero along the North American east coast (Valdes and Hoskins, 1989). Both orography and diabatic heating and cooling contribute to the formation of the westerly jet maximum near the Asian coast. The latent heat from tropical convection in the western Pacific is much stronger than anywhere else; consequently, the Asian jet predominates. Both orographic and thermal effects are easily seen in the barotropic vorticity equation. That equation is a useful description at an "equivalent barotropic" level (where the height field equals its vertical mean value).

The orographic effect is described in more detail in §7.2.1. For straight westerly flow encountering a mountain ridge oriented north-south, a trough is created on the downstream side. The mechanism is seen in the equivalent barotropic vorticity equation (EBVE). The EBVE is derived by integrating the vorticity equation in the vertical. The divergence term becomes the vertical velocity evaluated at the top and bottom of the domain. Topography enters the problem through the bottom boundary condition that vertical velocity there equal the slope flow up and down the topography. Mathematically,

$$\frac{\partial}{\partial t} \overbrace{\zeta_a} + \cdots = -\mathbf{V} \cdot \nabla \mathbf{H}$$

where ζ_a is the vertical component of absolute vorticity. $\mathbf{H}$ is the local height of the orography. The right-hand side of the equation is greater than zero when westerlies blow down the east slope of the Tibetan plateau. In that situation, the vorticity tendency is greater than zero, so positive vorticity is being created. Positive vorticity creates a trough in the stream function field downstream of the mountains. The orographic effect is magnified by the distribution of diabatic heating and cooling.

Exceptionally strong latent heating occurs in the western Pacific. Latent heating may be estimated from surface rainfall (Figure 5.15). Even though the maximum is centered in the Southern Hemisphere during January (10 S; 175 E), the divergent flow created by the convection affects the stream function field well to the north. When coupled with the strong diabatic cooling over the continent of eastern Asia during winter, the locations of the jet entrance and exit regions can be faithfully simulated. Sardeshmukh and Hoskins (1988) show the diabatic heating effects quite clearly in a linear barotropic vorticity equation model. The thermal effect is easy to see mathematically if one first separates the *advecting* velocities into rotational and divergent parts. Subscript ψ indicates the former velocity while subscript χ indicates the latter.

$$\frac{\partial \zeta_a}{\partial t} + \mathbf{V}_\psi \cdot \nabla \zeta_a = -\mathbf{V}_\chi \cdot \nabla \zeta_a - \zeta_a D \equiv S$$

where D is the divergence. S has been called the "Rossby-wave source" (Sardesh-mukh and Hoskins, 1988) because this equation shows that S equals the total rate of change of absolute vorticity by adiabatic processes; Rossby waves are vorticity-driven waves. The observed $\mathbf{V}_\chi$ field during July was shown earlier (Figure 5.25b); a corresponding diagram for January is given in Figure 6.22. Contours of velocity potential χ are again plotted. Schematic vectors of $\mathbf{V}_\chi$ are included in Figure 6.21. The divergent flow emanates from the source region near New Guinea toward the sink near northeastern China. In between, the subtropical jet is accelerated because a stream function ridge is built in the western Pacific at 15 N (near location "**A**" in Figure 6.21).

The ridge at location **A** is easily understood by looking at the term S. The diver-gence (D) in region **A** is approximately zero since the divergent wind is maximum there. The gradient of absolute vorticity is positive there (both planetary and shear vorticity increase with latitude). Consequently, S is strongly negative at location **A**. Decreasing vorticity means higher heights and a maximum in the stream function field ψ at **A**. A ridge is also produced in the opposite hemisphere during the op-posite season. The diabatic heating enhances the jet stream over Australia during their winter by building a ridge along 10 S between Africa and 170 E. The domi-nance of the latent heating over southeast Asia out to the western Pacific explains a feature noted in Figure 5.7: the winter subtropical jet is maximum at a similar longitude in both hemispheres. Sardeshmukh and Hoskins (1988) derive an integral constraint by integrating the vorticity equation along a closed ψ contour and applying the divergence theorem; the constraint shows that the ψ maximum must be displaced from the χ maximum. The displacement is seen in Figure 6.21; it was also seen in Figure 5.25.

The simplest conceptual picture is formed by thinking of the area of strong latent heating as creating a dome-shaped high at location **A** in Figure 6.21. To the northwest are lower heights due to both diabatic cooling mixed through a deep layer of the atmosphere and due to orography. (Valdes and Hoskins (1989) use a linear stationary wave model and estimate the long-wave created by orgraphy alone to be about 30% of the observed long-wave amplitude.) Air is forced between the orographic trough and the latent heating ridge and must be accelerated.

The long-wave circulation produced by the localized maxima of heating and cooling is illustrated in Figure 6.23 from Sardeshmukh and Hoskins (1988). They specify a zonally symmetric stream function, an area of heating, and an area of cooling. The divergent wind field is similar to Figure 6.22 and the nonlinear, total stream function is very similar to the observed pattern over the Pacific. It is noteworthy that the ridge in the stream function over the western Pacific shows a curving mean jet path. The curving jet path is seen in observations of the vector wind (Arkin et al. , 1986); it can be seen in Figure 6.9 and Plate 1, and it is included in Figure 6.21. The curving path is present in the rotational part of the wind and is also *reinforced* by the ageostrophic wind (Figures 5.7 and 6.20).

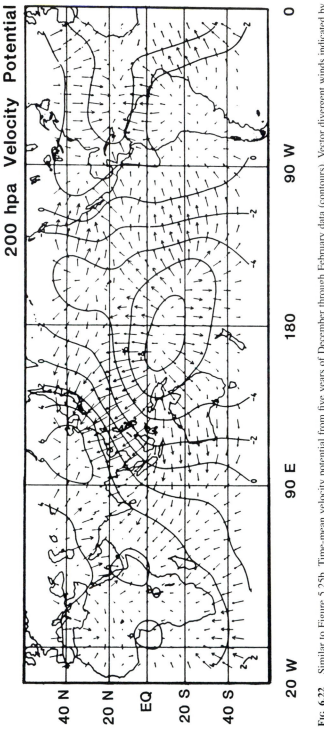

FIG. 6.22   Similar to Figure 5.25b. Time-mean velocity potential from five years of December through February data (contours). Vector divergent winds indicated by arrows; dashed lines in Figure 6.21 based on these arrows. From Arkin et al. (1986).

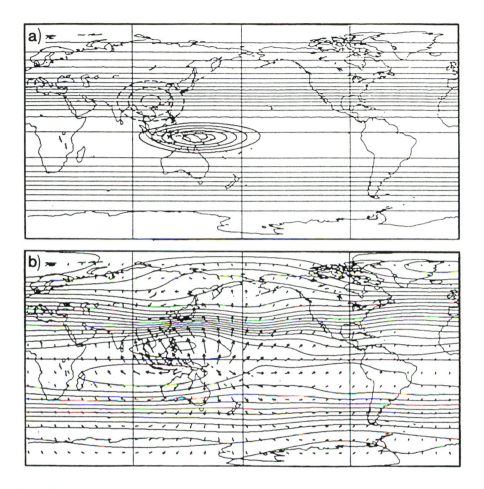

Fig. 6.23 (a) Observed zonal mean stream function at 150 hPa (straight lines), divergence region (solid contours centered on New Guinea), and convergence region (dashed contours). (b) Nonlinear response created by the fields in (a) and calculated by a barotropic vorticity equation model forced by this divergent flow. Total stream function (solid lines) and divergent wind vectors shown. From *J. Atmos. Sci.*, Sardeshmukh and Hoskins (1988) by permission of American Meteorological Society.

The jet exit regions appear in the solutions of Sardeshmukh and Hoskins (1988) based on diabatic heating not associated with frontal cyclones. The jet exit regions are consistent with the frontal cyclone eddy fluxes as well. The eddy momentum convergence is largest along the storm track, when the eddy reaches large amplitude. However, that location is well poleward of the subtropical jet location. Close inspection of Figure 6.19a shows that eddy momentum *divergence* must occur in the jet exit region because the momentum flux is largest to the north. The region of divergence is marked with hatching in Figure 6.21. Close inspection of Figure 6.19b shows (weak) meridional heat flux occuring here, too. Both fluxes act to weaken the jet, so both are consistent with thermal wind balance. These fluxes help move the time-mean jet even further equatorward, where the eddy fluxes are very small. The eddy forcing in the model of Valdes and Hoskins (1989) seems to show this effect.

6.5.4   *Time variability*

Eddies produce time and space variability in the flow; where eddies are large the jet appears more diffuse *on a time average*. A metaphor might be the wagging tail of a dog. Where the tail is attached to the dog might represent the jet entrance region; the jet exit region is the tip of the tail. As troughs and ridges amplify downstream of the jet entrance, large deflections ("wags") of the jet are created. Evidence for such excursions of the jet are plentiful on daily maps. Evidence can even be seen in Figure 6.21: the isotachs are wider downstream than upstream of the North American jet maximum. The jet entrance region stays comparatively fixed in part because it has a tropical source; the thermally direct circulation (the Hadley cell) is more constant over time. Orographic forcing also tends to be consistent over time, so the time and zonal average tends to place the subtropical jet at the east coast regions because the flow is less variable in those regions.

A semantic point should be addressed here. The term "Hadley cell" has been used in this section to describe a circulation having strength and poleward boundary that both vary zonally. Strictly speaking, that poleward edge is a "wave," but Hadley cell terminology has been used here to highlight the tropical origin of the ageostrophic motion in question. The "eddies" referred to here are principally the midlatitude frontal cyclones.

6.5.5   *Summary*

To summarize, the original view that the jet streams appear at the boundary between the Hadley and Ferrel cells is basically correct, though the situation is more complicated. The poleward, upper-level Hadley cell motion gains westerly speed due to angular momentum conservation. Regions where the subtropical jets are strong are areas most favorable for cyclone storm development by the baroclinic instability process. Baroclinic instability creates eddy heat fluxes that accomplish much of the poleward heat transport in middle latitudes. As the eddies evolve they are distorted by the jet flow, developing momentum flux convergences, too. The eddy momentum flux convergence tends to have the same sign around a latitude circle, so its zonal average is substantial. The eddy fluxes are proportional to the eddy amplitude squared, so the heat fluxes tend to be largest where the eddy has largest amplitude, even though the baroclinic instability process may not be efficient at that stage. The eddy momentum fluxes are strongest at latitudes where the eddy heat fluxes are large. From the Kuo-Eliassen equation, local "Ferrel cell" circulations are set up that largely cancel the changes brought about by both types of eddy fluxes. As the eddies evolve they also distort the jet flow. The large time variability of the jet location in the regions where the eddy fluxes are largest also contributes to the small values of the zonal mean jet at those latitudes. The Asian subtropical jet speeds are so large that the zonal mean subtropical jet location is largely determined by the Asian jet location. The Asian jet entrance region results from orographic and especially diabatic effects; both of these effects have relatively little daily variation compared with further downstream. Because of the predominance of the jet near the east coast of Asia, discussion of the zonal mean jet necessitated an introduction to the causes of the long waves. §7.2 discusses the maintenance of the long waves in greater depth. The jet exit region arises from a combination of diabatic circulation effects and eddy fluxes. The jet exit region is equatorward of the storm track and here both eddy heat flux and eddy momentum flux divergence decelerate the time-mean flow.

6.6 Simple zonally symmetric flow models

The first dynamical model in this chapter (§6.2.3.2) forecast $[u]$ when $[v]$ was specified (among other parameters). Next, the Kuo-Eliassen equation was developed; that equation derives $[v]$ while specifying $[u]$ (and other parameters). In this section, $[u]$ and $[v]$ are both allowed to adjust. The more advanced of the models discussed next allow other quantities, like $[\theta]$, to adjust as well. The analytic model of Schneider and Lindzen (1976) is derived in some detail. Descriptions of the other models are only sketched because their solutions are obtained numerically.

The work by Schneider and Lindzen (1976, 1977) and Schneider (1977) was originally meant to calculate basic states for stability studies including finite amplitude feedbacks. However, the models are useful simulations of the zonal average atmosphere as well. The system of equations used assumes steady-state conditions and ignores longitudinal variations. The flux forms of the u and v momentum equations, in spherical geometry, are

$$\frac{1}{R}\frac{\partial}{\partial \phi}\left(uv\cos\phi\right) + \frac{\partial}{\partial z}\left(uw\right) = fv + \frac{uv\tan\phi}{r} + F_\lambda \qquad (6.52)$$

and

$$\frac{1}{R}\frac{\partial}{\partial \phi}\left(v^2\cos\phi\right) + \frac{\partial}{\partial z}\left(vw\right) = -fu + \frac{u^2\tan\phi}{r} + F_\phi - \frac{1}{\rho_0 r}\frac{\partial P}{\partial \phi} \qquad (6.53)$$

where ϕ is latitude, r is the earth's radius, $R = r\cos\phi$, and ρ_0 is a horizontal mean density. Internal viscous damping is modelled using a nonlinear, second-order diffusion term in each equation. Internal diffusion, surface drag, and cumulus friction can all be included in the terms F_λ and F_ϕ. The thermodynamic equation is written

$$\frac{1}{R}\frac{\partial}{\partial \phi}\left(v\theta\cos\phi\right) + \frac{\partial}{\partial z}\left(w\theta\right) = -\tau^{-1}\left(\theta - \theta_c\right) + H \qquad (6.54)$$

where diffusion, latent heat release, and other diabatic processes may be included in term H. The Newtonian cooling term in (6.54) forces the model solutions back to a prescribed climatology given by θ_c. The Newtonian cooling radiative relaxation time is given by τ. The continuity equation is simply

$$\frac{1}{R}\frac{\partial}{\partial\phi}\left(v\cos\phi\right) + \frac{\partial w}{\partial z} = 0 \tag{6.55}$$

which is consistent with the Boussinesq approximation. That approximation is also employed to obtain the hydrostatic relation

$$\frac{1}{\rho_0}\frac{\partial P}{\partial z} = g\frac{\theta}{\theta_0} \tag{6.56}$$

where θ_0 is the horizontal mean of the prescribed, climatological potential temperature θ_c. This form of the hydrostatic relation is derived in Holton (1979, pp. 161–162), for example. The above system (6.52) through (6.56) differs from that employed by Schneider and Lindzen in order to simplify the derivation of the system of equations. The main differences are that they use T instead of θ as a dependent variable, and that they use minus the logarithm of the ratio of pressure over surface pressure as the vertical coordinate. These differences do not materially alter the discussion that follows.

To begin with, some further simplifications of (6.52) through (6.56) are made. In (6.52) and (6.53) one might assume that the advective terms and the geometric terms are small, leaving

$$fv = -F_\lambda = -\nu\frac{\partial^2 u}{\partial z^2} \tag{6.57}$$

and

$$fu = -\frac{1}{\rho_0 r}\frac{\partial P}{\partial\phi} + F_\phi = -\frac{1}{\rho_0 r}\frac{\partial P}{\partial\phi} + \nu\frac{\partial^2 v}{\partial z^2} \tag{6.58}$$

A simple second-order diffusion is used above to express the friction term, with ν a constant.

The pressure term in (6.58) is eliminated by taking a vertical derivative of (6.58) and a latitudinal derivative of (6.56). That obtains a thermal wind relation, modified by a diffusion term.

$$f\frac{\partial u}{\partial z} = -\frac{g}{r\theta_0}\frac{\partial\theta}{\partial\phi} + \nu\frac{\partial^3 v}{\partial z^3} \tag{6.59}$$

The dependent variable θ might be replaced by choosing $\theta = \theta_c$ as was done by Charney (1973, pp. 128–136). The solution to the resultant equations is a thermally direct circulation where the meridional transport is confined mainly to Ekman boundary layers. Held and Hou (1980) point out that this model has a serious flaw when $\nu \to 0$, namely that an angular momentum constraint cannot be satisfied.

A less restrictive assumption is to keep the potential temperature as an unknown, but to simplify the thermodynamic equation (6.54). Keeping the Newtonian cooling as the only nonadiabatic process, then (6.54) reduces to

$$\frac{1}{R}\frac{\partial}{\partial \phi}\left(v\theta \cos\phi\right) + \frac{\partial}{\partial z}\left(w\theta\right) = -\tau^{-1}\left(\theta - \theta_c\right)$$

Next one may use the continuity equation to obtain

$$\frac{v}{R}\frac{\partial}{\partial \phi}\left(\theta \cos\phi\right) + w\frac{\partial}{\partial z}\left(\theta\right) = -\tau^{-1}\left(\theta - \theta_c\right)$$

Two further assumptions are that horizontal variations of θ are negligible and that the vertical derivative (proportional to static stability) is constant. These assumptions obtain

$$\theta = \theta_c - \tau \kappa w \tag{6.60}$$

where $\kappa = \partial\theta/\partial z$, is now a constant. Substituting the definition of θ from (6.60) into (6.59),

$$f\frac{\partial u}{\partial z} = \frac{-g}{r\theta_0}\left(\frac{\partial \theta_c}{\partial \phi} - \tau\kappa\frac{\partial w}{\partial \phi}\right) + \nu\frac{\partial^3 v}{\partial z^3} \tag{6.61}$$

By taking the vertical integral of (6.57), one can eliminate the zonal wind from (6.61).

$$f\int_0^z v\,dz = \frac{g}{r\theta_0}\left(\frac{\partial \theta_c}{\partial \phi} - \tau\kappa\frac{\partial w}{\partial \phi}\right) - \nu\frac{\partial^3 v}{\partial z^3} \tag{6.62}$$

Equation (6.62) can be cast as one equation in one unknown by defining a stream function ψ where

$$v = \frac{-1}{\cos\phi}\frac{\partial \psi}{\partial z} \qquad \text{and} \qquad w = \frac{1}{R}\frac{\partial \psi}{\partial \phi} \tag{6.63}$$

Substitution of (6.63) into (6.62) yields

$$\frac{f}{\cos\phi}\psi - \frac{\tau\kappa}{R}\frac{\partial \psi}{\partial \phi} - \frac{\nu}{\cos\phi}\frac{\partial^4 \psi}{\partial z^4} = \frac{g}{r\theta_0}\frac{\partial \theta_c}{\partial \phi} \tag{6.64}$$

If the $(\partial\psi/\partial\phi)$ term is neglected, then (6.64) is essentially the same as an equation analyzed in Schneider and Lindzen (1976) and Held and Hou (1980).

Schneider and Lindzen (1976) obtain a solution by assuming separation of variables and radiative forcing (the θ_0 term) independent of $\cos\phi$. The result is an eigenvalue problem for the latitudinal ($\cos\phi$) structure and a heterogeneous vertical structure equation that is solved using an eigenfunction decomposition. A result they show is a hemispheric Hadley cell that has largest upward motion at 5 N and largest sinking at 45 N.

In a later paper, Schneider and Lindzen (1977) examine a linearized form of (6.52) through (6.56) that incorporates cumulus friction.

$$F_\lambda \sim \frac{\partial}{\partial z}\left(\nu \frac{\partial u}{\partial z} + M_c(\phi, z)\big(u - u_c(\phi, z)\big)\right) \tag{6.65}$$

The cumulus friction only enters into the zonal momentum equation. It models the drag thought to be exerted by trade wind cumulus clouds that extend into upper level westerlies but have cloud bases in the low-level easterlies. u_c is the zonal velocity of the cloud so that (6.65) is rather like a Rayleigh friction term for clouds; u_c was chosen to be the zonal wind at cloud base. M_c is the mass flux inside the cumulus clouds. Lindzen and Schneider also include latent heating in the form

$$H \sim M_c \frac{\partial \theta}{\partial z} \tag{6.66}$$

As pointed out in §6.3.2.4, laboratory simulations and numerical forecast models (Krishnamurti et al. , 1989) show that convection and friction have opposite effects: convection that is tilted along the direction of the shear feeds energy into the mean flow shear. (Eddy momentum fluxes analogously build shear; recall Figure 4.9.) Observations of tropical cumuli (LeMone et al. , 1984) confirm the "negative friction" along the direction of the cloud motion (but not perpendicular to the motion). Some others (Held and Hoskins, 1985) dispute the actual significance of cumulus friction. Despite these problems, including the cumulus friction term makes an important change to the model because zonal velocity now appears explicitly in the equations. Terms involving meridional derivatives are still ignored in Schneider and Lindzen (1977), but in a follow-up paper (Schneider, 1977) they are included.

An example solution from Schneider and Lindzen (1977) is shown in Figure 6.24. A zonal mean heating is placed asymmetrically about the equator, with a maximum near 5 N. The heating looks similar to the rainfall pattern (solid curve in Figure 3.27) except that the observed secondary midlatitude maxima are ignored and values poleward of 20° approach zero asymptotically as they move toward 50°. Their model develops a pair of Hadley cells having less meridional extent than in their earlier study. Their model still forms large [u] at the poleward boundary of each cell, so the jet stream axis moves closer to the equator. The primary cause of the change may be the cumulus friction parameterization. The model retains some curious features: westerly winds that occur everywhere (Figure 6.24a) and small thermally direct cells near the ground (equator to 20° and below 800 hPa in Figure 6.24b).

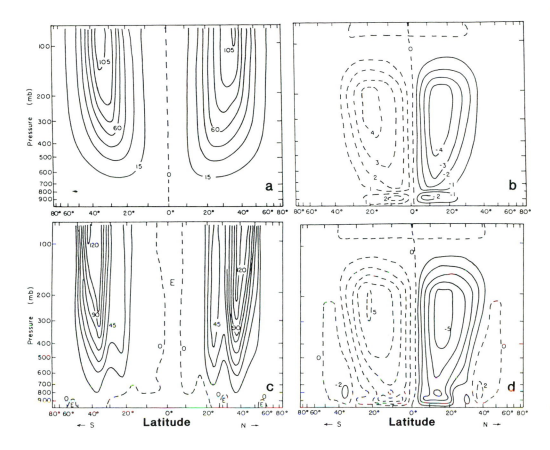

FIG. 6.24   Meridional cross-sections of $[u]$ in (a) and (c) using a 15 m/s interval; stream function ψ in (b) and (d) using a 10^{10} kg/s interval. (a) and (b) from Schneider and Lindzen (1977). (c) and (d) from Schneider (1977) using a similar model but incorporating meridional derivative (advection) terms. Both solutions are steady state. The circulation is clockwise around a negative maximum (solid contours) in (b) and (d). Figures reproduced from *J. Atmos. Sci.*, by permission of American Meteorological Society.

The follow-up paper by Schneider (1977) shows changes wrought by keeping the meridional derivative terms. A major improvement is that this "nonlinear" model allows nonzero surface winds. As a result, surface easterlies develop (Figure 6.24c) in the tropics. In addition, a thermally indirect, "Ferrel" cell is created in each hemisphere (Figure 6.24d). Forming a Ferrel cell is quite remarkable since there are neither eddy fluxes nor midlatitude diabatic heating (also from eddies) in this model. Results in §6.3 stress that the Ferrel cell is inseparable from the eddies. Schneider felt that the indirect cell appears in his solution because surface high pressure in the subtropics drives ageostrophic motion both equatorward and poleward. The heating centered off the equator creates a small amount of asymmetry; the circulation in the Southern Hemisphere is weaker. If seasonal changes are considered, one expects an asymmetric heating pattern to create a much stronger winter hemisphere Hadley cell. This point is a focus of studies by Lindzen and Hou (1988) and Hack et al. (1989).

Lindzen and Hou (1988) use the primitive equations with second-order interior viscosity, Rayleigh friction at the bottom, and Newtonian cooling. Similar to Held and Hou (1980), their equations are integrated numerically in time using an iterative, damping time scheme in order to reach a steady state. The method uses uncertain computational diffusion, and it does not always converge. An example solution is given in Figure 6.25. In contrast to the work done with Schneider, Lindzen and Hou (1988) find strong sensitivity to the location of the heating maximum. When heating is centered on the equator, the hemispheres are symmetric. In Figure 6.25 the heating is greatest at 6 N and the Southern Hemisphere Hadley cell is far the stronger. Though the summer Hadley cell is nearly vanquished, the winter cell is more than twice as great compared to the case where heating is centered at the equator. From this result, Lindzen and Hou (1988) argue that seasonally varying heating fields give larger *annual* average Hadley circulations than does the annual average heating field.

While Lindzen and Hou (1988) place much emphasis upon the latitude of the heating, Hack et al. (1989) raise an alternative view of the process. Instead of searching for time-invariant solutions, Hack et al. (1989) stress the time variation. The ICZ is an evolving structure in their model. Heating with vertical and latitudinal structure is imposed upon an atmosphere at rest. They integrate a balanced set of equations ahead in time until the temperature perturbation exceeds 2.5° C. The limit is achieved within a few days, consequently the air parcels do not travel far and angular momentum conservation does not have time to create large $[u]$. A sample result is presented in Figure 6.26, where the heating is centered at +10° latitude. The winter hemisphere (negative latitudes) cell is stronger and is greatest when the heating is 12 degrees off the equator. Subsidence is much stronger in the winter cell; they conclude that the surface boundary layer could be depressed (shallower) on the winter side and they cite supporting observational evidence in Ramage et al. (1981). A point they stress is that the heating profile simulating deep convection leads to a sign change of potential vorticity near the ICZ. The change in sign is unstable to the formation of easterly waves, which deform the ICZ as they develop, eventually breaking up the ICZ into a series of eddies. The inertial instability identified by Hack et al. (1989) has a length scale of roughly 1000 km. The easterly waves and the ICZ are both comprised of much smaller scale convection, the presence of which is explained in the next section.

6.7 Intertropical convergence zone structure and CISK

This final section examines more closely the ascending branch of the Hadley circulation. One expects the ascending branch to coincide with the so-called intertropical convergence zone (ICZ) of enhanced moist convection, but it was shown in §3.3 that the ascent could not be broad in scale due to the minimum in moist static energy near 700 hPa. Instead, the upward motion must be concentrated in strong updrafts

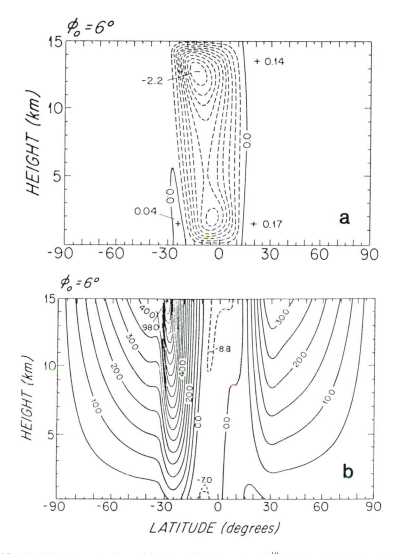

Fig. 6.25 Meridional cross-sections of (a) stream function using 10^{10} kg/s interval and (b) zonal wind using 5 m/s interval. Diabatic heating is largest at $+6°$ latitude. The solutions are asymptotically approached over time using a damping numerical scheme. From *J. Atmos. Sci.*, Lindzen and Hou (1988) by permission of American Meteorological Society.

imbedded deep within thunderstorms. Those updrafts insulate rising air from the debilitating effects of entraining dry mid-tropospheric air (Riehl and Malkus, 1958). Thus, a small-scale phenomenon (convection) is required to maintain a large-scale phenomenon (the Hadley cell). The model in this section shows why convection is chosen as a preferred scale of vertical motion in the tropics. The mechanism that explains this connection between large and small scales has come to be known as conditional instability of the second kind, or CISK.

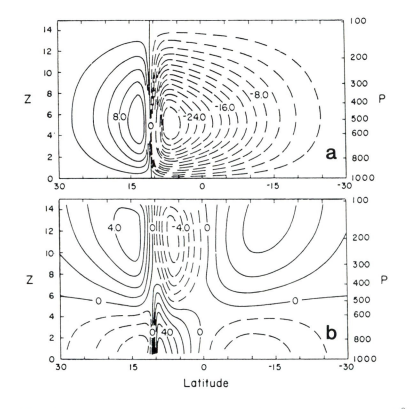

FIG. 6.26 Meridional cross-section of (a) effective stream function ($\psi r \cos \phi$) using 2×10^9 kg s^{-1} rad^{-1} interval and (b) [u] using 0.5 m/s interval. In contrast to solutions shown in the two previous figures, the solution is not steady state. Instead, the flow field after four days is shown after it has spun up from a rest state. From *J. Atmos. Sci.*, Hack et al. (1989) by permission of American Meteorological Society.

Before discussing CISK it is useful to briefly describe conditional instability (the "first" kind). Simply put, the atmosphere is conditionally stable when its lapse rate lies between the dry adiabatic and moist adiabatic lapse rates. Such a temperature profile is unstable for saturated air but stable otherwise. A related concept is potential instability. One definition states that the atmosphere is potentially unstable when the dewpoint lapse rate is larger than the moist adiabatic lapse rate. If a potentially unstable layer of air is raised, it becomes both saturated and unstable. Thus a small amount of lifting leads to vigorous cumulus convection. Commonly, potential instability is created when very dry air overlies very moist air. This is often the case in the tropics. A typical sounding has a very moist boundary layer (1 to 2 km thick) with a lapse rate similar to the moist adiabatic; this is usually topped by a trade wind inversion and much drier air above. It is easy to see how the sounding has a minimum in moist static energy near the inversion (Figure 3.20). The drier air above is created by large-scale subsidence, as might be expected for a Hadley cell.

Conditional instability of the second kind, hereafter referred to as CISK, operates as follows. In general terms the small-scale convective motions and the synoptic scale wind patterns reinforce each other by means of frictional convergence in the boundary layer. More specifically, a surface wind field with cyclonic vorticity will develop low-level convergence if friction is significant. In midlatitudes this could be so-called Ekman balance; in the tropics, a similar mechanism may apply at the equator even though geostrophy breaks down. The upward motion over regions of cyclonic vorticity is labelled Ekman pumping. This pumping forces very moist air to rise, reach saturation, and become convectively unstable. The great quantities of latent heat released encourage greater upward motion, stretch the vertical vortex tubes, and hence further increase the cyclonic vorticity. The released heat will lead to surface pressure falls (in the manner of Figure 3.22), which also increase the surface cyclonic vorticity. A positive feedback loop is formed because the greater cyclonic vorticity leads to more vigorous convection, etc. Charney and Eliassen (1964) first proposed this mechanism to explain the development of hurricanes.

The CISK mechanism can also be applied to the larger-scale intertropical convergence zone, or ICZ. The following derivation is presented in Charney (1971a, 1973). The system of equations neglects advection terms, all longitudinal variations, and *interior* friction. Ekman pumping, dependent upon frictional convergence in a planetary boundary layer, is kept.

$$\frac{\partial u}{\partial t} - fv = 0 \tag{6.67}$$

$$\frac{\partial v}{\partial t} + fu = \frac{-1}{\rho_0}\frac{\partial P}{\partial y} \tag{6.68}$$

$$\frac{1}{\rho_0}\frac{\partial P}{\partial z} = g\frac{\theta}{\theta_0} \tag{6.69}$$

$$\frac{\partial \rho}{\partial t} + \rho_0\frac{\partial v}{\partial y} + \frac{\partial}{\partial z}(\rho_0 w) = 0 \tag{6.70}$$

$$\frac{1}{\theta_0}\frac{\partial \theta}{\partial t} + \frac{N^2}{g}w = \frac{Q}{C_p T_0} \tag{6.71}$$

where the zero subscript denotes a horizontal average and

$$N^2 = g\frac{\partial}{\partial z}(\ln \theta)$$

The hydrostatic equation (6.69) is expressed in the form derived, for example, in Holton (1979, pp. 161–162).

The boundary layer meridional and vertical velocities are scaled by the Ekman number, $E = \nu/(fH^2)$ where ν is a viscosity coefficient (as in equations 6.57 and 6.58) and H is a scale height $H = R\overset{\frown}{T}/g$. Then

$$(u, v, w) = (Uu, E^{1/2}Uv, \frac{H}{B}UE^{1/2}w)$$

where B is a horizontal length scale in the y direction. Charney scales pressure and density by geostrophic and hydrostatic relations.

$$\left(\frac{P}{\rho_0}, \frac{\rho}{\rho_0}\right) = \left(fUB\psi, \frac{fUB}{gH}\mu\right)$$

Time is scaled by $t = tE^{1/2}f^{-1}$. With these scalings, the equations (6.67) through (6.71) become as follows.

$$\frac{\partial u}{\partial t} - v = 0 \qquad (6.72)$$

$$E\frac{\partial v}{\partial t} + u = -\frac{\partial \psi}{\partial y} \qquad (6.73)$$

$$\frac{f^2 B^2}{gH}\frac{\partial \mu}{\partial t} + \frac{\partial v}{\partial y} + \frac{1}{\rho_0}\frac{\partial}{\partial z}(\rho_0 w) = 0 \qquad (6.74)$$

The thermodynamic equation is manipulated as follows. The hydrostatic relation (6.69) can be approximated by

$$g\frac{\theta}{\theta_0} = \frac{\partial}{\partial z}\left(\frac{P}{\rho_0}\right) - \frac{\partial \ln \theta_0}{\partial z}\left(\frac{P}{\rho_0}\right) \simeq \frac{\partial}{\partial z}\left(\frac{P}{\rho_0}\right)$$

where z is dimensional in the last equation. Using this equation, multiplying (6.71) by g, and nondimensionalizing leaves

$$\frac{\partial^2 \psi}{\partial t \partial z} + \frac{N^2 H^2}{f^2 B^2}w = \frac{gH}{Uf^2 BE^{1/2}}\frac{Q}{C_p T_0} \qquad (6.75)$$

All dependent variables are now nondimensional. The "Rossby radius of deformation," B_R, is the chosen length scale; it depends upon the atmospheric state and is defined by $N^2 H^2/(f^2 B_R^2) = 1$. Since the square root of the Ekman number is order 10^{-2}, then to first order (6.73) and (6.74) reduce to

$$u = -\frac{\partial \psi}{\partial y} \qquad (6.76)$$

$$\frac{\partial v}{\partial y} + \frac{1}{\rho_0}\frac{\partial}{\partial z}(\rho_0 w) = 0 \qquad (6.77)$$

Charney now assumes that all moisture pumped out of the boundary layer condenses and releases latent heat. The formula for the heating rate is derived as follows. By definition,

$$Q = \frac{d\theta}{dt} \approx \frac{-Ldq_s}{dt} \approx -Lw\frac{\partial q_s}{\partial z} \approx -Lw_e\frac{\Delta q_s}{\Delta z} \approx w_e L\frac{q_s}{H}$$

where q_s is the saturation mixing ratio at the top of the boundary layer and L is the latent heat of condensation per unit mass. The steps in the equation above proceed by linking the heating rate to the change of moisture due to condensation, then to vertical advection, then to the (dimensional) Ekman pumping velocity and then to the assumption that $q_s \to 0$ at scale height H (all moisture condenses out). After nondimensionalizing, Q can be expressed as

$$Q = \lambda^2 \frac{W_E L q_s g}{C_p T_0 H N^2} = \lambda^2 W_E F(z)\eta \qquad (6.78)$$

where

$$\lambda^2 = \frac{N^2 H^2}{f^2 B^2} \qquad (6.79)$$

$$\eta = \frac{L q_s}{C_p T_0} \left(\frac{\partial \ln \theta_0}{\partial z} \right)^{-1} \qquad (6.80)$$

and the height derivative has been scaled by H. The vertical Ekman pumping velocity W_E is proportional to the vorticity at the top of the boundary layer.

$$W_E = \frac{-1}{\sqrt{2}} \frac{\partial u}{\partial y} = \frac{1}{\sqrt{2}} \frac{\partial^2 \psi}{\partial y^2} \qquad (6.81)$$

Finally, F is a vertical weighting function describing how the condensational heating is distributed in the vertical; its vertical average is unity when F is nonzero. Charney uses $F = 1$ for upward W_E, $F = 0$ for $W_E < 0$.

Given the approximate nature of this description, Charney chose to simplify the mathematics by formulating a two-layer version of the model. Thus (6.72), (6.74), and (6.76) are evaluated at the midpoint of layers 1 and 2 while (6.75) is evaluated at the interface between the layers (Figure 6.27). Given that $z \to \infty$ for $P \to 0$, he transformed his vertical coordinate from z to P. In a layer model the vertical derivatives are evaluated by finite differences. Thus, the vertical derivative in (6.77) is written

$$\frac{\partial v}{\partial y} - gH \left(\frac{\rho_b W_b - \rho_t W_t}{P_b - P_t} \right)$$

where the subscript b refers to the value at the bottom of the layer and the subscritp t refers to the value at the top of the layer. The scale height in the equation above comes from the coordinate transform and from an assumption that the static state be isothermal. W_i denotes the vertical velocity at the interface, W_E denotes the vertical velocity at the bottom, while the top is a rigid lid. Charney assumes that the two layers are equally thick and that the midpoints of the two layers have pressures

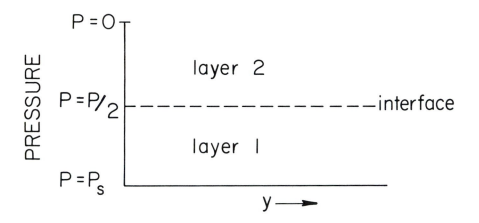

FIG. 6.27   Geometry used for CISK model. Variables evaluated at each midpoint of layers 1 and 2 are indicated by subscripts 1 and 2. Vertical velocity, W_i is evaluated at the interface.

$P_1 = 0.75P_s$ and $P_2 = 0.25P_s$. These assumptions produce the following equations at the midpoints of the layers.

$$\frac{\partial u_1}{\partial t} - v_1 = 0 \qquad\qquad \frac{\partial u_2}{\partial t} - v_2 = 0 \qquad\qquad (6.82)$$

$$u_1 = -\frac{\partial \phi_1}{\partial y} \qquad\qquad u_2 = -\frac{\partial \phi_2}{\partial y} \qquad\qquad (6.83)$$

$$\frac{\partial v_1}{\partial y} + W_i - 2W_E = 0 \qquad\qquad \frac{\partial v_2}{\partial y} - W_i = 0 \qquad\qquad (6.84)$$

where W_E is defined using ϕ_1 only. The ϕ's are nondimensional geopotentials. The thermodynamic equation is evaluated at the interface. Using the nondimensional hydrostatic relation to make the coordinate transform of (6.75) and using (6.78) through (6.81) yields

$$\frac{-\rho_i g H}{P_s} \frac{\partial \phi}{\partial t \partial P} + \lambda^2 W_i = Q = \lambda^2 W_E F \eta$$

From Charney's assumptions that the two layers be equally thick and the atmosphere isothermal, then the density at the interface is half that at the bottom surface. The thermodynamic equation becomes

$$\frac{\partial}{\partial t}\left(\phi_2 - \phi_1\right) + \lambda^2 \left(W_i - F\eta W_E\right) = 0 \qquad\qquad (6.85)$$

Charney (1973) chooses an exponential time dependence for all variables that is proportional to $\exp(\sigma t/\sqrt{2})$. Therefore, the most unstable solution is that with largest positive real σ. That substitution into (6.82) and (6.85) leaves

$$\sigma u_1 - \sqrt{2}v_1 = 0 \qquad\qquad \sigma u_2 - \sqrt{2}v_2 = 0 \qquad\qquad (6.86)$$

$$\sigma(\phi_2 - \phi_1) + \sqrt{2}\lambda^2(W_i - F\eta W_E) = 0 \tag{6.87}$$

The eventual goal of the derivation is to combine (6.83) and (6.84) with (6.86) and (6.87) to obtain one equation for v at one level. Substituting (6.83) into (6.87) eliminates u_1 and u_2.

$$\frac{\sqrt{2}}{\sigma}v_1 = -\frac{\partial\phi_1}{\partial y} \tag{6.88}$$

$$\frac{\sqrt{2}}{\sigma}v_2 = -\frac{\partial\phi_2}{\partial y} \tag{6.89}$$

Rewriting the two parts of (6.84),

$$W_i = 2W_E - \frac{\partial v_1}{\partial y} \tag{6.90}$$

$$W_i = \frac{\partial v_2}{\partial y} \tag{6.91}$$

Inserting (6.91) and (6.81) into (6.87), taking the y derivative, and substituting (6.88) and (6.89), and taking another y derivative, gives

$$\left(\frac{\partial v_1}{\partial y} - \frac{\partial v_2}{\partial y}\right) + \lambda^2\left(\frac{\partial^3 v_2}{\partial y^3} + \frac{F\eta}{\sigma}\frac{\partial^3 v_1}{\partial y^3}\right) = 0 \tag{6.92}$$

The v_1 in (6.92) is eliminated by combining (6.90) and (6.91).

$$\frac{\partial v_2}{\partial y} = 2W_E - \frac{\partial v_1}{\partial y} = -\left(\frac{2}{\sigma} + 1\right)\frac{\partial v_1}{\partial y} \tag{6.93}$$

where (6.81) and (6.88) have been used to express W_E in terms of v_1.

Substituting (6.93) into (6.92) and multiplying by $-(\sigma + 2)$ obtains

$$\frac{\partial}{\partial y}\left\{v_2(2\sigma + 2) - \lambda^2(2 + \sigma - F\eta)\frac{\partial^2 v_2}{\partial y^2}\right\} = 0 \tag{6.94}$$

The terms within the { } equal a constant that Charney assumes to be zero, leaving

$$\frac{\partial^2 v_2}{\partial y^2} - \frac{1}{\lambda^2}\left(\frac{2\sigma + 2}{2 + \sigma - F\eta}\right)v_2 = 0 \tag{6.95}$$

Charney (1973) solves (6.95) by assuming that there is rising motion ($F = 1$) for $-a \leq y \leq a$ and sinking motion ($F = 0$) elsewhere. For $-a \leq y \leq a$, (6.95) can have oscillatory solutions of the form

$$v_2 = A\sin\left(\frac{y}{\lambda_u}\right)\left(\sin\left(\frac{a}{\lambda_u}\right)\right)^{-1} \tag{6.96}$$

where

$$\lambda_u = \left(\frac{\eta - 2 - \sigma}{2 + 2\sigma} \right)^{1/2} \lambda$$

Elsewhere, (6.95) has exponential solutions. The exponential solution with positive argument is ignored since it is unbounded as $y \to \infty$. That leaves

$$v_2 = A \exp\left(\frac{a - y}{\lambda_d} \right) \tag{6.97}$$

for $y > a$ and

$$v_2 = -A \exp\left(\frac{a + y}{\lambda_d} \right) \tag{6.98}$$

for $y < -a$. Above,

$$\lambda_d = \left(\frac{2 + \sigma}{2 + 2\sigma} \right)^{1/2} \lambda$$

A dispersion relation can be derived from mass continuity across the matching points $y = \pm a$. Pressure, and thus geopotential ϕ, must be continuous across each boundary.

$$\phi_2(a^+) = \phi_2(a^-) \qquad \text{and} \qquad \phi_2(-a^-) = \phi_2(-a^+)$$

where the left-hand side of each equation above would use formulas based on (6.97) or (6.98), while the right-hand sides use (6.96). The matching is done by integrating (6.89) at both matching points.

$$0 = \frac{\sqrt{2}}{\sigma} \int_{-a^-}^{-a^+} v_2 dy = \phi_2(-a^-) - \phi_2(-a^+)$$

$$0 = \frac{\sqrt{2}}{\sigma} \int_{a^-}^{a^+} v_2 dy = \phi_2(a^-) - \phi_2(a^+) \tag{6.99}$$

Dividing (6.95) by the coefficient of v_2 then integrating across each of the matching points allows (6.99) to be used. The integral of the v_2 term in (6.95) vanishes from (6.99); for example, at $y = -a$,

$$0 = \lambda^2 \int_{-a^-}^{-a^+} \left(\frac{2 + \sigma - F\eta}{2\sigma + 2} \right) \frac{\partial^2 v_2}{\partial y^2} dy = \lambda^2 \left(\frac{2 + \sigma - F\eta}{2\sigma + 2} \right) \frac{\partial v_2}{\partial y} \bigg|_{-a^-}^{-a^+}$$

$$= \lambda^2 \left(\frac{2 + \sigma - \eta}{2\sigma + 2} \right) \frac{\partial v_2}{\partial y} \bigg|_{-a^+} + \lambda^2 \left(\frac{2 + \sigma}{2\sigma + 2} \right) \frac{\partial v_2}{\partial y} \bigg|_{-a^-} \tag{6.100}$$

Substituting (6.96) and (6.98) into (6.100) yields

$$\frac{1}{\lambda_d} = \frac{1}{\lambda_u} \tan\left(\frac{a}{\lambda_u} \right) \tag{6.101}$$

Charney states that σ increases monotonically from $\sigma = 0$ for length scales

$$\frac{a}{\lambda} = \left(\frac{\eta}{2} - 1\right)^{1/2} \tan\left(\frac{\eta}{2} - 1\right)^{1/2} \tag{6.102}$$

to a maximum of

$$\sigma = \frac{\eta}{2} - 1 \tag{6.103}$$

at $a/\lambda = 0$.

Instability is possible only for scales less than a certain maximum scale, given by (6.102). It is clear from (6.102) that $\eta > 2$ is needed for the expression to make sense. Substituting realistic values into (6.80) yields $\eta \sim 2$; it becomes progressively more difficult to get $\eta > 2$ as the equator is approached, since B_R rapidly increases. Using $\eta = 2.1$, $q_s = 0.03$, and latitude 4 N, the maximum a is about 400 km. This length scale is comparable to the width of the rain area in tropical storms and to the width of the ICZ. Scales smaller than this are unstable. From (6.103) the most unstable scale is one that has a width that is infinitesimally small!

One anticipates that diffusion would preclude the development of infinitesimally wide convective cells that extend through the depth of the tropical troposphere. However, diffusion cannot be expected to give the observed scales of 10 to 100 kilometers. One might simply expect the width of the convection to be comparable to the depth of the fluid. If so, the CISK solution may explain why the observed ICZ convection is organized into rainbands that each have 20 to 50 km width. In Chapter 5 (Figure 5.13), the time-mean ICZ over the tropical oceans appears as a narrow, east-west oriented cloud band. However, that is not how the convective clouds appear at any given moment (Figure 5.12). Instead, the convection is organized into mesoscale "cloud clusters" that are spaced along the ICZ. Between the cloud clusters the air is relatively clear. The problem of maximum instability at shortest scale can be avoided by at least two means.

One solution (Chang and Williams, 1974) points out that the layer model formulation here does not include damping caused by the Ekman layer. By comparing the layer model with a continuous model, they show that the Ekman damping may be approximated by modifying (6.81). The Ekman pumping velocity would not depend upon the lowest layer ϕ_1, but would be formulated in terms of geopotential at the top of the Ekman layer (ϕ_E). Since a new variable is introduced, a new equation is needed, and the thermodynamic equation evaluated at the interface between the Ekman and lowest layers is used.

$$\frac{\partial}{\partial t}\left(\phi_1 - \phi_E\right) + \lambda^2 W_E = 0 \tag{6.104}$$

The similarity to (6.85) should be clear. The Ekman pumping is defined using ϕ_E instead of ϕ_1.

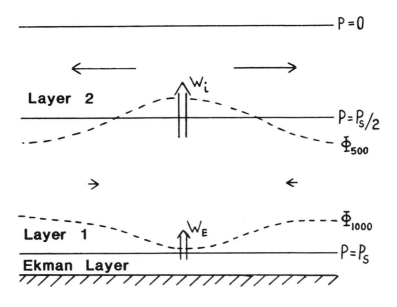

FIG. 6.28   Schematic diagram showing the motions in the two-layer CISK model. Neglectibly thin Ekman layer shown. Dashed lines show geopotential height surfaces for 1000 hPa and 500 hPa. A cyclonic circulation is placed in the middle. Motions in upper layer are much stronger than in the lower layer.

$$W_E = \frac{1}{\sqrt{2}} \frac{\partial^2 \phi_E}{\partial y^2} \qquad (6.105)$$

Chang and Williams (1974) report that this formulation of the Ekman damping has a different wavenumber k dependence than the continuous solution (k^2 rather than k dependence). Nonetheless, (6.104) and (6.105) preserve the feature that the Ekman damping increases for smaller scales. The competing effects of the CISK and Ekman damping should lead to a most unstable wavelength that differs from zero given a favorable environment for CISK. A dynamical explanation for Ekman damping may be inferred from Figure 6.28. P is the vertical coordinate, so geopotential, ϕ, must do the adjusting. A warm core low would have geopotential height contours similar to the dashed lines in the figure. The vorticity is cyclonic, so W_E is upward. The vertical motion out of the Ekman layer adiabatically cools the air in layer 1, causing the geopotential contours to flatten, reducing W_E, etc. The flattening would be stronger for smaller scales because smaller scales have proportionally larger W_E due to the k^2 dependence in (6.105).

A second approach is to refine the heating function definition (6.78). As described in Mak (1986, 1982) the relation between the convection and the heating must change with the size of the convection. Small cumulus clouds would not be insulated from their environment. These clouds would mainly increase the moisture content of the lower troposphere, but would not lead to net condensational heating because they do not precipitate. Larger cloud masses would form precipitation, so Q would be

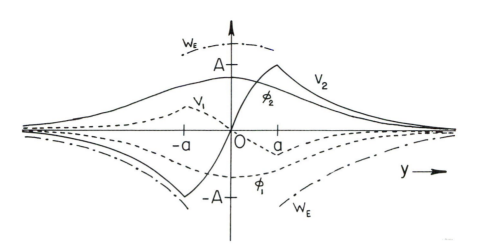

FIG. 6.29   Selected schematic solution structures from the two-layer CISK model. $[v]$ in the upper layer of the CISK model clearly shows upper-level divergence centered over the heating. Heating occurs between $-a \leq y \leq a$. $[u_2]$ (not shown) similar to $[v_2]$ but much larger due to the scaling.

activated only for such large scales. Mak (1982) applies a formulation where Q is tied to the large-scale vorticity field in a baroclinic instability model (§7.1.1). Instead of an infinitely small scale, he finds a finite scale of maximum instability. An adiabatic form of the model is described at the beginning of Chapter 7.

Another peculiarity is that the upper layer solution is much larger than the lower layer solution. Figure 6.28 also illustrates how this happens, because the W_E term in (6.93) is much larger than the $\partial v_1 / \partial y$ term. Other aspects of the solution are given in Figure 6.29.

In Chapter 3 (Figure 3.27) the maximum precipitation occurs about five degrees north of the equator. Even if account is taken of the asymmetry caused by the Asian summer monsoon and only oceanic areas are examined, the ICZ still lies some distance from the equator (e.g., the eastern Pacific in Figure 5.13.). Charney (1971a) sought to explain this offset from the equator by modifying the above analysis made on an f-plane. If f is allowed to vary only in the dispersion relations, then Charney finds that the Ekman pumping vanishes at the equator and the CISK mechanism is turned off. Schneider and Lindzen (1977) dispute this *a posteriori* inclusion of a variable f. They point out that the B_R (deformation length scale) becomes infinite at the equator, which shuts down the Ekman pumping; their own analysis on a sphere finds that the deformation radius approaches a finite number. Consequently they find largest upward circulation at the equator when the heating and surface temperatures are largest there. Schneider (1977) found that the rising branch occurs where the sea surface temperatures are greatest, which is not necessarily at the equator. A comparison of Figures 5.13 and 5.3 seems to verify his conclusion, especially over the eastern Pacific ocean.

7

THEORIES FOR NONZONAL FIELDS

But seeing that so great Continents do interpose and break the continuity of the
Oceans, regard must be had to the Nature of the Soil, and the Position of the
high Mountains, which I suppose [are] the two principal Causes of the several
Variations of the Winds.

Halley, 1686

The nonzonal motions are wave-like for at least four reasons. One expects waves
for a fluid such as the atmosphere because waves are "smooth." One expects waves
in response to stable oscillations, like the lee waves that form in statically stable air
downwind of mountain ridges. A similar concept applies in the horizontal dimension,
where the stabilizing factor is the rotation of the earth ($\partial f / \partial y$), which is described in
§7.2.1 employing potential vorticity conservation. One expects very strong vertical
and horizontal shears to develop (§6.6) if only meridional circulations are allowed.
The horizontal shears can initiate the growth of waves by barotropic instability. The
vertical shears can lead to growing waves by baroclinic instability (§7.1.1). A good
model for observed developing cyclones appears to be a combination of these two
special cases of instability. Finally, one expects waves from the prevalence of quasi-
geostrophy. Either there are waves or there must be strong ageostrophic velocities.
Beyond the maintenance of $[u]$ (§6.2.2), waves are needed for heat transport. If air
motion is geostrophic, one must have waves since $[v][T] = 0$, except possibly in
valleys between north-south oriented mountain ranges.

In the previous chapter, eddies enter the discussion in several places as needed
to explain various features of the zonal mean circulation. This chapter focuses upon
explaining the properties of eddies. The first question that could be asked is: why must
eddies form in atmosphere's general circulation? After all, the Hadley circulation is
capable of performing the necessary poleward heat transport, and it can do so without
violating mass, angular momentum, and other balance requirements. The deflection
of air as it encounters a mountain range is one obvious explanation for the existence
of eddies. A second explanation lies in the differing thermal properties of ocean

and land. Both of these mechanisms were proposed by Halley (1686) as reasons for the lack of constancy of the winds where there are monsoons (in Africa and Asia) and in the Northern Hemisphere middle latitudes. Halley admits to having little knowledge of the winds in the Southern Hemisphere middle latitudes. If Halley had known, then he would have needed a third mechanism to generate eddies, since those latitudes are nearly uniformly covered by ocean. The third mechanism is more subtle, and the basic theory to support it was laid down two hundred years later. The third mechanism is the instability of flows with large vertical and horizontal shears.

Each of these aspects of eddy motion is discussed in this chapter. First to be shown is baroclinic instability; strong but realistic zonal mean winds are unstable to the formation of eddies. Next, topography and land–sea contrasts are shown to play a major role in the formation of the long waves. Then various properties of the traveling midlatitude wave-cyclones are described. In particular, the life–cycles of these storms are detailed using observations, models, and analysis tools such as "Eliassen-Palm" fluxes.

In a subject as vast as the general circulation, one must necessarily exclude some topics. For one example, unusual, persistent, weather patterns that are large-scale but longitudinally variable are not discussed. The reader is referred to other recent books (e.g., Benzi et al. , 1986) for discussion of "blocks" and other anomalous flows.

7.1   The instability of zonal mean fields

7.1.1   *A simple analytic model*

A great deal of evidence supports the view that frontal cyclones form as a dynamical instability of the zonal mean flow. The earliest linear theories are by Charney (1947) and Eady (1949) for baroclinic growth and Kuo (1949) for barotropic growth. Later studies combine the baroclinic and barotropic mechanisms; these studies demonstrate, for example, that the two mechanisms are not independent because the optimal wave-cyclone structure for baroclinic growth conflicts with the optimal structure for barotropic growth (Grotjahn, 1979). These and many other linear studies are remarkably successful at reproducing the observed structure and energetics of the wave-cyclones. While details in the results vary, these studies clearly demonstrate that vertical and to some extent horizontal shears observed in the atmosphere are unstable to cyclonic storm development. More recently, a few studies examine the nonlinear development of eddies; such studies have some success at reproducing the properties of these cyclones at later stages in their life cycle. The nonlinear studies are discussed at some length in §7.3.

The following analysis from Grotjahn (1979) illustrates the instability of typical atmospheric flows. The mathematical formulation is analogous to Eady (1949) except that compressibility and linear variation of the Coriolis parameter are allowed. The model is the quasi-geostrophic potential vorticity equation in Cartesian coordinates. (Derivations of (7.1) are in Grotjahn, 1979, and Holton, 1979, p. 180.)

$$\left(\frac{\partial}{\partial t} - \frac{\partial \Psi}{\partial y} \frac{\partial}{\partial x} + \frac{\partial \Psi}{\partial x} \frac{\partial}{\partial y} \right) Q = 0 \tag{7.1}$$

where

$$Q = \nabla^2 \Psi + f + S\epsilon \frac{\partial \Psi}{\partial z} + \epsilon \frac{\partial^2 \Psi}{\partial z^2} \tag{7.2}$$

$$S = \frac{\partial \ln \rho_s}{\partial z} \qquad \kappa = D \frac{\partial \ln \theta_s}{\partial z}$$

and

$$\epsilon = \frac{f^2 L^2}{g \kappa D}$$

The subscript s refers to a horizontal average static state that is prescribed. The equations have been nondimensionalized and scaled with values of L and D typical of midlatitude wave-cyclones.

Next, Ψ and Q are partitioned into a prescribed basic state (indicated by square brackets) and a perturbation (indicated by a lowercase letter).

$$Q(x, y, z, t) = [Q(y, z)] + q(x, y, z, t) \tag{7.3}$$

and

$$\Psi(x, y, z, t) = [\Psi(y, z)] + \psi(x, y, z, t) \tag{7.4}$$

Here the basic state stream function $[\Psi]$ (nondimensional pressure field) is assumed to be a linear function of Cartesian latitude (y).

Equations (7.2), (7.3), and (7.4) are substituted into (7.1). The resulting equation is linearized by neglecting all terms that include the perturbation solution multiplying itself. Because $[\Psi]$ is chosen to be a linear function of y, the resulting equation has the perturbation appearing just once in each term. The resulting eigenvalue problem for ψ becomes

$$\left(\frac{\partial}{\partial t} + [U] \frac{\partial}{\partial x} \right) q + \frac{\partial \psi}{\partial x} \frac{\partial [Q]}{\partial y} = 0 \tag{7.5}$$

Here

$$\frac{\partial [Q]}{\partial y} = \beta - S\epsilon \frac{\partial [U]}{\partial z} - \epsilon \frac{\partial^2 [U]}{\partial z^2} \tag{7.6}$$

is the meridional gradient of basic state potential vorticity. The geostrophic, basic state velocity is $[U] = -\partial [\Psi]/\partial y$, and it is only a function of height. The first meridional derivative of the Coriolis parameter, β is assumed to be constant.

To solve (7.5) requires boundary conditions. A channel is used that is periodic in the x direction and that has impermeable walls along the northern and southern boundaries. The latter condition reduces to vanishing northward geostrophic velocity ($\partial \psi / \partial x = 0$) at the walls. The basic state automatically satisfies both of these boundary conditions. A rigid lid and a flat, rigid bottom are assumed for the other z boundary conditions. The z boundary conditions are obtainable from the adiabatic equation

$$\frac{d\theta}{dt} = 0 \tag{7.7}$$

The quasi–geostrophic version of (7.7) is

$$\left(\frac{\partial}{\partial t} - \frac{\partial \Psi}{\partial y}\frac{\partial}{\partial x} + \frac{\partial \Psi}{\partial x}\frac{\partial}{\partial y} \right) \frac{\partial \Psi}{\partial z} + \frac{W}{\epsilon} = 0 \tag{7.8}$$

where W is the vertical velocity. By setting $W = 0$ in (7.8) one can express the z boundary conditions in terms of a single unknown ψ. The next step is to linearize (7.8) at the top and bottom ($z = 0, 1$).

$$\left(\frac{\partial}{\partial t} + [U]\frac{\partial}{\partial x} \right) \frac{\partial \psi}{\partial z} - \frac{\partial \psi}{\partial x}\frac{\partial [U]}{\partial z} = 0 \tag{7.9}$$

at $z = 0, 1$.

The perturbation is assumed to be a propagating wave in the x (longitudinal) direction having a simple y structure.

$$\psi = \sin(ky)\Re\{\phi(z)\exp(ik(x - ct))\} \tag{7.10}$$

"$\Re$" denotes taking the real part; k is both the x and the y wavenumber. This y structure satisfies the boundary conditions $\partial \phi / \partial x = 0$ at $y = 0, B$ where $B = \pi / k$.

Substituting (7.10) into (7.5) yields

$$([U] - c)\left(-2k^2\phi + S\epsilon\frac{\partial \phi}{\partial z} + \epsilon\frac{\partial^2 \phi}{\partial z^2} \right) + \phi\frac{\partial [Q]}{\partial y} = 0 \tag{7.11}$$

The common factor $ik\sin(ky)\exp\{ik(x - ct)\}$ is factored out of (7.11).

The basic state velocity is chosen to be

$$[U(z)] = \Lambda\left(\exp(-Sz) - 1\right) + \frac{\beta z}{S\epsilon} \tag{7.12}$$

This velocity profile is zero at the bottom and maximum at the top, and is chosen so that $\partial [Q]/\partial y = 0$ everywhere in the interior, thus simplifying (7.11). For the solutions that are being sought, one may assume that $[U] - c \neq 0$ and (7.11) reduces to

$$\frac{\partial^2 \phi}{\partial z^2} + S \frac{\partial \phi}{\partial z} - \alpha^2 \phi = 0 \tag{7.13}$$

where $\alpha^2 = 2k^2/\epsilon$ is a scaled wavenumber.

Solutions to (7.13) have the form

$$\phi(z) = \exp(r_2 z)\{\cosh(r_1 z) + A \sinh(r_1 z)\} \tag{7.14}$$

where

$$r_1 = \frac{1}{2}(S^2 + 4\alpha^2)^{1/2} \qquad \text{and} \qquad r_2 = \frac{-S}{2}$$

What remains to be found are the value of the complex phase speed c and the complex constant A. These are found from the boundary conditions (7.9).

Substituting (7.10) into (7.9) yields

$$([U] - c)\frac{\partial \phi}{\partial z} - \phi\frac{\partial [U]}{\partial z} = 0 \tag{7.15}$$

at $z = 0, 1$. At $z = 0$, after substituting (7.14) into (7.15), one obtains

$$A = \frac{-1}{c\,r_1}\frac{\partial [U]}{\partial z} + \frac{S}{2r_1}$$

This relation is inserted into (7.14) and the result is substituted into (7.15) at $z = 1$. One finally obtains

$$c = \frac{-b \pm (b^2 - 4ae)^{1/2}}{2a} \tag{7.16}$$

where

$$a = \left(\frac{r_2^2}{r_1} - r_1\right)\sinh(r_1)$$

$$b = \left(\frac{\partial [U(0)]}{\partial z} - \frac{\partial [U(1)]}{\partial z}\right)\cosh(r_1)$$

$$+ \left([U(1)]\left(r_1 - \frac{r_2^2}{r_1}\right) + \frac{r_2}{r_1}\left(\frac{\partial [U(0)]}{\partial z} + \frac{\partial [U(1)]}{\partial z}\right)\right)\sinh(r_1)$$

and

$$e = -\frac{\partial [U(0)]}{\partial z}[U(1)]\cosh(r_1)$$

$$- \frac{1}{r_1}\left(\frac{\partial [U(0)]}{\partial z}\left(r_2[U(1)] - \frac{\partial [U(1)]}{\partial z}\right)\right)\sinh(r_1)$$

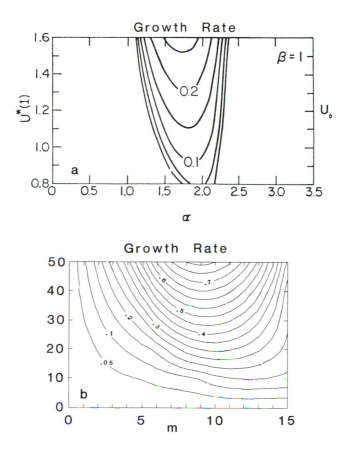

FIG. 7.1 (a) Growth rate spectra for $\beta = 1$ versus $U_0(1)$ for the Cartesian coordinate model solution (7.16). Only Λ is varied, $S = -1$. Since $U_0(0) = 0$, $U_0(1)$ is a crude measure of the vertical shear. In general, the growth rate increases with $U_0(1)$. As the vertical shear increasingly dominates the β-effect, the long wave cut-off shifts to longer wavelengths and the short wave cut-off approaches the value found for $\beta = 0$. (b) Growth rate as a function of zonal wavenumber (m) and maximum jet speed U_0 for a midlatitude jet on a sphere. Quasi-geostrophic linear solution. (b) from Grotjahn (1991b).

The imaginary part of (7.16) equals the growth rate when multiplied by zonal wavenumber, k. An Eady (1949) solution is recovered if (7.12) is replaced by $[U] = z$, $S = 0$ (incompressible flow), and $\beta = 0$. In this case, the basic state vertical shear equals one, $r_2 = 0$, $r_1 = \alpha$, and the cosh term in b vanishes. With these modifications it is easy to see that instability increases with vertical shear (the cosh term in e dominates). Figure 7.1a illustrates this property by plotting the growth rate as a function of the magnitude (Λ) of the basic state wind. It can also be shown analytically (Grotjahn and Lai, 1991) that lower static stability increases the instability (by increasing ϵ, which leads to larger k for a given α, and thus larger growth rate, $k\Im(c)$). Eliassen shows a similar result in the book by Petterssen (1956; p. 316).

The derivation began by nondimensionalizing the equations. Some dimensional values are now inserted. For midlatitudes, one expects $f \approx 10^{-4} \text{ s}^{-1}$, $k \approx .13$ for the troposphere, $D = 10$ km, and $g = 10 \text{ m s}^{-2}$. By setting $\epsilon = 1$, one obtains a Rossby radius of deformation length scale $L = 1100$ km. For the troposphere, and for this scaling, $S \approx -1$. $\beta \approx 1$ is appropriate for this scaling in midlatitudes. As a velocity scale, 22 m/s is selected so that the Rossby number (= 0.2) is small. These numeric choices suggest a reasonable value for the basic state velocity at 10 km elevation to be $[U(1)] = 30.8$ m/s. From Figure 7.1 the growth rate is largest at $\alpha = 1.7$ and equals 0.2. The dimensional values are about 5700 km for the most unstable wavelength and a minimum doubling time of about 2 days. The length scale and growth rate are comparable to those observed for atmospheric wave cyclones.

Figure 7.1 shows that as the shear increases ($[U(1)]$ increasing) the growth rates are also increasing. In addition, the long waves (small α) and the short waves (large α) are both stable in this result. Non-amplifying long and short waves are also found in the classic studies using twolayer models (Phillips, 1956). Here, however, stable long and short waves are an artifact of the rather special choice of basic state velocity (7.12). That choice of $[U]$ caused $\partial[Q]/\partial y = 0$ and that allowed (7.5) to be significantly altered into the form (7.13). A well known condition for instability is that $\partial[Q]/\partial y$ change sign in the domain (Charney and Stern, 1962). Despite vanishing in the interior, the basic state potential vorticity gradient still changes sign due to boundary contributions, which appear like delta functions of opposite sign at the two z boundaries (Bretherton, 1966).

It is not difficult to formulate this model using a more realistic definition of $[U]$. Grotjahn (1980) made several improvements to the model discussed here; for example, he used a $[U]$ profile that reached a maximum at the tropopause level so that the basic state potential vorticity gradient no longer vanished. In that more general situation essentially all the wavelengths become unstable. In general terms, the maximum instability remains at the same wave-cyclone scale found here and has about the same doubling time. Expressing the problem in spherical geometry is not difficult either. Corresponding growth rate curves are shown in Figure 7.1b for a model using spherical geometry and a general midlatitude jet where $\partial[Q]/\partial y \neq 0$ (Grotjahn, 1991b). The patterns in Figures 7.1a and 7.1b are similar, implying that the simple model results derived here are quite general. The principle disagreement is that short wavelength modes are neutral in the simple model, but not in the fancier model.

In reality the source of the growth may be more subtle than imposing temperature gradients at two z boundaries. In a paper on potential vorticity (PV), Hoskins et al. (1985) discuss how temperature gradient at the surface and PV contrast between troposphere and stratosphere act like the two boundary sources of instability. When these are superimposed (as when an upper-level vorticity source overrides a low-level thermal front) then amplification can occur. The superposition just described is well known to synopticians (e.g., Petterssen, 1956; p. 335).

It is quite important to understand how results from a linear study relate to the real atmosphere. Linear instability results are not intended to apply to the real atmosphere in a literal sense because there are many features of the atmosphere not incorporated into such simple linear models. Some approximations used in the formulation are obvious: linearity, geometry, scaling, discretization, adiabatic restrictions, and simplistic basic flow. The linearization means that the results of this section are strictly valid for small amplitude perturbation solutions. Nonlinearity makes important changes that are discussed in §7.3. The simplicity of the mean flow is worthy of further comment since there has been some confusion about this issue in the recent literature. Whereas a linear instability study searches for an optimal wave structure, in reality many different waves are present throughout the atmosphere and are propagating at various rates. What linear studies tell us is that when the observed waves come together in a way that *approximates* the unstable linear solution structure, then there will be growth. For example, all the baroclinically unstable modes have the trough axis tilted upstream with height; most developing frontal cyclones have the same structure. The linear studies explain why the growth occurs and under what types of situations it is more effective, not the exact shape of the waves to be observed.

The type of situations studied go far beyond the simple model derived here. These studies cover a wide range of factors: cyclone wavelength; basic flow vertical shear, horizontal shear, curvature, turning, downstream variation, and other properties; static stability; latitude; topographic interaction; surface interaction and diabatic effects; various geometries; etc. The number of papers on baroclinic and barotropic instability exceeds one hundred, too many to review adequately in this book. Instead, a list of general results, in addition to the items mentioned earlier, would include the following. The most unstable solutions tend to be square; the zonal and meridional scales tend to be similar. Their shape and size respond to the mean flows; narrower jets select waves with smaller meridional scale. Smaller static stability gives larger growth rate and vice versa. A lower latitude location (due to larger β) inhibits instability. In cases where the mean flow shear varies along the track, eddies in areas of stronger shear (similar to observed jet maxima) grow faster than elsewhere. Spherical geometry causes a discrete spectrum of solutions that seems to inhibit instability. Topography can increase or decrease the growth rate depending upon how the storm track and basic state flow are changed by the topography.

Most linear instability studies assume a propagating wave solution similar to (7.10) and solve the resulting eigenvalue problem for complex phase speed c given real k. Alternative approaches have been introduced. One example is to solve the problem as an initial value problem (e.g., Pedlosky, 1964).

Another possibility is to specify a real frequency ω, and solve for complex k (Gaster, 1962). This approach was introduced to the study of cyclogenesis by Merkine (1977). Imaginary components of wavenumber (k_i) can be found that grow exponentially. These "spatially unstable" modes are linked to "temporally unstable" modes, such as (7.16), in Peng and Williams (1986, 1987) for barotropic and baroclinic flows. The issue of spatial instability is closely allied with the notions of "absolute" instability and "convective" (advective) instability. (The term "convective" refers to advection; it does not refer to convection as meteorologists use the term.) Absolute and convective instability differ from each other in terms of events at a fixed point: the amplitude increases continuously at a fixed point for an absolutely unstable mode. A convectively unstable wave packet moves downstream fast enough so that the amplitude at a fixed point may increase, then decrease as the packet passes by.

Absolutely unstable modes are confined to a localized region, whereas convective modes propagate away from their initial location. Not surprisingly, the existence of absolute instability in a two-layer model requires weak vertical-average flow compared with the vertical shear. Mathematically, if a wave packet is to remain stationary, then absolute instability requires vanishing complex group velocity, i.e., $\partial \omega / \partial k = 0$. Merkine (1977) shows that certain instances of spatial instability are examples of absolute instability. Both types of modes could be localized in space, unlike the form (7.10).

A further distinction lies between "local" and "global" modes. Solutions derived above are examples of global modes because of the form (7.10) and because the growth rate is invariant to a Galilean transform. Global modes require periodicity to be bounded in space. In addition, adding a constant to $[U]$ does not change the imaginary part of (7.16). (The reader should be careful to note that a more general form of (7.16) is needed that does not assume $[U(0)] = 0$.) An absolutely unstable mode's growth rate is not invariant to a Galilean transform, and therefore must be local. Pierrehumbert (1984) questioned the relevance of global modes. His critique was based on the periodicity requirement and the very long time needed for global eigenmodes to develop in certain parameter ranges of his two-layer model. In models that more realistically described the atmosphere (e.g., more levels) the local modes have less importance and global modes are, at the least, of comparable import. Cai and Mak (1990a) found similar growth rates for global and local modes. In later work, Pierrehumbert (1986) studied spatial instability in a continous, Charney (1947) model and found that *no* absolute instability is present when $[U(0)] \geq 0$. This suggests that convective instability may be more relevant for atmospheric conditions.

Temporally unstable modes in an eigenvalue problem are usually constructed from "global" functions that extend uniformly around a latitude circle. Eigenfunctions may be localized by an amplitude envelope that modulates the global functions; the localization arises as the modes distinguish the stronger growth in areas of stronger shear even though the growth rate is constant over the domain. The amplitude envelope remains fixed in space. Such envelopes typically have largest amplitude just downstream from regions of stronger shear (e.g., Frederiksen, 1983). Amplitude envelopes are also found for localized topographic features in a zonal mean flow (e.g., Grotjahn and Wang, 1990).

This lag between the locations of maximum amplitude and maximum vertical shear is analyzed in Peng and Williams (1987). When absolute instability is present,

Pierrehumbert (1984) shows that the growth rate of a local mode in the domain is almost the same as the *absolute* instability growth rate at the point of maximum vertical shear. The constancy of the absolute instability frequency can determine a complex wavenumber. The amplitude envelope of the spatially unstable mode would be given by the imaginary part of the wavenumber, (k_i). Since k_i is maximum at the point of maximum shear, the amplitude of the wave packet cannot be largest there. By definition, the wave packet is maximum or minimum where k_i goes to zero. For the situation of a zonally varying current, the amplitude envelope is maximum downstream from where the spatial instability (or local absolute instability) vanishes.

One could view the storm track as either like an absolutely unstable wave packet or as an amplitude envelope modulating a global function. In either case, the individual highs and lows one sees on weather maps would correspond to the peaks and valleys of the carrier wave. An individual peak grows and decays as it traverses the wave packet.

7.1.2   *Laboratory experiments*

Laboratory experiments also confirm the view that a simple Hadley cell circulation will break down into wave-like motions if the rotation rate or heating differential from equator to pole are within certain bounds. These studies most commonly use a rotating cylinder or annulus-shaped tank of water to represent the atmosphere. While it is difficult to directly compare these experiments with the earth's atmosphere, the results are fascinating. Laboratory models have been used to study more phenomenona than just instability of the flow, the subject of this subsection. Other example applications of laboratory models are given elsewhere in this book. Reviews of such laboratory models have been made by Fultz et al. (1959), Fowlis and Hide (1965), and Hide and Mason (1975).

The early experiments focused upon the types of circulations that would appear for various ranges of parameters such as size of the annulus, rotation rate, temperature contrast, and so forth. The major results of the early experiments are usually summarized by means of a "regime diagram." A regime diagram displays in a compact form the type of circulations observed for two nondimensional parameter ranges. The nondimensional parameters express innate properties of the fluid and experimental setup. The power of such parameters is that the same behavior should be seen for the same parameter values, even though a variety of experimental devices may be compared.

An example of a regime diagram is Figure 7.2a. This figure elegantly displays how the character of the circulation varies as one alters rotation rate, temperature differential between the annulus side walls, etc. The axes are proportional to nondimensional "thermal Rossby" and Taylor numbers. The former is proportional to the thermal forcing and the latter to the mechanical forcing. A thermal Rossby number

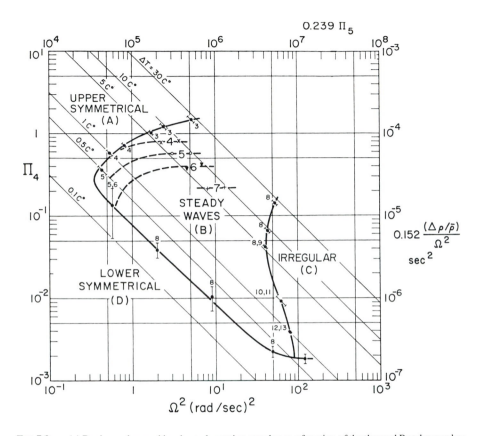

FIG. 7.2a   (a) Regimes observed in a heated, rotating annulus as a function of the thermal Rossby number (Π_4) and the rotation rate (Ω). To the right of the heaviest line the flow typically develops waves; to the left the circulation is axis-symmetric (sinking along the inner wall, rising along the outer wall). For some of the parameter values near the top of the diagram, the flow that develops depends upon the past state of the experiment: it is wavy if rotation has been decreasing, symmetric if rotation has been increasing to reach that point, for example. Since Π_4 depends upon the temperature gradient and the rotation rate, increasing rotation rate for fixed temperature difference (ΔT) is equivalent to moving diagonally downward parallel to the thin lines labelled with ΔT values. From *J. Atmos. Sci.*, Fowlis and Hide (1965) by permission of American Meteorological Society.

can be defined as

$$\Pi_4 = \frac{gH(\Delta\rho/\rho)}{\Omega^2(b-a)^2} \simeq \frac{\mu gH\Delta T}{\Omega^2(b-a)^2}$$

Here b is the radius of the outside boundary, a is the radius of the inside boundary, ΔT is the imposed temperature difference across the distance $(b-a)$, μ is the fluid's thermal expansion coefficient, H is the depth of the fluid and Ω the rotation rate of the annulus. The thermal Rossby number is thus proportional to the temperature differential and inversely proportional to the rotation rate. The abcissa in Figure 7.2 is the rotation rate squared. Thus, lines of constant temperature are diagonals on such a chart. More typically, the Taylor number is used, since the size of the annulus and the kinematic viscosity (ν) are included in the Taylor number.

$$T_a = \frac{4\Omega^2(b-a)^4}{\nu^2}$$

In practice, a series of experiments is run where the temperature difference (ΔT) and rotation rate are varied. Since the geometry of the apparatus and the fluid are unchanged, the changes in T_a are due to rotation rate changes in most sequences of experiments. The value in using T_a is that different sequences by different apparati can be more directly compared.

Characteristic types of circulations occur for certain ranges of the nondimensional parameters. In Figure 7.2a, heavy lines indicate the midpoints of transition regions between various circulation types; the "error bars" indicate the width of the transition region. Values of temperature differential and rotation rate that fall to the left and below the longer, heavy line in Figure 7.2a lead to symmetric, "Hadley cell" type overturning. Fluid rises on the hotter outer wall and sinks along the cooler inner wall (or fluid center if there is no inner wall). The overturning is "symmetric" in that it has no variation in the "longitudinal" (constant radius) direction. To the right of that heavy line several flow patterns are possible; they are characterized by zonal wavenumbers as indicated by the integers. The locations where an indicated wavenumber occurred most frequently are specified by the dashed lines. In the upper right portion of the figure, waves may or may not develop depending upon how one arrives at that location on the diagram. (For example, to arrive at some point the experiment must start from a prior state, such as an initial state of rest and no temperature difference).

If one keeps rotation rate fixed and slowly increases the temperature differential, then fluid parameters are changing so as to move upwards on the diagram. As one increases the temperature differential, there is a point (at the heavy line) where a Hadley cell no longer handles the heat transport and the flow breaks down into small waves. For further increases in the temperature differential, progressively longer waves are destabilized. This experimental result was later reproduced by a mathematical analysis in Lorenz (1962). One caveat is that the most prominent wavenumber observed is less than that predicted by linear theory (Hart, 1981). Hart explains this discrepancy as follows: at finite amplitude, the longer waves modify the mean flow more slowly than the shorter waves; consequently the shorter waves cut off the power source for their growth by more effectively reducing the vertical shear. In the lower symmetrical flow regime the temperature difference is small and the flow is weak, and viscous boundary layers are thought to keep waves from forming (e.g., Holton, 1979; p 278). For higher heating rates the eddies redistribute enough heat to significantly alter the static stability of the fluid. Static stability increases as temperature difference increases until baroclinic eddies are stabilized and only the symmetric overturning can accomplish the heat transport.

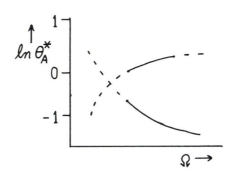

FIG. 7.2b   (b) Approximate solutions, derived from theoretical calculations by Lorenz (1962, 1963), illustrating the location of the main transitions between symmetric and asymmetric modes on a regime diagram. The shapes of the two solid lines approximate the heavy solid line in (a) for certain nondimensional parameter ranges.

For a given temperature differential, if one slowly increases the rotation rate, then the fluid moves through parameter space along a diagonal line from upper left to lower right. (Illustrative diagonal lines are drawn for various ΔT.) If the fluid is in the region characterized by waves, then this diagonal path on the diagram leads to smaller and smaller wavenumbers characterizing the flow. Eventually, a second heavy line (Figure 7.2a) is crossed beyond which the flow is very irregular.

The general shape of the transition curve between symmetric and wavy flows can be approximated by examining the analytic results in Lorenz (1962, 1963). The heavy line in Figure 7.2a is similar to the critical curve for instability of a wave on a zonally symmetric flow. Lorenz finds an equation governing that critical curve to have the form

$$\Delta\Theta = \Omega^{-1}G$$

where

$$\Delta\Theta^\star = \Delta\Theta + \left(\Delta\Theta\right)^3$$

and where $\Delta\Theta^\star = \Delta T$ is the imposed temperature difference across the annulus. G contains a polynomial in both the numerator and denominator. Both polynomials are sixth-order in the temperature difference across the domain. Some progress can be made by examining certain axis ranges in the regime diagram and using parameter values given in Lorenz (1962). For $\Delta\Theta < 0.5$, $\Delta\Theta \simeq \Delta\Theta^\star$ and $\Delta\Theta \approx \Omega^{-1}$. This curve has a shape like that plotted as the lower heavy line in Figure 7.2b. For $\Delta\Theta \simeq 1$ then $\Delta\Theta^\star \simeq 2\Delta\Theta$ and

$$\Delta\Theta \approx \left(\Omega^2 + \Lambda\right)^{-\frac{1}{2}}$$

where Λ is a constant determined by the wavenumbers of the wave involved. This curve has a shape similar to the upper curve plotted in Figure 7.2b.

The types of flows encountered as one increases the rotation rate are illustrated in Figure 7.3. Experiments by Buzyna et al. (1984) are indicated by dots in Figure 7.3a. The numbers in parentheses refer to the number label of each experiment; the number to the right is the principal wavenumber observed. Figure 7.3b shows typical snapshots of the temperature field at mid-depth. Experiments 75 and 28 are in a regime where the flow undergoes increasingly large wave "amplitude vacillation" as the rotation rate increases. Amplitude vacillation is a regular, cyclic change in eddy amplitude over time; the corresponding plots in Figure 7.3b are at the midpoint of the cycle. At higher rotation rates, the wave energy has reached an approximate maximum and the *shape*, more than amplitude, varies with time. The wave amplitude is similar in experiments 76 and 82; but the isotherms become more concentrated near the sidewalls as rotation rate increases. As a result, $[u]$ develops maxima near each sidewall instead of in the middle of the channel. For high rotation rates the flow becomes progressively more turbulent. The isotherms in experiment 80 not only show peak amplitude in a higher wavenumber, but the energy is spread more broadly across the wavenumbers. In this regime the circulation is chaotic; wavenumber amplitudes vary irregularly in space and time.

Much recent work with laboratory models focused upon interactions with bottom topography. As a link to the following section, some effects of topography upon baroclinic waves are listed below. Most studies used a sinusoidal bottom surface that has amplitude up to about ten percent of the total fluid depth; the topography forces a long wave. Leach (1981) found that the forced wave was confined to the bottom for symmetric circulations and extended through greater depth when baroclinic waves were present. Jonas (1981) found that topography increased the symmetric regime, as if the bold solid line in Figure 7.3a were slid diagonally downward along the isotherms. He also found that the remaining unstable waves have higher wavenumber when topography is added: wavenumber increases of 1 or 2 were typical. Li et al. (1986) examined a wavenumber 2 bottom topography that forced a long wave whose ridge was centered upstream of each ridge top. The baroclinic eddies tended to concentrate downstream from the ridge. Increasing the rotation rate (holding ΔT fixed) caused the eddy amplitude to increase (up to a maximum). As a consequence, $[u]$ decreased and so did the eddy phase speed. The same effects were seen in a follow-up report (Pfeffer et al. , 1989) where particular attention was paid to the long wave; the long wave was displaced upstream further (asymptotically) as the rotation rate increased.

7.2 Forced planetary waves

A prominent long wave pattern is identifiable in observed variables of the Northern Hemisphere (§5.2, §5.3, §5.4). Given that these planetary waves are present after time averaging, one can regard them as stationary, at least compared to the shorter wavelength, traveling cyclonic storms. Since long waves show up on a climate mean, one would anticipate that these waves are forced by phenomena that vary slowly in

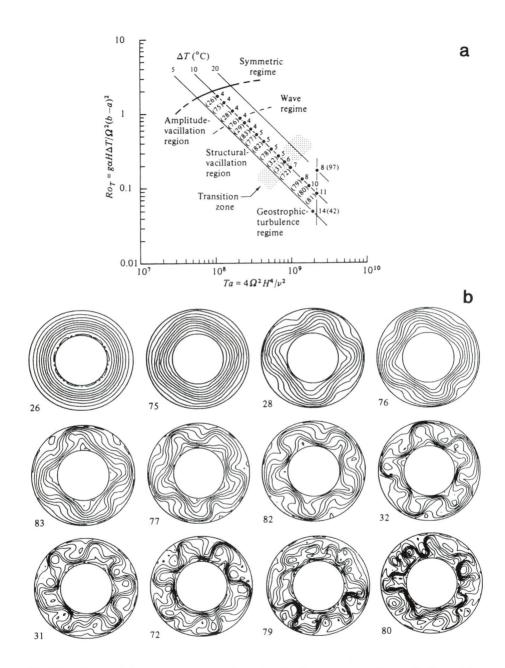

FIG. 7.3 Characteristic structures encountered at various locations on a regime diagram like Figure 7.2a. (a) Regime diagram showing locations in parameter space of experiments (indicated in parentheses) and the principal wavenumber seen in the given experiment. Rotation rate is varied, $\Delta T = 10\,\text{K}$. Ro_T is the thermal Rossby number; T_a the Taylor number. (b) Typical structures illustrated by isotherms at mid-depth in the annulus. Numbers refer to experiments indicated in (a). From Buzyna et al. (1984).

time, if at all. This line of reasoning suggests two prime candidates for maintaining the planetary waves. The first possibility is that the long waves are deflections in the flow that are caused by large-scale mountain ranges. Topography was the first mechanism to be examined theoretically. The seminal papers are by Charney and Eliassen (1949) and Bolin (1950). The other prime candidate is the contrasting thermodynamical properties of the earth's surface between land and ocean. The key paper for the second mechanism is that by Smagorinsky (1953). On the time scales under consideration here, these two mechanisms vary slowly in time.

The maintenance of the long waves by these two mechanisms, orographic and thermal forcing, turns out to be quite complex. For example, one expects there to be nonlinear feedbacks among the long waves and such surface properties as sea surface temperature, snow or ice coverage of land and ocean, surface roughness, soil wetness and vegetation, to name some obvious properties. In addition, the vertical distribution of heating could play an important, nonlinear role, especially in maintaining the upper-level amplitude of the planetary waves. Our knowledge and models of how the radiative and latent heating behave may be crude, according to Held (1983, p. 128). Of course, the shape of the large-scale topography is known quite well. Yet, the atmospheric *response* to orographic forcing can be complex.

The complexity of the orographic problem can be illustrated in two ways. First, a model that includes moisture is likely to generate nonzonal sources of heating when air condenses as it is forced over mountain ranges. Is that a topographic or a diabatic forcing? Second, while the topography certainly does not change with the season, seasonal variations in the stability, celerity, and direction of the flow encountering the mountains can elicit quite different atmospheric responses. The Northern Hemisphere long-wave pattern seasonal changes might be explainable in terms of such atmospheric changes. However, the proximity of the upper-level trough (Figure 5.9) to eastern North American during summer suggests that the Baffin and Greenland Islands' ice sheets play causal roles. If so, is this a diabatic process forcing a planetary wave, or does the high elevation of the Greenland plateau play some role?

A third maintenance mechanism is based on the notion that long waves clearly influence and are influenced by the traveling wave-cyclone storm tracks. Cyclones prefer certain regions for development. Where mature, the transient eddies have magnitude comparable to the long waves. The mature, large amplitude stage in the life cycle of the traveling wave-cyclones has only recently been considered theoretically, and much remains to be understood about the dynamics and thermodynamics of storms at this stage.

Back in 1983, Held concluded that orographic forcing was the dominant mechanism maintaining the long waves, in part because the diabatic heating distribution was poorly known. However, he also suggested (p. 160) that the surface heating would have little effect above 850 hPa. One might conclude that the thermal forcing would have difficulty maintaining the planetary waves since the waves have large amplitudes in the upper troposphere. However, that conclusion would contradict some other studies (e.g., Saltzman, 1968). In more recent papers, thermal forcing has regained some importance (e.g., Held and Ting, 1990; if the low-level flow is weak). The current view (e.g., Valdes and Hoskins, 1989) is that topography, diabatic heating, and eddy feedbacks play important roles in the maintenance of the stationary long wave pattern. While one mechanism may have greater effect in one instance, all three need to be discussed.

The seminal papers on orographic and thermal forcing are detailed in the next two subsections. The third mechanism, transient eddy forcing, is discussed briefly. The fourth subsection weighs the relative contributions by the three mechanisms. Finally, issues of energy propagation raised in the discussions of the mechanisms are touched upon.

7.2.1   An early study of topographic forcing

Perhaps the simplest model to predict the distortion of a flow caused by a mountain barrier is one that conserves potential vorticity. Potential vorticity, Q, can be defined in many ways depending upon the approximations used. A very simple definition of potential vorticity, Q_{sw}, arises in the shallow water equations. (A derivation and related discussion are in Holton, 1979, pp. 87–92.) One could also use an adiabatic form of the potential vorticity, Q_a.

$$Q_{sw} = \frac{\zeta + f}{\Delta P}$$

$$Q_a = \zeta_a \cdot \nabla\theta \approx \zeta_3 \frac{\partial\theta}{\partial P} \approx \Delta\theta \frac{\zeta + f}{\Delta P_\theta}$$

(7.17)

where ζ is relative vorticity vertical component, f is the Coriolis parameter ("planetary" vorticity) and ΔP is the thickness of a shallow layer of fluid. The vector absolute vorticity is ζ_a, and ΔP_θ is the thickness between two theta surfaces. For adiabatic flow, a column of air will remain between two theta surfaces, and one easily sees the similarity between Q_{sw} and Q_a in (7.17).

Conservation of Q requires a distortion of a straight flow when that flow encounters a mountain barrier. Figure 7.4 illustrates the result for westerly flow encountering a north-south mountain ridge. Initially, at point A, $f = f_1$ and thickness equals ΔP_1; the flow is straight so $\zeta_1 = 0$. As the air flows up the ridge the column compresses; to balance a shrinking ΔP, the absolute vorticity must decrease; in fact $f < f_1$ and $\zeta < 0$. At the ridge top (point B), $\zeta_2 = 0$, $f = f_2$, and $\Delta P = \Delta P_2$. It is easy to see why ζ must vanish at the ridge top by considering what happens east of the ridge line. As air flows down the slope, ΔP increases but f is still decreasing since the motion retains a southward component; hence ζ must be greater than zero to balance changes in ΔP and f. Different ridge widths and heights give different curving paths, but all paths have this same characteristic shape. While ascending the mountain, the

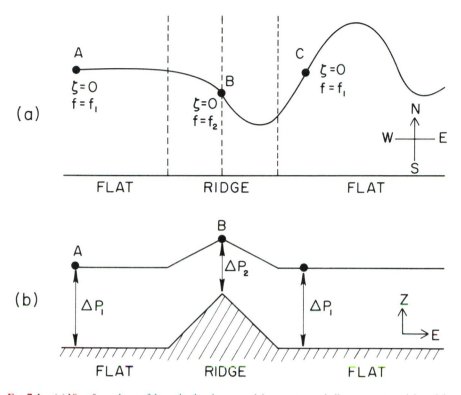

FIG. 7.4 (a) View from above of the path taken by a parcel that conserves shallow water potential vorticity in the Northern Hemisphere. The parcel moves from west to east over a ridge. (b) View of the path, looking north, as a vertical cross-section.

curvature cannot be so great as to cause flow due south because then the decreasing f could not be balanced by a constant ΔP, implying that $\zeta > 0$, which would turn the flow back eastward. As the air flows down the east slope, ΔP increases while f continues to decrease. At some point the cyclonic curvature is great enough so that the trajectory path begins to head northward. That point, where f is a minimum along the trajectory, marks the axis of a trough to the east of the ridge. The actual location of the trough axis varies with the properties of the flow, the variation of f, etc. The trough axis need not lie at the junction of the ridge and flat land, it need only be to the east of the ridge line. Eventually, the air crosses over its initial latitude (point C) where $\Delta P = \Delta P_1$ again, $f = f_1$ again, and $\zeta = 0$ again. However, the trajectory has a northward motion and the parcel overshoots its original latitude. This sets the stage for a trajectory path that oscillates about the original latitude as it moves downstream. In the figure, the oscillation drawn has been damped in order to model the effect of internal friction (which would make Q unconserved). Such damping improves the simulation of the atmosphere in models like the one by Charney and Eliassen (1949).

The situation is different if one assumes that the long waves have infinite meridional extent. This assumption has the effect of eliminating the geostrophic zonal wind associated with the long wave. That case has somewhat different properties than the situation worked out in Figure 7.4. Bannon (1991) presents a conceptual analysis for this more restrictive case. He also includes the additional effect of vertical shear in the mean flow. In Bannon's model the long waves only have meridional motion, so only $\zeta = \partial v/\partial x$ can balance the changes in f and ΔP. In this case $\zeta < 0$ to balance both increasing f and decreasing ΔP as the column ascends the ridge. Consequently, the flow is simpler than in Figure 7.4. High pressure is centered symmetrically on the ridge. This high pressure decreases with height. Bannon shows how westerly shear stretches the column, since the top of the column moves faster than the bottom. The shear stretching partially cancels the compression; since ΔP is not changed as much by traversing the ridge, the high is weaker. The results described by Bannon have a further restriction because they apply only as long as the relative vorticity dominates over the planetary vorticity. An "ultra long" wave might have negligible relative vorticity; since only changes in f could balance the changes in ΔP, then a trough is centered over the topographic ridge. It will be shown that the restrictive case described by Bannon has some relevance to less restrictive flows on a sphere because the periodic domain leads to curving flow, not straight flow, encountering the mountain.

The classic study of topographic forcing of long waves is by Charney and Eliassen (1949). They use the "equivalent barotropic" form of the vorticity equation. That form allows them to model the divergence term in the vorticity equation without needing to solve for a second variable. The divergence term includes a contribution by vertical velocity created when horizontal winds blow up and down topography. Topography appears only through this divergence term. The other contributions to the divergence correction help control excessive retrograde phase speeds that occur for the barotropic vorticity equation. The Rossby wave phase speed C for the barotropic vorticity equation (in one dimension) is simply

$$C = U - \frac{\beta}{k^2} \tag{7.18}$$

where U is the mean flow speed, k is the wavenumber of the Rossby wave, and β is the meridional derivative of f. The analogous Rossby wave phase speed for the equivalent barotropic model (with a flat bottom) is

$$C = \frac{U - \frac{\beta}{k^2}}{1 + \frac{f^2 A_0}{gHk^2}} \tag{7.19}$$

where A_0 is the surface value of a vertical structure function $A(p)$ used for the wind. H is the scale height, which is assumed constant. From (7.18) it is clear that C becomes very large and negative (westward propagation) as $k \to 0$ (for long waves). These large retrograde velocities do not develop in the equivalent barotropic model because k^{-2} also appears in the denominator of (7.19).

The derivation of the Charney and Eliassen model begins with the quasi-geostrophic vorticity equation.

$$\frac{\partial \zeta}{\partial t} + \mathbf{V} \cdot \nabla \zeta + \beta v = f \frac{\partial \omega}{\partial p} \qquad (7.20)$$

where it has been assumed that $\zeta \ll f$ in the divergence term. All vectors are evaluated on constant pressure surfaces. Integrating (7.20) in the vertical and choosing a structure function for the vector velocity obtains

$$\mathbf{V}(x, y, p, t) = A(p)\mathbf{V}(x, y, t)$$

such that the vertical average (indicated here by a tilde) is normalized ($\widetilde{A} \equiv \tilde{A} = 1$). That obtains

$$\frac{\partial \tilde{\zeta}}{\partial t} + \tilde{A}^2 \tilde{\mathbf{V}} \cdot \nabla \tilde{\zeta} + \beta \tilde{v} = \frac{f_0 \omega_0}{p_0} \qquad (7.21)$$

where ω_0 and p_0 are the surface values of pressure velocity and pressure. If a level $p^\star$ is defined at which

$$\mathbf{V}^\star \equiv \mathbf{V}(x, y, p^\star) = \tilde{A}^2 \tilde{V} \qquad (7.22)$$

then multiplying (7.21) by $\tilde{A}^2$ yields

$$\frac{\partial \zeta^\star}{\partial t} + \mathbf{V}^\star \cdot \nabla \zeta^\star + \beta v^\star = \frac{\tilde{A}^2 f_0 \omega_0}{p_0} \qquad (7.23)$$

where the $\star$ superscript refers to the value at the special pressure level and not to deviations from a vertical average as defined with an asterisk in the Appendix. The vertical velocity W is related to pressure velocity ω by

$$gW = \frac{d}{dt}(gZ) \equiv \frac{d\phi}{dt} = \frac{\partial \phi}{\partial t} + \mathbf{V} \cdot \nabla \phi + \omega \frac{\partial \phi}{\partial p} \qquad (7.24)$$

Using the hydrostatic equation, one obtains an expression for ω in terms of ϕ and W.

$$\omega = \rho \left(\frac{\partial \phi}{\partial t} + \mathbf{V} \cdot \nabla \phi - gW \right) \qquad (7.25)$$

The relationship (7.25) is valid for the interior of the domain, but at the surface several other contributors to W are possible. Two main contributors could be Ekman suction or pumping, and flow up and down sloping topography. The former describes the vertical motion at the top of a frictional boundary layer caused by convergence and divergence within that layer. The velocity is cross-isobaric from higher to lower pressure for a three-way balance between diffusion, Coriolis, and pressure gradient terms. The ageostrophic flow causes surface convergence around a low pressure center. Since the vorticity is positive for a low pressure center, one expects upward velocity to be proportional to geostrophic vorticity. Ekman pumping is used in §6.7. For flows up and down slopes, the vertical component of motion is the dot product of the horizontal wind vector with the gradient of the topographic height H_T. Finally, for quasi-geostrophic motion, $\mathbf{V}_0 \cdot \nabla \phi = 0$.

Therefore

$$\omega_0 = \rho \left(\frac{\partial \phi_0}{\partial t} - g(\mathbf{V_0} \cdot \nabla H_T) - \frac{gHr}{f_0} \zeta_0 \right) \tag{7.26}$$

where zero subscripts denote values at the surface and r is a viscosity coefficient. The Coriolis parameter f_0 is constant.

Substituting (7.26) into (7.23) and setting all time derivatives equal to zero (because stationary solutions are sought) yields

$$\mathbf{V^\star} \cdot \nabla \zeta^\star + \beta v^\star = \frac{-\tilde{A}^2 f_0 g}{p_0} \left(\mathbf{V_0} \cdot \nabla H_T + \frac{Hr}{f_0} \zeta_0 \right) \tag{7.27}$$

Assuming that the values of $[\mathbf{V}]$ and ζ at the ground are some fraction, $A_0/\tilde{A}^2$ of their value at the $p^\star$ level, then (7.27) is written

$$\mathbf{V^\star} \cdot \nabla \zeta^\star + \beta v^\star = \frac{-A_0 f_0 g}{p_0} \left(\mathbf{V^\star} \cdot \nabla H_T + \frac{Hr}{f_0} \zeta^\star \right) \tag{7.28}$$

Next, (7.28) is linearized about a basic state zonal mean zonal wind $[u]$ that is constant within the channel domain. Periodic boundary conditions in x (longitude) are assumed. The linearized form of (7.28) becomes

$$[u] \frac{\partial \zeta^\star}{\partial x} + \beta v^\star = -\lambda \big([u] + u^\star \big) \frac{\partial H_T}{\partial x} - \lambda v^\star \frac{\partial H_T}{\partial y} - S \zeta^\star \tag{7.29}$$

It is assumed that the topography varies in y at a rate comparable to or slower than its rate of x variation. It is further assumed that the perturbation zonal velocities ($u^\star$) are small compared to $[u]$.

$$[u] \frac{\partial \zeta^\star}{\partial x} + \beta v^\star = -\lambda [u] \frac{\partial H_T}{\partial x} - S \zeta^\star \tag{7.30}$$

where

$$\lambda = \frac{f_0 g A_0}{p_0} \quad \text{and} \quad S = \frac{Hr\lambda}{f_0}$$

To simplify the solution of (7.30), Charney and Eliassen make the reasonable assumption that the stream function ψ and H_T have the same sinusoidal variation in y.

$$\psi = \Re\{\eta(x)\sin(\ell y)\} \tag{7.31}$$

where

$$\nabla^2 \psi = \zeta^\star \qquad \text{and} \qquad (u^\star, v^\star) = \left(-\frac{\partial \psi}{\partial y}, \frac{\partial \psi}{\partial x}\right) \tag{7.32}$$

Substituting (7.31) and (7.32) into (7.30) yields a nonhomogeneous ordinary differential equation for η.

$$[u]\frac{d^3\eta}{dx^3} + S\frac{d^2\eta}{dx^2} + (\beta - [u]\ell^2)\frac{d\eta}{dx} - S\ell^2\eta = -\lambda[u]\frac{\partial H_T}{\partial x} \tag{7.33}$$

Equation (7.33) can be written as

$$L(\eta) + \epsilon\eta = F(x) \tag{7.34}$$

where the homogeneous form of (7.34) is an eigenvalue problem, where $\epsilon = -S\ell^2$ and $F(x) = -\lambda[u]dH_T/dx$. Homogeneous solutions of (7.34) are labelled ϕ_n and ϵ_n. The structure function η is expanded in terms of the eigenfunctions.

$$\eta(x) = \sum_{n=-\infty}^{\infty} c_n\phi_n \tag{7.35}$$

When (7.35) is substituted into (7.34), the result is

$$\sum c_n(\epsilon - \epsilon_n)\phi_n = F(x) \equiv \sum b_n\phi_n \tag{7.36}$$

where the forcing F has been expanded in terms of the same eigenfunctions ϕ_n. Formally, the expansion is correct if one uses solutions to the self-adjoint form of the problem. From (7.36) it is clear that

$$c_n = \frac{b_n}{\epsilon - \epsilon_n} \tag{7.37}$$

as long as ϵ is not an eigenvalue. The case when ϵ equals a particular eigenvalue can be handled if b_n equals zero. The b_h are determined from F by an inversion such as

$$b_n = (1/R)\int_\alpha^\beta F(x)\hat{\phi}_n(x)dx \tag{7.38}$$

where $\hat{\phi}$ is the adjoint eigenvector and R is the range ($R = \beta - \alpha$). Substitution of (7.37) into (7.34) and interchanging the sum and integral gives

$$\eta(x) = \int_\alpha^\beta \sum_{n=-\infty}^\infty \frac{F(\xi)}{R} \left\{ \frac{\phi_n(x)\hat{\phi}_n(\xi)}{\epsilon - \epsilon_n} \right\} d\xi \qquad (7.39)$$

where ξ is introduced to avoid confusion with x. The quantity inside the $\{\}$ brackets is a Green's function. In Charney and Eliassen (1949), the eigenvalue ϵ_n is labelled $-\gamma$ and the range is $R = 2\pi$. A Fourier exponential series solution is used to find the eigenvalue. Returning to the homogeneous forms of (7.33) and (7.34) one obtains

$$\sum_{n=-\infty}^\infty \left(-i[u]n^3 - Sn^2 + i(\beta - [u]\ell^2)n \right) b_n \exp(inx)$$

$$= - \sum_{n=-\infty}^\infty \gamma b_n \exp(inx)$$

From the orthogonality condition, the eigenvalue must be

$$\gamma = in^3[u] + Sn^2 - i(\beta - [u]\ell^2)n \qquad (7.40)$$

The specific form of (7.39) can be written

$$\eta(x) = \frac{1}{2\pi} \int_0^{2\pi} \sum_{n=-\infty}^\infty \frac{in\lambda[u]H_T(\alpha)}{\gamma - S\ell^2} \exp(in(x+\alpha)) d\alpha \qquad (7.41)$$

with γ given by (7.40). A factor of in has appeared from evaluating the derivative of H_T. It is understood that the real part of the right-hand side of (7.41) is desired.

The atmospheric response to the earth's topography around the latitude circle 45 N is shown in Figure 7.5a. There are several curves plotted for different values of σ, where $\sigma = S\ell/[u]$. Charney and Eliassen (1949) used $\beta = 2\pi$, $[u] = 17°$ longitude per day, $\lambda = 0.4$, and $\ell^2 \simeq 15$. (The value of ℓ corresponds to a channel width of 33 degrees latitude.) The observed time-mean geopotential height considered by Charney and Eliassen is plotted as the dark solid line in Figure 7.5a. The agreement between this model and observations is really rather remarkable given the simplifications used in this model. Figure 7.4a shows two damped wavetrains extending eastward from the two major mountain ranges. The two wavetrains are more easily seen as the Ekman pumping (σ) increases. Considering the $\sigma = 0.5$ curve, the major troughs are located downstream from the Rockies and the Himalayas. One could anticipate this pattern from the simple potential vorticity arguments that began this section. However, some care is necessary on this point.

In Figure 7.4 it is assumed that the flow upstream of the mountain is straight and zonal. Yet on a sphere the downstream wavetrain could extend entirely around

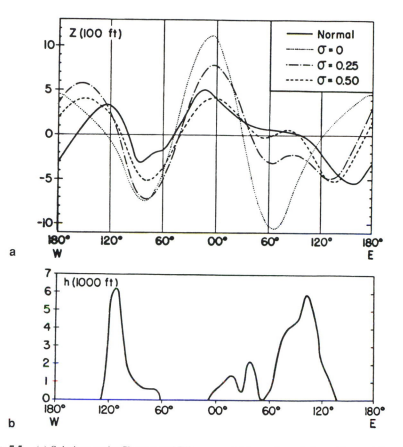

FIG. 7.5 (a) Solutions to the Charney and Eliassen model for various choices of internal viscosity (σ). $[u] = 17°$ longitude day^{-1}. (b) Terrestrial topography at 45 N. From Smith (1979).

a latitude circle and thus the flow impinging upon the mountain could have nonzero vorticity. Consequently, the higher viscosity (larger σ) curves shown in Figure 7.5 agree best with simplistic potential vorticity arguments.

The smaller σ values allow more and more interference between the wavetrains generated by the Rockies and Himalayas. An intuitive feeling for this interference is gained by examining the inviscid version of the Charney and Eliassen model. The denominator of (7.41) reduces to

$$in\left(n^2[u] - \beta + \ell^2[u]\right) \tag{7.42}$$

when viscosity is set equal to zero. Defining an absolute wavenumber k, such that $k^2 = n^2 + \ell^2$, then the sign of (7.42) will be different depending upon whether k^2 is greater or less than $\beta/[u]$. Waves satisfying $k^2 = \beta/[u]$ will be resonant waves.

As pointed out by Smith (1979), it is possible to have "ultra long" waves $k^2 < \beta/[u]$ for which (7.42) is negative. From (7.41) this leads to a negative value of η

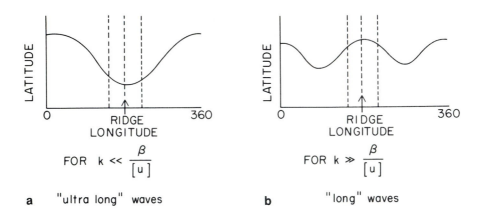

FIG. 7.6 Balanced solutions depend upon the wavelength. (a) For "ultra long" waves, a trough is located over the ridge line. (b) For waves that are merely "long," a high is located over the ridge line.

where H_T is positive; that is, to lower pressure centered over the ridge. For merely "long" waves $k^2 > \beta/[u]$ then (7.42) is positive and high pressure is centered over the ridge. The long wave pattern is superficially similar to the flow analyzed by Bannon (1991). Both patterns are schematically illustrated in Figure 7.6. Absolute vorticity still decreases in order to balance the compression as the air flows up the mountain, but for the ultra long waves this is accomplished mainly by changing the Coriolis parameter since ζ^* for these waves is so small. Hence the relative vorticity of ultra long waves is a positive maximum at the ridge top. Almost all of the change in absolute vorticity is caused by changing f. On the other hand, for the long waves, the relative vorticity change exceeds β. Therefore the decrease in absolute vorticity for the long waves is accomplished mainly by reducing ζ^*. To do this, ζ^* develops a negative maximum at the ridge top.

The demarcation between long and ultra long waves depends upon $[u]$ and by implicit assumption on H_T (since H_T depends upon ℓ). It is likely then that observed planetary waves will exhibit phase shifts that intermix the long and ultra long solutions described here. The observed time-mean 500 hPa height field (Figure 5.9) has a ridge over the Rockies in summer but further west during winter; Figure 7.7 shows this shift. The observed ridge north of the Tibetan plateau shifts the same way as indicated in Figure 7.7, but the opposite way from that shown in the corresponding figure in Smagorinsky (1953). In summer $[u]$ is weaker, so $\beta/[u]$ is larger, implying that some "long" wavelengths become "ultra long" making η negative over the Rockies. This analysis fits the sea-level pressure changes in Figure 7.7 better than at 500 hPa; a ridge is over the topography in winter, a trough in summer.

So, orographic forcing can have seasonal variation that is partly consistent with observations. However, showing the *ability* of a mechanism is far from demonstrating its actual preeminence. The work by Charney and Eliassen (1949) was attacked by

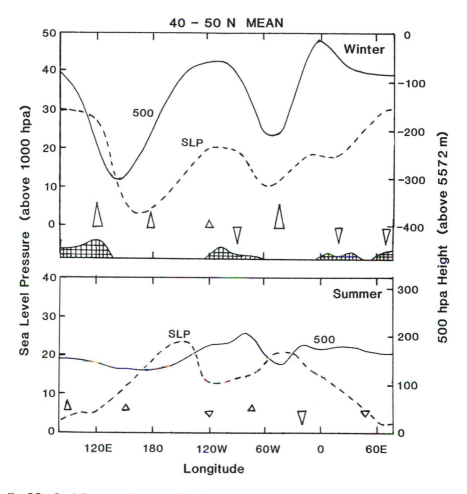

FIG. 7.7 Spatially averaged pattern of 500 hPa geopotential height (dashed lines) and sea level pressure (solid lines). Winter and summer periods used, where data from figures 5.8 and 5.9 have been averaged over 40 N to 50 N latitude. Shaded regions schematically show major topographic features. Upward pointing triangles show major areas of net diabatic heating, integrated from the surface to 50 hPa, as presented in Hoskins et al. (1989). Downward pointing arrows indicate cooling; triangle length is proportional to the magnitude of the heating. This figure updates and is comparable to a similar figure in Smagorinsky (1953).

Sutcliffe (1951), who strongly supported the thermal forcing view. However, even Sutcliffe had to concede that topographic barriers inhibit low-level advection leading to large gradients of moisture and temperature. In other words, orographic and thermal forcing are not cleanly separable mechanisms.

7.2.2 An early study of thermal forcing

The previous section ended by demonstrating that the atmospheric response to orographic forcing could be capable of explaining the seasonal changes in the planetary wave pattern. But the observations (§5.4) during summer show the North American trough centered over the large ice sheets covering most of Greenland and Baffin Islands and little evidence of a trough near east Asia. One might assume that the cold surface temperatures over the ice sheets are somehow mixed through most of the troposphere. And one might assume that the trough over east Asia withers away during summer due to melting of the snow cover over Siberia and consequent warming. It is easy to see how heating the surface can lead to a vertical redistribution of the heating since surface heating destabilizes the lapse rate. But strong surface cooling should *stabilize* the lapse rate and therefore inhibit vertical mixing.

One mechanism to overcome the stabilization might be frictional convergence around a surface low that is surrounded by cold air. This situation was identified during the decaying, "occluded" stage of the wave–cyclone (Figure 4.22). The surface convergence forces cold air to rise. Interestingly enough, the surface pressure pattern (Figure 5.8b) does show persistent low pressure centered near Baffin Island. Because the trough axis is vertical, the low is decaying by converting kinetic to potential energy, and one would need a mechanism to bring active lows into the area so that they might occlude there. This process could be viewed as an example of transient forcing; §7.2.3 pursues these points further. Next, the earliest theoretical study of thermal forcing of planetary waves is presented.

The earliest study is that by Smagorinsky (1953). His model is derived, some results shown, and more recent extensions and critiques are described. A vertical component quasi-geostrophic vorticity equation is used, similar to (7.20), but using height as the vertical coordinate.

$$\frac{\partial \zeta}{\partial t} + \mathbf{V} \cdot \nabla \zeta + \beta v = \frac{f}{\rho} \frac{\partial}{\partial z} (\rho w) \tag{7.43}$$

The dependence upon vertical velocity is eliminated by incorporating the thermodynamic equation (7.8) but with diabatic heating Q.

$$\left(\frac{\partial}{\partial t} - \frac{\partial \psi}{\partial y} \frac{\partial}{\partial x} + \frac{\partial \psi}{\partial x} \frac{\partial}{\partial y} \right) \frac{\partial \psi}{\partial z} + \frac{w}{\epsilon} = g \frac{Q}{f} \tag{7.44}$$

As in §7.1.1, $\epsilon = f^2 L^2/(g\kappa D)$ and κ is the horizontal mean static stability: $\kappa = \partial(\ln \theta)/\partial z$.

Smagorinsky examines the heating function

$$Q(x, y, z) = N \exp\left(\frac{-z}{h}\right) \sin\left(\frac{\pi z}{z_T}\right) \sin(kx) \sin(my) \tag{7.45}$$

where N is a normalization factor. From (7.45) it is clear that the heating vanishes at the surface ($z = 0$) and the tropopause ($z = z_T$). The heating is a maximum near the level h.

The vertical variation of the density is assumed to be exponential.

$$\rho = \rho(y)\exp(\frac{-z}{\tilde{H}}) \qquad (7.46)$$

where $\tilde{H} = \tilde{T}(g/R - \gamma)^{-1}$ and γ is a prescribed, constant lapse rate. $\gamma = 6.5$ K km^{-1} is used, and $\tilde{T} = 260$ K is the mean temperature.

Substituting (7.46) into (7.43) and substituting for w from (7.44) yields

$$\left(\frac{\partial}{\partial t} + \mathbf{V}\cdot\nabla\right)\left\{\zeta - \frac{f\epsilon}{\tilde{H}}\frac{\partial\psi}{\partial z} + \epsilon f\frac{\partial^2\psi}{\partial z^2}\right\} + \beta v = \epsilon g\frac{\partial Q}{\partial z} - \frac{\epsilon g}{\tilde{H}}Q$$

A linearization is made about a zonal current $U = U(0) + \Lambda z$ that is a linear function of z. Steady state solutions ψ are sought. Therefore,

$$U\frac{\partial}{\partial x}\left(\nabla^2\psi - \frac{f\epsilon}{\tilde{H}}\frac{\partial\psi}{\partial z} + f\epsilon\frac{\partial^2\psi}{\partial z^2}\right) + \frac{\Lambda f\epsilon}{\tilde{H}}\frac{\partial\psi}{\partial x} + \beta\frac{\partial\psi}{\partial x}$$
$$= g\epsilon\frac{\partial Q}{\partial z} - \frac{g\epsilon}{\tilde{H}}Q \qquad (7.47)$$

Consistent with the variation assumed for Q, Smagorinsky assumes a similar variation for the solution

$$\psi = \sin(my)\,\Re\{\phi(z)\exp(ikx)\} \qquad (7.48)$$

This solution reduces (7.47) to the form

$$Uik\left(-(k^2 + m^2)\psi - \frac{f\epsilon}{\tilde{H}}\frac{\partial\psi}{\partial z} + f\epsilon\frac{\partial^2\psi}{\partial z^2}\right) + \left(ik\beta + \frac{ik\Lambda f\epsilon}{\tilde{H}}\right)\psi$$
$$= g\epsilon\frac{\partial Q}{\partial z} - \frac{g\epsilon}{\tilde{H}}Q \equiv F(z)\sin(kx)\sin(my) \qquad (7.49)$$

where (7.49) defines the net forcing F.

The interior equation (7.49) is a nonhomogeneous, second-order ordinary differential equation for ψ. The interior equation can be further simplified. By choice, the heating distribution ($\sin\{my\}$) term can be factored out. Second, (7.49) can be separated into expressions for the real (ϕ_r) and imaginary (ϕ_i) parts of ϕ. ϕ_r and ϕ_i are real-valued functions of height. Making this substitution into (7.49) yields terms that include either $\sin(kx)$ or $\cos(kx)$. By the orthogonality of sines and cosines, separate equations for the real and imaginary coefficients are obtained. In addition, only one of these equations includes the diabatic heating. The resulting two equations are as follows.

For the real part of ϕ,

$$
\begin{aligned}
&- \epsilon fUk\frac{\partial^2 \phi_r}{\partial z^2} + \frac{\epsilon fUk}{\tilde{H}}\frac{\partial \phi_r}{\partial z} + \\
&\left(kU\left(k^2 + m^2\right) + \frac{k\Lambda f\epsilon}{\tilde{H}} - k\beta \right)\phi_r = F(z)
\end{aligned}
\tag{7.50}
$$

where the common factor of $\sin(kx)$ has been divided out of both sides of (7.50). For the imaginary part of ϕ,

$$
\begin{aligned}
&- \epsilon fUk\frac{\partial^2 \phi_i}{\partial z^2} + \frac{\epsilon fUk}{\tilde{H}}\frac{\partial \phi_i}{\partial z} + \\
&\left(kU\left(k^2 + m^2\right) + \frac{k\Lambda f\epsilon}{\tilde{H}} - k\beta \right)\phi_i = 0
\end{aligned}
\tag{7.51}
$$

In (7.51) the common factor of $\cos(kx)$ has been eliminated. Both the x and y dependencies have been removed; only dependence upon z remains. The real and imaginary parts decouple: ϕ_r appears only in (7.50) and ϕ_i is only in (7.51). The equations are now much easier to solve analytically. Since (7.50) and (7.51) remain second-order differential equations, two boundary conditions are needed to obtain a solution.

The boundary conditions could take several forms. The bottom boundary condition could include Ekman suction and pumping, flow up and down orographic slopes, and heat transfer across the earth's surface. The thermodynamic equation could specify the conditions by setting w equal to the Ekman velocity plus the slope flow. Linearizing and assuming steady-state conditions obtains

$$
\begin{aligned}
Uik\frac{\partial \psi}{\partial z} + \Bigg\{ & \frac{\epsilon f^2}{g}\left(ik\frac{\partial H_T}{\partial y} - m\cot(my)\frac{\partial H_T}{\partial x} \right) \\
& - \frac{fD}{\epsilon}\left(k^2 + m^2 \right) - ik\Lambda \Bigg\}\psi = \frac{gQ}{f} - \frac{\epsilon f^2 U}{g}\frac{\partial H_T}{\partial x}
\end{aligned}
\tag{7.52}
$$

where H_T is the topography and D is the Ekman layer depth. Equation (7.52) is similar to the form Roads (1982) uses. Smagorinsky uses a flat bottom ($H_T = 0$) in his 1953 study. The top boundary condition Smagorinsky uses is of two types: in one case he places a rigid lid at $z = z_T$; in the other case he includes a "stratosphere" in the upper part of his domain with an unspecified higher static stability. (Static stability enters into this problem via the definition of ϵ.) In Smagorinsky's cases the top boundary condition has the same mathematical form as the bottom. Roads (1982) reexamines the Smagorinsky model and concludes that a radiation condition

$$
\frac{\partial \psi}{\partial z} = 0
\tag{7.53}
$$

as $z \to \infty$ is more appropriate than the rigid lid.

At this point, several choices are available for solving (7.49) or (7.50) and (7.51) subject to the boundary conditions. Smagorinsky (1953) uses a formulation similar to (7.50) and (7.51). However, instead of evaluating the common x derivative, he uses geostrophic meridional velocity on the left-hand side of (7.50) and (7.51). He also splits the boundary condition (7.52) into separate real and imaginary equations. The homogeneous form of his equations is analogous to the system solved by Charney (1947) in a classic study of baroclinic instability. The resulting solution is a pair of hypergeometric series.

Roads (1982) chooses a different approach so as to improve the interpretation of the analytical result. He transforms (7.49) into a form that could be solved using a WKB approximation. The solution involves Airy functions whose integrals are best solved numerically using the Langer uniform approximation. (Further details are in the article by Roads.)

A representative solution found by Smagorinsky is shown in Figure 7.8. The solution is a wave with longitudinal wavelength of 140° and meridional wavelength of 53.9°. The heating function Q is a positive maximum at the +40° longitude and negative minimum at the −40° longitude. The upper level trough lags the surface trough by about 25° longitude. He concludes that this agrees well with the climatological locations of the surface (Aleutian and Icelandic) lows relative to the upper-level troughs (over the east coasts of Asia and North America; Figures 5.8 and 5.9, respectively). This phase tilt can also be seen in Figure 7.7a, if allowances are made for the orientation of the Icelandic low in Figure 5.8.

To understand the tilts of the pressure axes and vertical velocity, one must include the Ekman pumping and suction as well as the stream function definition. Figure 7.8c illustrates the physical mechanism. The heating is a maximum at Q_m. Intuitively, one expects a source of heating to cause expansion of the air as indicated by the double-shafted arrows. At the top of the Ekman layer ($z = 0$) upward motion is centered over the surface trough (**L**). East of the trough (point **A**) vertical velocity should increase with height from the continuity equation because $\partial u / \partial x$ is negative. The lengthening vertical arrows show this. Below the maximum heating at Q_m one expects the diabatic expansion to oppose the Ekman pumping, making $\partial w / \partial z$ negative; hence $\partial u / \partial x$ must be positive; that is consistent with point **B** being west of the trough. Point **C** is located far enough away from the surface low for Ekman pumping to be cancelled by the thermal expansion. Point **D** is located where the diabatic heating crosses through zero and where the upper level w is greatest. If the diabatic heating contributes only to *horizontal* divergence ($\partial w / \partial z = 0$) above 2 km (e.g., at point **E**), then the vertical velocity must be greatest to the east of the trough. In Smagorinsky's stratosphere case shown, this is at point **D**.

The triangles in Figure 7.7 show locations of heat sources and sinks. Smagorinsky considers three possible types: latent heat release (e.g., west coast of North America), sensible heat transfer from the oceanic western boundary currents, and diabatic heating over the continents. He deduced the diabatic heating from geostrophic temperature

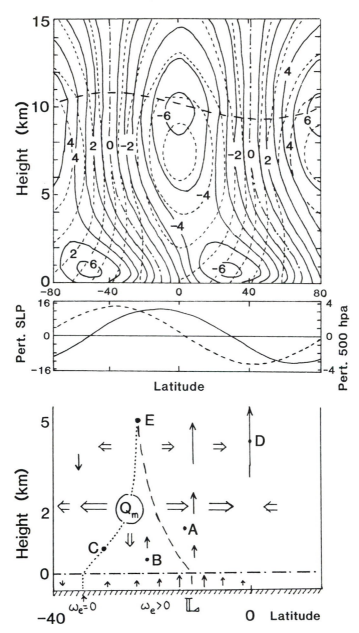

FIG. 7.8   (a) East-west cross-section through the "stratosphere" case solution shown by Smagorinsky (1953). Solid contours are northward velocity (v); dashed contours are vertical velocity (positive is upward motion). (b) 500 hPa height (dashed) and sea level pressure (solid) patterns. As in (a), this diagram is redrawn from Smagorinsky (1953). The heating is a maximum at almost 2 km elevation and longitude +40°. Heating is a minimum at −40°. (c) Schematic explanation of the tilts of the w and ψ axes seen in (a). Single-shafted arrows show vertical motion with length proportional to magnitude. Double-shafted arrows show divergence postulated to result from the diabatic heat source (maximum at "Q_m"). Points **A** through **E** are discussed in the text. The top of the Ekman boundary layer ($z = 0$) is the dot-dashed line. The axis of $w = 0$ is dotted; the trough axis is a dashed line. The Ekman pumping velocity at $z = 0$ is w_e. "**L**" denotes the location of the surface low pressure center.

advection which may explain the differences between his Figure 7 and the diabatic heating shown in Figure 7.7 here. (The principal differences are at 20 E and 120 E in winter and 150 E in summer.) The observed principal troughs are located off the east coasts of the continents, a location of prominent (sensible) heating associated with the western boundary currents. Smagorinsky argues that the oceanic heat fluxes are consistent with his model: the upper trough is above the maximum heating. In addition, the surface trough is about 25 degrees east both in observations (his Figure 7) and in his model (Figure 7.8). The agreement for other diabatic heating is not as good, however. Smagorinsky notes that heating in the middle troposphere would largely result from latent heat release. The horizontal distribution of latent heating should be essentially the same as the pattern of precipitation. Climatologically, there is a relative maximum in rainfall along the northwest coast of North America. However, the diabatic heating deduced from more recent observations has a stronger maximum along the storm tracks (Figure 5.19b) than along the west coast of North America. The largest net diabatic heating is maximum well upstream, near 140 E (Hoskins et al., 1989). Such a heating field improves the simulation by the Smagorinsky model.

Smagorinsky seeks to avoid solutions that are too close to resonance. (He incorporates surface friction in the hope of avoiding resonance, but Dickinson (1980) and Roads (1982) point out that resonance is not necessarily avoided. If $U(0)$ goes to zero, then the surface friction does little to dampen the resonance.) Smagorinsky finds that the further his solution is from resonance the less upper-level amplitude it has. Yet planetary waves have large upper-level amplitude (e.g., Figure 7.7). From the hydrostatic equation an approximate conversion may be derived at the surface that 1 hPa = 8 m; clearly, the planetary wave amplitude is much larger at 500 hPa. From these results Smagorinsky concludes that the thermal forcing might play a role but that orographic forcing was likely to be more important.

Shutts (1987) argues that the problem is not resonance but the specification of the diabatic heating. Shutts concludes that a forced resonant wave or a nonlinearly equilibrated wave (such as a "modon"; McWilliams, 1980) could be in *balance* with the diabatic heating. The diabatic heating is then thought of as creating a thermal equilibrium state where the wave-like response is part of that state. This differs from Smagorinsky's more common approach where only the *difference* from thermal equilibrium is specified as the diabatic heating. Shutts gives a simple example where the heating is interactive with the wave (by using Newtonian cooling as suggested by Döös, 1962; e.g., Equation (6.54). When near resonance, the wave amplitude is much larger but bounded (unlike Smagorinsky's specification) and the net amount of diabatic heating is zero. This type of approach, where equilibration between heating and response is possible, has been used in Mitchell and Derome (1983), for example. Shutts further speculates that orographic studies could have a similar problem when a specified flow is forced over, rather than around a mountain; in that case the orographic response might be overestimated. On the other hand, the flow around the mountain would be an example of "orographic equilibration" because that non-zonal flow around the mountain appears in the stationary eddy field (at low levels) but does not appear in a forcing term (because the slope flow vanishes).

7.2.3   Transient eddy forcing

In the previous two subsections, compelling evidence has been developed to link orographic and diabatic processes to the long waves. However, the story cannot end there. Otherwise, it would contradict our synoptic experience that frontal cyclones form in favored regions and subsequently reach large amplitude in favored regions. This subsection explores the idea that transient eddies contribute to the climatological long waves. One can imagine several ways in which eddies can contribute. A high frequency of occurrence of large amplitude lows in specific locations would appear as a time-mean trough. Transient eddies tend to stall when they reach large amplitude and often merge with the "climatological" lows. The nonlinear, turbulent energy cascade from cyclone-scale waves goes both to longer and shorter wavelengths. The eddy fluxes of momentum, if not heat, may force lower heights in the region of a long wave trough.

Nonlinear interactions between waves cause energy to be transferred between wavenumbers. One way to show the energy cascade is to examine a one-dimensional, barotropic flow having three wavenumbers: k_1, k_2, and k_3. The cascade can be easily shown if one makes the reasonable assumption that enstrophy (V) and kinetic energy (E) are conserved. The enstrophy of a given wavenumber is proportional to the kinetic energy of that wavenumber.

$$\left(\nabla^2 \psi_i\right)^2 \equiv V_i = k_i^2 E_i \equiv k_i^2 \left(\nabla \psi_i \cdot \nabla \psi_i\right) \tag{7.54}$$

where ψ_i is the stream function for wave k_i. The conservation of enstrophy and energy for the three-wave system becomes

$$k_1^2 \frac{\partial E_1}{\partial t} + k_2^2 \frac{\partial E_2}{\partial t} + k_3^2 \frac{\partial E_3}{\partial t} = 0 \tag{7.55}$$

$$\frac{\partial E_1}{\partial t} + \frac{\partial E_2}{\partial t} + \frac{\partial E_3}{\partial t} = 0 \tag{7.56}$$

It is straightforward to combine these two equations to obtain relations between the energy tendencies for each wavenumber. For example, one eliminates E_1 by combining (7.55) and (7.56) and by then combining that with the result of eliminating E_2 from the same equations. Dividing by $(k_1^2 - k_3^2)(k_2^2 - k_1^2)$ yields

$$\underbrace{\frac{1}{k_3^2 - k_2^2} \frac{\partial E_1}{\partial t}}_{>0} = \underbrace{\frac{1}{k_1^2 - k_3^2} \frac{\partial E_2}{\partial t}}_{<0} = \underbrace{\frac{1}{k_2^2 - k_1^2} \frac{\partial E_3}{\partial t}}_{>0} \tag{7.57}$$

It should be clear that a change in energy by the middle wavelength is balanced by opposite changes to the shorter and longer wavelengths. The amounts of the energy transfers depend on the specific wavenumber combinations, but the important point is that the cascade goes both ways. Fjortoft (1953) was the first person to arrive at this conclusion. A similar conclusion follows for a baroclinic system of equations (e.g., Pedlosky, 1987; §3.27).

If one accepts that midlatitude cyclones grow by baroclinic instability, which selects most unstable wavenumbers around 5 to 8, then one would expect to see the cyclones transferring some of this energy up scale to the long waves. Shepherd (1987) finds that stationary long waves draw energy from transient, shorter waves in observational data (§5.9). Cai and Mak (1990b) have used a Cartesian geometry, channel model with linearly varying f to investigate this connection further. They find that energy transfers from a diabatically-forced zonal mean flow to a baroclinically unstable "synoptic" wave that in turn transfers energy (mainly barotropically) to a longer "planetary" wave.

Gall et al. (1979) find up-scale energy transfers in a GCM that has neither orography nor diabatic heating. Nonlinear interactions are still present, and they find long waves to be forced by shorter, unstable waves. The presence of the long wave causes zonal variations in the baroclinic instability, leading to a spectrum of moderate and intense vortices. One would expect "long waves" to appear in a Fourier decomposition of the height field around a latitude circle when there are just a few, intense, localized vortices. Two other factors lead to the appearance of longer waves. First, some lows merge (leading to a larger low). Second, most lows migrate to a higher latitude; if the dimensional size does not change, then the "wavenumber" of the low will decrease. Observations support these properties. Contrary to the impression created by the band-passed height variance used in §6.5 (Figure 6.19c), the storm tracks are not straight. Instead, the tracks are often shaped like a shepherd's hook—they tend to start out straight (roughly following the time-mean flow downstream from the long wave trough) and then bend poleward, sometimes westward, as the cyclone gathers large amplitude (e.g., Grotjahn and Lai, 1989). Not all storms do this, and the median path varies from year to year, but many storms do.

Observational studies of transient forcing of long waves initially focused upon fluxes of momentum and heat by the rotational part of the flow. All the early studies examined Northern Hemisphere winter patterns. Youngblut and Sasamori (1980) considered the potential vorticity equation and stressed that nonlinear advection by the transient and stationary eddies must be included along with orography and thermal forcing. They found that eddies tended to *dissipate* the long waves on the whole, though in some regions the eddy fluxes may reinforce the long wave pattern. When just the momentum fluxes were considered, Holopainen and Oort (1981) found that the eddies have a more positive role. On a global average the transient eddies build the zonal mean flow but destroy the standing eddies. However, upon closer examination of the horizontal patterns, Holopainen and Oort (1981) found transient fluxes (using the $\bar{\zeta}$ equation) reinforcing the Asian trough and to a lesser extent the North American trough. However, in a follow–up study Holopainen et al. (1982) found a result more like Youngblut and Sasamori. Lau and Holopainen (1984) subdivided the eddy contributions into low ($>$ 10 days) and band-pass (2.5 to 6 day) frequency ranges. The low frequency eddies exceed the band-pass forcing of the time-mean geopotential field, especially at the eastern end of each storm track: deep upper-level lows were generated near the climatological Aleutian and Icelandic lows, consistent with transient waves that slow down, have trough axes tilted more vertically and merge (at the surface). The higher frequency transients induce a ridge near the start of the storm track, consistent with the baroclinic instability draining off long wave energy. They find that the high and lower frequency changes to the time-mean geopotential partially cancel, as did Holopainen (1984).

The eddy heat fluxes tended to dominate the eddy momentum fluxes; since the heat fluxes tend to be down the gradient (Figure 5.4), the long wave is diminished. However, as discussed in §6.5 (and later in §7.3) the eddy heat fluxes would diminish as the eddy passes maximum amplitude and decays. The baroclinic conversion could become negative (§4.5.2; Figure 4.22) and the merging of the decaying low with the parent long wave trough could supply vorticity to reinvigorate the long wave. Trough merging is frequently seen on weather maps; it appears in at least one diagnostic, the E vector.

The E vector has roots in several early studies; it is extensively analyzed in Hoskins et al. (1983). The E vector is not a true vector, but it does have an advantage over calculating the transient eddy terms in the potential vorticity equations (as done by Holopainen et al. , 1982). Those eddy terms contain many derivatives and consequently are quite noisy. The derivation of the E vector is easily seen from the vorticity equation (e.g., Holton, 1979; p. 128). The flux form of horizontal advection is divided into time mean and transient contributors.

$$\frac{\partial \zeta}{\partial t} = -\nabla \cdot \left(\overline{\mathbf{V}} \, \bar{\zeta} \right) - \underbrace{\nabla \cdot \left(\mathbf{V}' \zeta' \right)}_{(A)} + \left(\frac{\partial u}{\partial p} \frac{\partial \omega}{\partial y} - \frac{\partial v}{\partial p} \frac{\partial \omega}{\partial x} \right) - \omega \frac{\partial \zeta}{\partial p} \qquad (7.58)$$

Hoskins et al. focus on term A and neglect the other terms. Term A includes both the rotational ($\mathbf{V}'_{\psi}$) and divergent ($\mathbf{V}'_{\chi}$) parts of the transient wind. After applying a time average, the time derivative in (7.58) vanishes and the time mean vorticity is maintained (against friction, say) by term A.

$$\begin{aligned} \overline{A} &= \nabla \cdot \left(\overline{\mathbf{V}'_{\psi} \zeta'_{\psi}} \right) + \nabla \cdot \left(\overline{\mathbf{V}'_{\chi} \zeta'_{\chi}} \right) \\ &= \frac{\partial^2}{\partial x^2} \left(\overline{u'v'} \right) + \frac{\partial}{\partial x} \left(\overline{v' \frac{\partial u'}{\partial x}} \right) - \frac{1}{2} \frac{\partial^2 \overline{u'^2}}{\partial x \partial y} + \frac{1}{2} \frac{\partial^2 \overline{v'^2}}{\partial x \partial y} \\ &\quad + \frac{\partial}{\partial y} \left(\overline{u' \frac{\partial v'}{\partial y}} \right) - \frac{\partial^2 \overline{u'v'}}{\partial y^2} + \nabla \cdot \left(\overline{\mathbf{V}'_{\chi} \zeta'_{\chi}} \right) \end{aligned} \qquad (7.59)$$

where $\mathbf{V}'_\psi = (u', v')$ are the x and y components of the transient, *rotational* wind. Defining quantities

$$N \equiv \overline{u'v'} \qquad \text{and} \qquad M \equiv \frac{1}{2}\left(\overline{u'^2 - v'^2}\right) \tag{7.60}$$

then (7.59) can be written

$$\overline{A} = \frac{\partial^2 N}{\partial x^2} - \frac{\partial^2 N}{\partial y^2} - \frac{\partial^2 M}{\partial x \partial y} + \frac{\partial}{\partial y}\left(\overline{u'\frac{\partial v'}{\partial y}}\right) + \frac{\partial}{\partial x}\left(\overline{v'\frac{\partial u'}{\partial x}}\right) \\ + \nabla \cdot \left(\overline{\mathbf{V}'_\chi \zeta'_\chi}\right) \tag{7.61}$$

Since the rotational part of the flow is nondivergent, then $\partial u/\partial x = \partial v/\partial y$ and (7.61) reduces to

$$\overline{A} = \frac{\partial^2 N}{\partial x^2} - \frac{\partial^2 N}{\partial y^2} - 2\frac{\partial^2 M}{\partial x \partial y} + \nabla \cdot \left(\overline{\mathbf{V}'_\chi \zeta'_\chi}\right) \\ \simeq -2\frac{\partial^2 M}{\partial x \partial y} - \frac{\partial^2 N}{\partial y^2} + \nabla \cdot \left(\overline{\mathbf{V}'_\chi \zeta'_\chi}\right) = \frac{\partial}{\partial y}\left(\nabla \cdot E\right) + \nabla \cdot \left(\overline{\mathbf{V}'_\chi \zeta'_\chi}\right) \tag{7.62}$$

In (7.62) the forcing has been expressed in terms of the meridional derivative of the divergence of E, given by

$$E = \left(-2M, -N\right) \tag{7.63}$$

The $\partial^2 N/\partial x^2$ term is neglected in (7.62) by Hoskins et al. by assuming that the longitudinal scale of $u'v'$ is much longer than the meridional scale. That assumption is not unreasonable, based upon data in Figure 6.19a. Some studies ignored the divergent wind term in (7.61), an error of approximately 20% (Hoskins et al., 1983). Plumb (1990) relaxed this restriction a bit, while assuming that the mean flow and eddy variances vary slowly in space. Plumb found a time-mean flow to be balanced when transient eddies have small amplitude, a "non-acceleration" condition. In the same journal issue, Andrews (1990) relaxes the small amplitude constraint.

A stated goal in using E is to avoid taking many derivatives; but of course one cannot eliminate them. Instead, the derivatives are mentally applied after a diagnostic quantity E has been calculated. Hoskins et al. state that the divergence of E vectors can be regarded as a westerly momentum flux by this reasoning: divergence of E is analogous to the zonal wind in the vorticity equation in that a meridional derivative is applied to both.

Inspection of (7.61) and (7.63) reveals that the x component of E describes the shape of the transient feature. Eastward pointing means that $\overline{v'^2} > \overline{u'^2}$ and the low would tend to be elongated in the meridional direction. Westward E implies zonal elongation. Hoskins et al. (1983) find high-frequency E vectors pointing downstream, mainly in the Asian jet exit region (just upstream, their divergence is consistent with acceleration of the time-mean flow by momentum fluxes). The lower-frequency E vectors point upstream and have greatest convergence near the jet maxima. The low- and high-frequency patterns largely cancel each other along and on the poleward side of the storm track. At lower latitudes, the low-frequency E vectors seem to have greater divergence, implying possible acceleration of the mean flow there. Figure 7.9 illustrates the upper-level synoptic situation. A small-amplitude cyclone begins to develop on the east side of a long wave trough (Figure 7.9a). The warm and cold air advection caused by the cyclone displaces the height contours (builds the fronts) in a manner indicated by the dashed lines, intensifying the trough and increasing the jet speed at points **A** and **C**. As the cyclone passes maturity, it separates from the surface front and migrates into the cold air; this decreases the winds at **A** but could increase them at **B** (Figure 7.9b). Separated from the front and located in weak background flow, the cyclone slows down, presumably becoming a "low-frequency" phenomenon. The long-wave trough is west of the climatological surface low. The *surface* lows merge but the cyclone's upper-level trough stretches the upper-level trough to the east. The cyclone has large amplitude now, so the low-frequency E vectors point upstream. The merging feeds some transient vorticity into the time-mean wave. The cancellation between the heat and momentum fluxes is consistent with a cancellation between high and low pass E vectors since the eddy momentum fluxes develop late in the cyclone's life cycle; the result is a decrease in the vertical shear and a simultaneous increase in the vertical average wind speed.

7.2.4   *Which forcing dominates?*

Transient eddies are unlikely to be the dominant forcing mechanism. While long waves and synoptic waves can exist in consonance, the long wave that results would not be stationary except in remarkable circumstances. Accordingly, one must return to orographic and thermal forcing to find a fixed source of long wave forcing. At best, the eddies play a supporting, not dominant role. This subsection examines recent studies that include more than one type of forcing. The discussion commences with reformulations of the classical models of §7.2.1 and §7.2.2.

Saltzman (1963) extended Smagorinsky's (1953) model to include bottom to-pography. He worked out an example and concluded that the solution combining both orography and heating was better than solutions that use either alone. When Roads (1982) reconsidered the study by Smagorinsky (1953), he also incorporated orography, so he could consider the relative importance of orographic and thermal forcing. However, Roads did not attempt to model the observed structure, presumably because of the increased mathematical difficulty in solving that problem and presum-ably because the structure of Q (7.44) is not well known. Roads chose to emphasize solutions that are close to resonance. He did so in part because the surface friction did not necessarily avoid resonance, as mentioned earlier. He also examined solutions

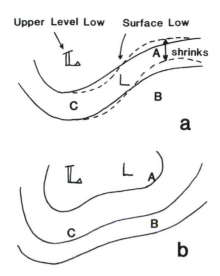

FIG. 7.9   Schematic diagrams showing typical evolution of the upper level geopotential height field (solid contours). Fixed geographic locations labelled **A**, **B**, and **C**. (a) Situation for a developing, small amplitude, short wave with surface low located midway between **C** and **A**. A quasi-stationary long wave is present with trough axis intersecting **C**. Dashed lines indicate changes to heights caused by short wave heat fluxes. (b) Situation 1 to 3 days later when short wave has greatly amplified while moving downstream. Surface low is near **A**. The upper-level lows have merged into an elongated low; the jet is further south (closer to **B** than to **A**).

near resonance because he felt that such a near-resonance state could be sustained. Indeed, one theory of persistent, "blocking" patterns states that such patterns may arise from near-resonance waves (e.g., Charney et al. , 1981). Roads concluded that either the orographic forcing exceeds the thermal forcing or that the atmosphere is near resonance, and that both contribute to the maintenance of the planetary waves.

When Chen and Trenberth (1988) reconsidered the study by Charney and Eliassen, they chose a forced eigenvalue problem resulting from linearized balance equations (Charney, 1962). Whereas Charney and Eliassen (1949) included the slope flow from $[u]$, Chen and Trenberth added the eddy slope flow as well. They found the $[u]$ and eddy slope flows to have much cancellation. Intriguingly, the strong upslope flow over the Himalayas given by $[u]$ is nearly canceled by easterly u' flow (creating a Siberian High and a low over India). The thermal and orographic forcing interact in the model because of this boundary condition, and they find the two to be comparable in the troposphere. Orography appears to be like a governor, reducing the sensitivity of the response to changing the heating.

The question of whether orographic or thermal effects dominate has been addressed in other ways. A recent theoretical study by Donner and Kuo (1984) reexamines the relative magnitudes of various forcing mechanisms. Their study includes a rather careful treatment of clouds and radiative effects. The dynamics are simplified: meridional variation is neglected; zonal wavenumbers 1,2,3, and 4 are treated individually; a potential vorticity equation is linearized about zonal mean temperature and velocity profiles (based upon Oort and Rasmussen, 1971, for pressure exceeding 50 hPa). A nonhomogeneous, second-order, ordinary differential equation with variable (in height) coefficients results where boundary conditions at the bottom incorporate slope flow. The nonhomogeneous term comes from diabatic heating: radiation (clouds, water vapor, carbon dioxide, ozone), latent heat release, and surface sensible heating. The cloud distribution is estimated from an atlas of oceanic cloud cover (Hahn et al., 1982).

Donner and Kuo (1984) find the main thermal effects to be due to radiative effects of clouds and water vapor in their model. Overall, radiative heating amplifies wavenumber 1 and tends to diminish wavenumber 2 by counteracting topographic and other thermal forcing. When thermal processes are included, the amplitude and phase of zonal wavenumbers 1 and 2 are markedly better than when the sole forcing is by topography. Hence they conclude that topography is *not* important in determining the long wave (tropospheric) structure. In this regard they concur with a similar study by Bates (1977). However, their simulation when all forcing terms (topographic and thermal) are included still has significant discrepancies with observations. Donner and Kuo's solutions are least accurate in the lower troposphere. It may be that the temperature and velocity profiles chosen by Donner and Kuo for their basic state inhibit the ability of the surface properties (topography and heat flux) to affect deep planetary waves. By default, the middle atmospheric radiative processes would have a greater impact. The vertical propagation of energy is incorporated into the next subsection (§7.2.5).

The roles of orographic and thermal forcing have been tested in general circulation models too. Manabe and Terpstra (1974) examine two long period integrations of a sophisticated GCM, one with mountains present, one without. Judging by Figure 7.10, the long waves are considerably stronger when mountains are present. However, even the simulation with mountains underpredicts some critical features such as the ridge over the northeast Pacific. One is tempted to conclude that orography is dominant and that thermal effects are responsible mainly for the seasonal shift of surface pressure. However, the incorporation of mountains entails additional diabatic heating, and thus the thermal and orographic effects are not cleanly separated in these GCM experiments. This point is demonstrated in Figure 5.15, where the precipitation maximum along the western North American coast originates largely from orography. Held (1983) presents maps of the diabatic heating at 850 hPa for these GCM integrations. Held felt that the heating is similar enough in the two cases to conclude that nonlinear interaction between the planetary wavetrains was more important than the interference between thermally and orographically forced waves. He also noticed that the diabatic heating has relative maxima along the wave-cyclone storm tracks (as well as along the west coast of North America). Held concluded that a thermal forcing mechanism would have to include knowledge of the tracks and nonlinear properties of wave-cyclones. Not all GCM results agree with Manabe and Terpstra (1974). For example, Kasahara and Washington (1971) concluded from their study that thermal

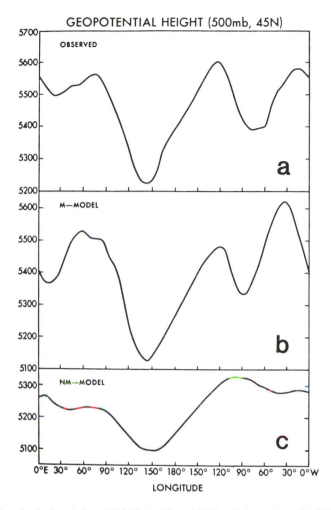

FIG. 7.10   Longitudinal variation of 500 hPa height at 45 N in (a) observations, (b) GFDL model with mountains, and (c) GFDL model without mountains. From Manabe and Terpstra (1974).

forcing is more important than orography. Lau (1986) examined 15 years of simulated data from a spectral GCM. He found the heating field along the Pacific storm tracks and in the vicinity of Indonesia to be much stronger when mountains were included. One might conclude that the question is not which process dominates, but how do the processes interact?

Nigam et al. (1988) tried to sort out the GCM responses to heating and topography. The zonal flow and heating generated by a GCM were input into a linear model that has strong Rayleigh friction in the tropics (to localize the response outside the tropics). Their calculations were then compared to the original GCM climatology. The individual contributors to the heating were input separately, including isolation of transient eddy heat release. In their model, orography contributed the following proportions of these fields: half of 300 hPa u'; two- thirds of 300 hPa Φ' (geopotential); and most of the 300 hPa v'. The heating was more important at lower levels, but their linear model did not match the GCM as well at low levels. (The linear model did not accurately match at upper levels over North America, either.) Each of these two contributors: (a) topography and (b) heating plus transient forcing, were greatest and of comparable magnitude near the east coast of Asia.

Opsteegh and Vernekar (1982) include topography and transient forcing of the steady-state, linear, two-level primitive equations. The eddy fluxes are five-year observations of eddy convergence of momentum and heat. The orographic (rather than the thermal) forcing creates a bigger response in their model. The transient eddy response is similar to the orographic pattern, but displaced eastward, suggesting a causal link (Figure 7.11). The combined response has troughs and ridges located eastward of their topographic positions. Several cautions are in order. First, Opsteegh and Vernekar use a zonal mean wind when defining the slope flow. Second, their diabatic heating seems too weak along the storm track and in the tropics, especially near Indonesia. That tropical heating plays a key role in the localization of the jets (§6.5), and one expects a similar role regarding long waves.

Sardeshmukh and Hoskins (1988) introduce an upper-level divergence source centered near Indonesia. The source simulates the response from deep convection below. When the source is added to the steady-state vorticity equation (initialized by a climatological $[u]$ for winter), a long wave pattern is created with jet maximum at the east coast of Asia, a result consistent with the next study. Valdes and Hoskins (1989) use a steady-state, multi-level model linearized about a zonal mean flow. Topography is incorporated into the vertical coordinate. The diabatic heating they use is estimated as a residual in the thermodynamic equation using ECMWF data, similar to Figure 6.16. Figure 7.12a shows the observed eddy stream function field during winter. The three contributors to the zonally varying stream function are given separately in the same figure. The total simulated response compares favorably with the observations with these exceptions: the simulation is too low in the western tropical Pacific and along the west coast of North America; it is too high over eastern Europe and western Russia. Valdes and Hoskins remark that the orographic response (Figure 7.12b) is about 30% of the total in midlatitudes. This estimate is about half the value stated by Nigam et al. (1988). Most of the orographic response occurs in midlatitudes; it tends to have the same sign throughout the troposphere. The diabatic heating response has greatest amplitude in the subtropics; it tends to reverse sign between the surface and the upper troposphere. As in a companion study, Valdes and Hoskins find that diabatic heating dominates where it builds a strong high over southern Asia and thus intensifies the Asian jet. The subtropical troughs created by diabatic heating are opposed by the transient eddies in mid-ocean, but reinforced in the eastern oceans. The heat and momentum fluxes by the eddies nearly cancel in middle latitudes, as mentioned before (§6.5, §7.2.3).

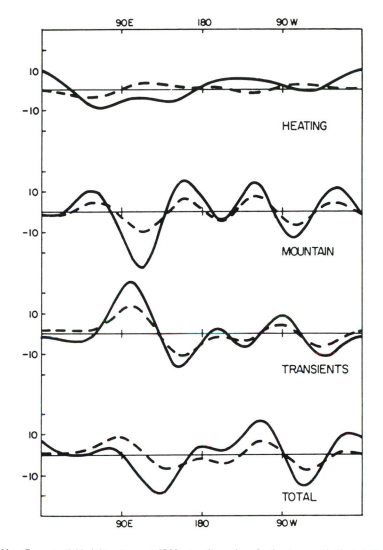

FIG. 7.11 Geopotential height patterns at 45 N set up by various forcing terms as indicated. The bottom pair of curves shows the sum of the three groups of forcing. Solid lines are solutions at 400 hPa; dashed lines show solutions at 850 hPa. From *J. Atmos. Sci.*, Opsteegh and Vernekar (1982) by permission of American Meteorological Society.

The results in Valdes and Hoskins (1989) can be compared with a related study by Schneider (1990). Some parameterizations, notably Newtonian cooling and Rayleigh friction, are avoided by Schneider. However, he analyzed GCM model output, not observations, and used a residual method to gauge the transient forcing. His orographic response is roughly similar to Figure 7.12b in middle latitudes. His diabatic heating response field is similar to Figure 7.12c except over North America (where both studies find a comparatively modest response). Except in the central Pacific and over

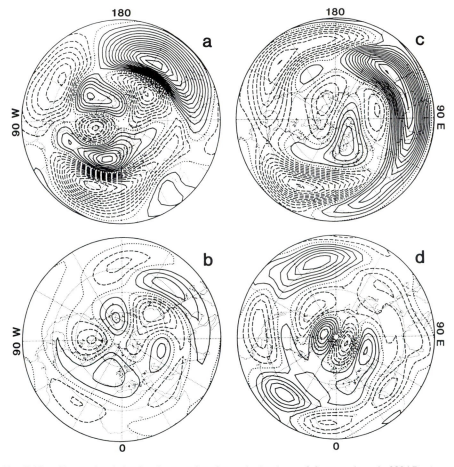

FIG. 7.12 Observed and simulated stream function at the level $\sigma = 0.2$, approximately 200 hPa. ($\sigma = p/p_s$ where p_s is the surface pressure.) Zonal mean pattern removed. (a) Observed eddy stream function. (b) Simulated stream function response to mountain forcing alone. (c) Simulated stream function response to observed diabatic heating forcing alone. (d) Simulated stream function response to observed transient eddy fluxes of heat and momentum. Contour interval is the same in all panels and is approximately 2.2×10^6 m^2 s^{-1}. From *J. Atmos. Sci.*, Valdes and Hoskins (1989) by permission of American Meteorological Society.

eastern Russia, the transient response is different in Schneider (1990). Curiously, large transient forcing values are found near the storm tracks.

In this subsection, three possible mechanisms for maintenance of the long waves have been described. It should now be clear that the question of which forcing dominates depends upon the location and variable of interest. Diabatic processes tend to dominate in the subtropics with significant feedback by eddy heat and momentum fluxes on the equatorial side of the storm tracks. Those fluxes tend to weaken the mean flow and shift the mid-ocean troughs eastward. In middle latitudes, orography is consistently important for large mountain ranges. Diabatic heating in the western

Pacific enhances the Asian trough, too. Developing cyclones appear to stretch the topographically forced trough eastward, particularly at higher latitudes.

7.2.5   Energy propagation

The energy in stationary wavetrains can be viewed as "propagating" downstream from locations of thermal or orographic forcing. The Charney and Eliassen (1949) model uses the equivalent barotropic vorticity equation in one dimension. Hence, their model has no vertical propagation, and wavetrains respond only with longitude structure. What happens when these two restrictions are lifted? The pattern of response seen in Figures 7.12b and 7.12c indicates that the downstream wavetrain is complex; the troughs and ridges are not restricted to lie along latitude circles. This subsection is specifically concerned with how the properties of the environment influence the downstream structure of the response to a local source. The discussion is divided into vertical and horizontal propagation.

Vertical propagation of Rossby waves is thoroughly examined by Charney and Drazin (1961). They examine the linearized quasi-geostrophic equations and combine the vorticity and thermodynamic equations to obtain the potential vorticity equation. Restricting attention to stationary waves and linear vertical shear, one can use the homogeneous form of (7.49) to illuminate the process. Further mathematical simplification is helpful for this discussion. If basic state velocity $U = U(0) + \Lambda z$ is chosen to be representative of the lower stratosphere, then $U(0)$ is large while Λ is small and negative. Neglecting the vertical variation of U, except where differentiated, gives a simpler equation to solve in the end; not making this approximation leaves a confluent hypergeometric equation instead of (7.64). Factoring out common functions of x and y leaves

$$\epsilon f \frac{\partial^2 \phi}{\partial z^2} - \frac{\epsilon f}{\tilde{H}} \frac{\partial \phi}{\partial z} + \left(\frac{\beta}{U} - \frac{\Lambda \epsilon f}{U \tilde{H}} - k^2 - m^2 \right) \phi = 0$$

where ϕ is a stream function defined by (7.48). Defining

$$A \equiv \frac{1}{\epsilon f} \left(\frac{\beta}{U} - \frac{\Lambda \epsilon f}{U \tilde{H}} - k^2 - m^2 \right)$$

then

$$\frac{\partial^2 \phi}{\partial z^2} - \frac{1}{\tilde{H}} \frac{\partial \phi}{\partial z} + A\phi = 0$$

Making the change of variables,

$$\varphi = \phi \exp\left(- \frac{z}{2\tilde{H}} \right)$$

yields

$$\frac{\partial^2 \varphi}{\partial z^2} + \nu^2 \varphi = 0 \tag{7.64}$$

where

$$\nu^2 = \frac{\beta}{U} - \frac{\Lambda}{U\tilde{H}} - \frac{k^2 + m^2}{\epsilon f} - \left(\frac{1}{2\tilde{H}}\right)^2 \qquad (7.65)$$

Charney and Drazin refer to ν as a refractive index based upon an analogy with wave mechanics. The structure of the solutions depends upon the sign of ν^2. For ν imaginary then the solutions to (7.64) are "trapped" modes that exponentially decay with height. In contrast, for ν purely real, one obtains propagating modes that can bring energy to the upper atmosphere. From (7.65) it is clear that when such propagating modes exist, they must be long waves with $k^2 + m^2$ less than some cut-off value. Therefore, if one imagines the stratospheric waves (§6.2.4) to be forced by a broad spectrum of tropospheric waves, then the long waves are the only modes capable of penetrating the middle stratosphere. Observations are consistent with this interpretation. The result may apply to the upper troposphere, too. Only the larger-scale components present in locally small-scale forcing by topography or heating might reach the upper atmosphere. From (7.65), ν can vary with height. For the example derived, a constant vertical shear, Λ is chosen. But a more general profile of the wind results in a dependence of ν upon the meridional gradient of the zonal mean flow potential vorticity (7.6) rather than simply the vertical shear. Similarly, the stratosphere's static stability is about three times that of the troposphere. Since ϵ is inversely proportional to static stability, then $k^2 + m^2$ must be two-thirds smaller in order for a wave to propagate into the stratosphere. (If ϵ or $\partial U/\partial z$ did vary with height, there would be additional terms in Equation 7.65 as illustrated in Grotjahn, 1979, but the qualitative discussion above would not be materially altered). The study by Charney and Drazin (1961) has other implications for the maintenance and structure of the planetary waves; Held (1983) discusses several of these.

The horizontal structure of the forced Rossby waves is examined by Grose and Hoskins (1979) and Hoskins and Karoly (1981). Grose and Hoskins find that when the equivalent barotropic model of Charney and Eliassen (1949) is formulated on a sphere, the amplitude of the wavetrain at 45 N is reduced by a factor of three. Part of this reduction could be accounted for by the too slow meridional variation of topography assumed by Charney and Eliassen. Part of the reduction is also due to how the Rossby wavetrains disperse in two dimensions. On the sphere, Hoskins and Karoly (1981) are able to show that the wavetrains emanating from a single region of thermal or orographic forcing follow great circle routes downstream (Figure 7.13a). For the complex multiplicity of sources on the Earth, the pattern still has large scale, but the great circle paths are less clear (Figure 7.12) due to interaction of wavetrains. These paths can undergo refraction, reflection, and absorption; Branstator (1983) examines these three influences upon the paths. By using the observed flow at 300 hPa as his basic state, Branstator identifies regions that act as wave guides (strong jet streams) or reflecting zones (equatorial easterlies). The tropical reflecting regions can enhance the resonant amplitude of midlatitude long waves. Observations by Wallace and Gutzler

(1981) seemed to confirm this resonance mechanism. Wallace and Gutzler (1981) sought to identify a connection between deflections of the midlatitude westerlies and anomalous heating (due to enhanced deep convection) over the equatorial western Pacific. "One point" correlation maps, such as Figure 7.13b, suggest this tropical-extratropical connection. These maps are constructed by correlating height values at one grid point with values at all other grid points.

7.3   Wave-cyclone life cycles

7.3.1   *Theoretical evidence*

Perhaps the first theoretical treatment of the eddy life cycle was by Brown in 1969. He used a simple quasi-geostrophic channel model. His prediction equation was the vorticity equation, and it required diagnostic solution of the "omega equation" for pressure velocity at each time step. He used separate equations for the zonal mean state and the zonally asymmetric eddies. He solved his equations by numerical integration from an initial condition. The initial conditions for the zonal mean and eddy fields were determined from the linear solution found for a specified mean flow. The eddy was initially given a very small amplitude because formally, the linear results are an excellent approximation to the nonlinear solution when the solution has small amplitude relative to the mean flow. He encountered nonlinear computational instability when he integrated the adiabatic form of the equations; to rectify that problem he included friction that more strongly damps the shortest waves. In order to conserve total energy he included a diagnostic heating function to balance the frictional losses. When Brown changed the initial eddy amplitude, the chaotic behavior at the start was a little different, but the same cyclic behavior was eventually reached. This means that the initial condition was not important in determining the asymptotic character of the solution. The period, amplitude, and other properties of the asymptotic oscillations were related to the diabatic input and output that were present.

Brown's solutions fluctuated wildly early in each time integration, but after one or two cycles of growth and decay a regular pattern emerged. The cyclic behavior can be characterized as follows (Figure 7.14). When the eddy reaches minimum amplitude, the baroclinic and barotropic mechanisms both convert energy from the mean flow to the eddy. (The former is an $A_Z \rightarrow A_E \rightarrow K_E$ conversion; the latter is $K_Z \rightarrow K_E$.) Soon, the baroclinic mechanism takes over most of this conversion. The barotropic conversion diminishes and eventually becomes negative. This slows the growth rate until the eddy reaches its maximum amplitude. At that time, the barotropic conversion is negative and just exceeds the baroclinic, which is positive but beginning to diminish. The amplitude of the eddy begins to decline. Eventually the barotropic (negative) conversion reduces in amplitude and finally changes sign when the eddy reaches its minimum amplitude. At this time the zonal mean flow becomes strongest and the baroclinic process restarts, repeating the cycle. In the simplest terms, the eddy grows

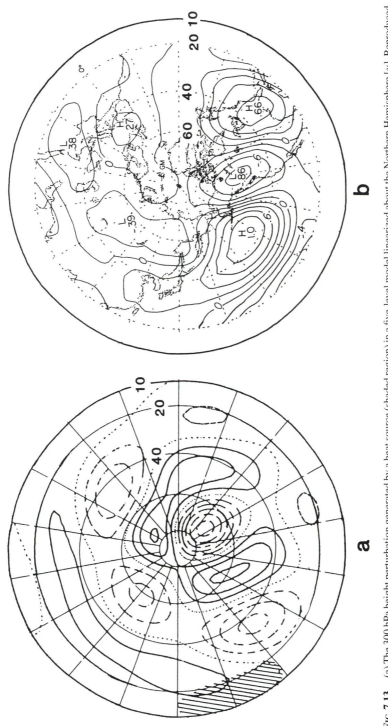

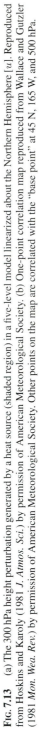

FIG. 7.13 (a) The 300 hPa height perturbation generated by a heat source (shaded region) in a five-level model linearized about the Northern Hemisphere [u]. Reproduced from Hoskins and Karoly (1981 *J. Atmos. Sci.*) by permission of American Meteorological Society. (b) One-point correlation map reproduced from Wallace and Gutzler (1981 *Mon. Wea. Rev.*) by permission of American Meteorological Society. Other points on the map are correlated with the "base point" at 45 N, 165 W, and 500 hPa.

344

mainly by the baroclinic process until it reaches maximum amplitude then decays by the barotropic mechanism until it reaches minimum amplitude. This description highlights another feature Brown found: the barotropic conversion changes sign strongly, whereas the baroclinic tends to remain positive.

Brown's result can be visualized in the following principal features of the eddy structure (Figure 7.15). During the growing stage the waves have an upstream tilt with height of the trough and ridge axes. This tilt arises because the temperature field lags behind the mass field at the surface by less than one-half wavelength. The lag allows heat fluxes that transport heat from warm to cold areas. Given the usual geometry of cyclones and anticyclones and their flows, the thermal trough is west of the pressure trough and the heat flux is poleward. This vertical tilt structurally identifies the baroclinic energy conversion. At the same time, the trough and ridge may have horizontal tilts that are required for a net convergence of eddy momentum. The barotropic energy conversion process is indicated in Figure 7.15a for initial barotropic (and baroclinic) growth. In the decaying stage, the vertical tilt has been lost, a net heat flux is no longer possible, and the baroclinic conversion has ceased. The barotropic mechanism is now removing energy from the eddy (and feeding it back to the mean flow), so the horizontal tilts are reversed, as in Figure 7.15c.

This description differs from that used for the closed system analog described in §4.5.2 (Figure 4.22). In the earlier discussion barotropic processes are ignored but surface frictional convergence (divergence) is allowed at the surface low (high). That convergence generates potential energy by the cold air rising above the surface low. Thus the kinetic energy of that system is decaying by a *negative* baroclinic conversion.

Brown does not include the Ekman layer frictional divergence (surface vertical velocity is zero), though he does include interior viscosity (to control the numerical instability). He is mainly left with the barotropic process to reduce his eddy amplitude. Which decay mechanism, barotropic or boundary-layer driven baroclinic conversion, is the correct description for the atmosphere? Both. The relative importance of each mechanism probably differs from storm to storm. However, the time-mean momentum convergence is strong (Figure 6.19), the positive areas of $\overline{\omega'T'}$ in Figure 5.18 are weak, and the time-average baroclinic energy conversion (Figure 5.23) is generally positive; judging from these observations, the negative barotropic conversion should be more significant. The observations just cited are suggestive; more careful composites of observations, discussed in the next subsection, confirm this conclusion.

Brown mentions that the conversions tend to minimize the changes in A_Z plus K_Z (where one is maximum the other is minimum). To some extent, this is built into the model by inputing G_Z to balance the total frictional loss. The phase of the G_Z variation is highly synchronized with the A_Z variation and the $A_Z \rightarrow K_Z$ conversion. In contrast, the G_Z generation is less synchronized with the $A_Z \rightarrow A_E$ conversion. The $A_Z \rightarrow A_E$ conversion is almost synchronized with the $A_E \rightarrow K_E$ conversion: the phase of the former leads the latter, but the magnitude is nearly identical; perfect synchrony would validate the "express bus lines" analogy used when the energy

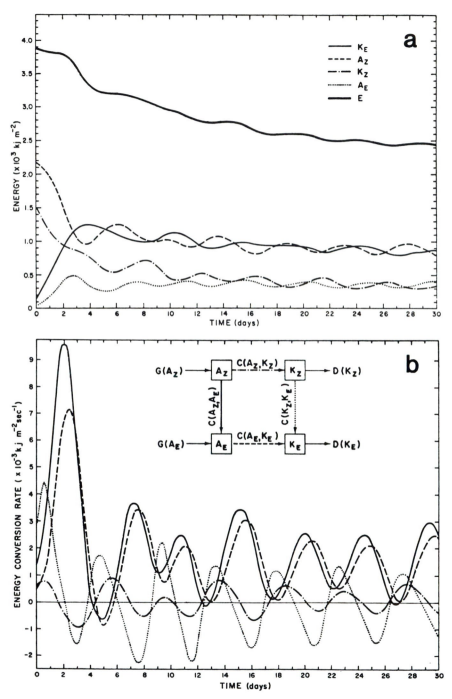

FIG. 7.14a and b Time series of various energetics quantities in a nonlinear channel model. The various quantities are identified on the energy "box" diagram in panel (b). (a) Energy amounts in various forms. (b) Energy conversions between the four atmospheric forms.

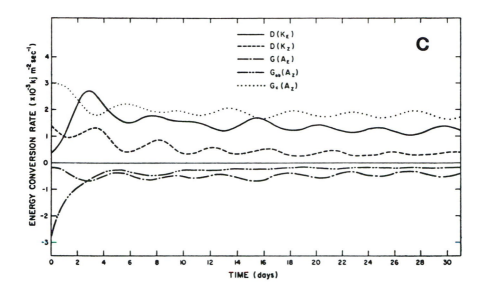

FIG. 7.14c Generation (G) and destruction (D) amounts. From *J. Atmos. Sci.*, Brown (1969) by permission of American Meteorological Society.

box diagrams were introduced in §4.6. The cycle proceeds as follows. Generation of zonal available potential energy goes into the meridional temperature gradient, but also immediately into the zonal wind field as well, from thermal wind balance. The zonal mean wind shear must build up sufficiently to initiate baroclinic instability. The maximum $A_E \rightarrow K_E$ conversion also awaits the growth of eddies to large amplitude. On the other hand, the barotropic conversion is initially positive when the eddy amplitude is small (hence the K_E maximum leads the A_E maximum).

Further work on the nonlinear life cycles can be subdivided into more sophisticated channel models and spherical coordinates models. The quasi-geostrophic system develops fronts symmetrically, yet observations clearly show isotherms to be concentrated on the cold air side of the surface trough. Introducing ageostrophic terms can create this type of asymmetry in linear (e.g., Grotjahn, 1979) and nonlinear (e.g., Mudrick, 1974) models. Polavarapu and Peltier (1990) use a *nonhydrostatic*, primitive equation model having very high resolution to simulate the life cycle. Polavarapu and Peltier emphasize the small-scale features of the occlusion process and the differences between assuming f to be either constant ("f-plane") or a linear function of latitude ("β-plane"). The β-plane approximation creates sharper fronts quicker and forms the occluded low further north. The nonhydrostatic terms create bands of vertical overturnings near the sharp fronts.

Other work upon the question of life cycles has been carried out in spherical geometry (Gall, 1976; Simmons and Hoskins, 1978, 1980; Frederiksen, 1981; Frederiksen and Puri, 1985; Grotjahn and Lai, 1991). These articles show graphs with

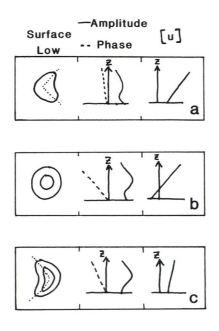

FIG. **7.15**   Schematic diagrams showing the eddy structure at various stages in the life cycle. Structures are deduced from Brown's channel model energetics shown in the previous figure. On the left side of each panel is a schematic illustration of the horizontal shape of the surface low (dashed lines indicate trough axes). The middle of each panel includes vertical profiles of amplitude (solid line) and phase (dashed line) of eddy geopotential. The right side of each panel shows the vertical profile of $[u]$. The approximate times are (a) Day 22.5, where eddy amplitude is minimum (b) day 24 where barotropic conversion vanishes, and (c) day 25 where eddy amplitude is maximum.

greater vertical and meridional detail than those published by Brown. The remainder of this subsection emphasizes the structure of the eddies.

Figure 7.16 shows the typical progression of eddy development found by these studies. Two levels are shown at four different days. In order to see the eddy structure one must remove some type of mean field. It is not a trivial matter to choose the appropriate mean field. Grotjahn and Lai (1991) choose the zonal mean at the *start* of the time integration. Other researchers use the *instantaneous* zonal mean. The latter choice is not an appropriate description during the late stages of the life cycle because the lows line up along a different latitude than the highs (Figure 7.16c); the instantaneous zonal mean underestimates the eddy strength. Using the initial zonal mean ignores the net effect of the eddy heat fluxes leading to the domain average positive values in Figure 7.16h. Fortunately, calculations by the author find the eddy fluxes and amplitude for these two methods of defining the "eddy" to be very similar when zonal and *time* means are taken. The migration of lows and highs to different latitudes (comparing Figures 7.16a and 7.16c) is commonly observed (Figure 5.20). Kuo (1951) predicted that cyclones and anticyclones would migrate toward latitudes of similar absolute vorticity. The eddies start off with small amplitude, vertical upstream tilts (baroclinic growth) and horizontal downstream tilts (barotropic decay). The upper-level eddy amplifies much more than the lower-level solution. Of particular note are the highs in Figure 7.16h. They have large amplitude and are located *south* of the initial position; these features are consistent with Eliassen-Palm fluxes discussed in §7.4.5.

Most later studies find results similar to Brown's results: baroclinic growth followed by barotropic damping. The energy conversions during the first cycle in the Simmons and Hoskins model are presented in Figure 7.17. Two differences from Brown are notable. First, the barotropic conversion is negative essentially throughout the cycle. Figure 7.15a may be contrasted with 7.16a. Second, they find the eddy to nearly disappear by the end of the cycle.

Simmons and Hoskins only show their solution over the first life cycle, so it is not clear how much their results are influenced by the initial conditions they choose. Some caution is advisable given the wild fluctuations during the first life cycle in Brown's model. Simulations by Brown (Figure 7.14a), by Frederiksen (1981), and by the author find that the period of initial growth is not followed by a period of complete decay. Instead, the eddy amplitude oscillates. The long-term behavior looks like an amplitude vacillation cycle for weakly nonlinear disturbances (Pedlosky, 1972). Frederiksen concludes that his results are different because Simmons and Hoskins use stronger diffusive damping. (Frederiksen uses a diffusion formulation very similar to that in an operational forecast model.) Neither study includes a boundary layer or even surface friction. Hence, the negative baroclinic conversion driven by convergence at the surface low (Figure 4.22) could not be included.

Returning to Simmons and Hoskins (1978), the major result of their study is that the upper-level eddy development is much greater than that predicted by linear theory (Figure 7.18). Averaged over the life cycle of the eddy, the eddy has much greater amplitude at high levels than predicted by linear theory (at least for the longer waves). Wavenumber 6 is used as the example. The upper-level intensification for wavenumber 9 (a short wave) was considerably less than that found for wavenumbers 6 or 3. The eddy fluxes are changed in a similar fashion: heat fluxes now have a high-level secondary maximum in the meridional transport in addition to the low-level maximum (Figure 7.19); only the low-level maximum was predicted by linear theory. For the momentum fluxes (Figure 7.20) the linear theory predicts surface and tropopause-level maxima in the transport. The life cycle average shows the tropopause-level transport to be the dominant one by far; the surface maximum is negligible on the life cycle average.

The linear solution has strong convergence of momentum with both equatorward and poleward transport, but when averaged over the whole nonlinear life cycle, the poleward momentum flux dominates. The nonlinear result is more like the observed distribution of zonal average eddy momentum flux (Figures 4.11, 5.16, or 6.19). The horizontal tilts of the linear and nonlinear solutions illustrate the changes in momentum flux. The lows and highs in the linear solution (Figures 7.16a, and 7.16e) are symmetric about the jet, but they develop a "triangular" shape (Figures 7.16c and

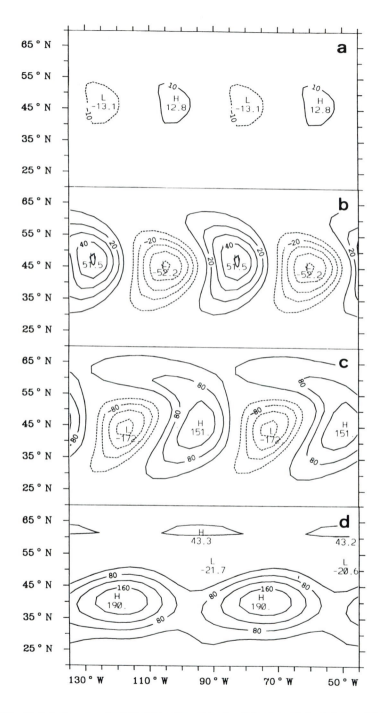

FIG. 7.16   Perturbation geopotential height patterns at two levels in a general circulation model simulation of nonlinear frontal cyclone life cycle. (a) through (d) Patterns at 950 hPa. (e) through (h) corresponding patterns at 500 hPa. The initial condition (a) and (e) is based on the most unstable linear solution for the

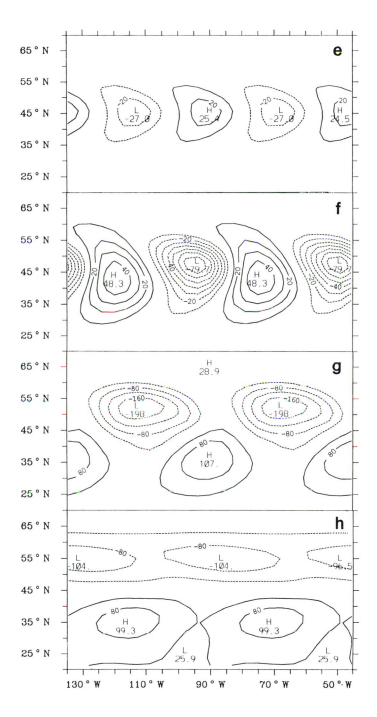

initial zonal mean state. Later solutions are day 2 (b) and (f) day 5 (c) and (g), and day 10 (d) and (h). The contour interval is the same in all panels. From Grotjahn and Lai (1991).

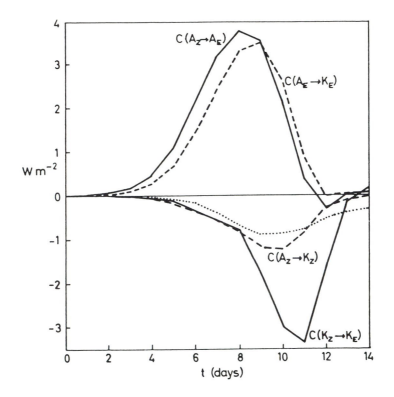

FIG. 7.17   Time series of energy conversions in the first life cycle in a nonlinear spherical coordinates model. The model is initialized with a zonal wavenumber 6 disturbance upon a $[u]$ jet initially at 45 N. Arrows in the labelled conversions indicate the direction for positive energy conversion. From *J. Atmos. Sci.*, Simmons and Hoskins (1978) by permission of American Meteorological Society.

7.16g) at large amplitude. The latter shape emphasizes the poleward transport of eddy westerly momentum and is similar to observed structures (e.g., Figures 5.8 and 6.8).

Most studies in spherical geometry find that during the course of the life cycle, the zonal mean jet position migrates to a higher latitude. Simmons and Hoskins (1978) show a case where the jet is initially at 45 N, and the final position at the end of the integration (after 13 days) is near 50 N. In Grotjahn and Lai (1991) the jet moves from 40 N to 45 N during the integration. Consistent with the energy conversions, Grotjahn and Lai find the vertical shear to decrease while the eddies grow (but the $[u]$ jet maximum stays roughly constant). The vertically averaged zonal mean zonal wind increases, especially as the eddies decay.

7.3.2 *Observational evidence*

Observations largely confirm the eddy life cycle theories. Evidence from three studies are presented in this subsection: one study isolates the life cycle stages by geographic preference, the other two carefully select periods in time.

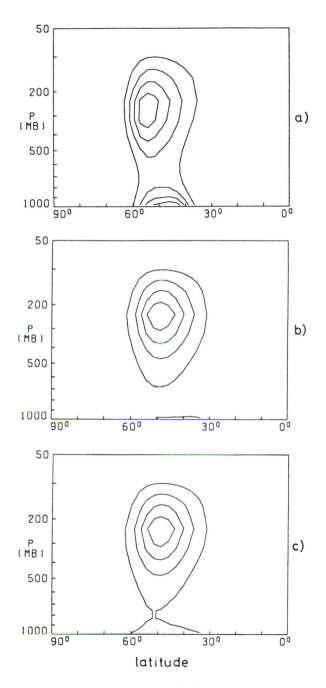

Fig. 7.18 Meridional cross-sections of zonal mean eddy kinetic energy. Patterns shown are at (a) day 0, (b) day 10, and (c) average from days 4 through 14. The contour interval is arbitrary and varies between plots. The initial condition comes from the linearized version of the model. Nonlinearity enhances upper-level development. From *J. Atmos. Sci.*, Simmons and Hoskins (1978) by permission of American Meteorological Society.

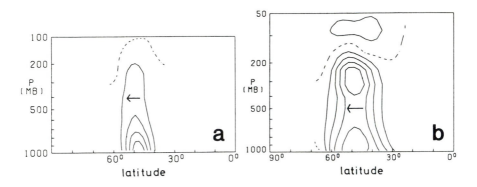

FIG. 7.19   Comparison of linear (a) and nonlinear time mean (b) solutions when the initial $[u]$ has a jet at 45 N. Meridional cross-sections of meridional eddy heat flux ($[v'T'] \cos \phi$) for the wavenumber 6 solution shown in Figure 7.18. From *J. Atmos. Sci.*, Simmons and Hoskins (1977, 1978) by permission of American Meteorological Society.

Lau (1978) partitioned the Northern Hemisphere into longitudinal sectors (Figure 7.21). The scheme assumes that particular sectors (EAS and ENA) sample developing cyclones while other sectors (WNA and EUR) sample decaying storms, with the mature stage sampled by sectors in between. This scheme takes advantage of stationary preferred regions of cyclogenesis in the Northern Hemisphere. A band-pass filter was used to isolate frontal cyclones having a 2 to 6 day period from long wave pattern.

The heat fluxes (Figure 7.22) found by Lau (1978) agree well with the theories. In the jet entrance regions, (i.e., EAS and ENA) the heat flux is largest at about the 850 hPa level. The profile looks like the linear model solution (Figure 7.19a). In the central oceanic areas (i.e., PAC and ATL) a second distinct maximum develops in the lower stratosphere. This double maximum in heat flux looks gratifyingly similar to the mature stage result found in the nonlinear models (Figure 7.19b). The low-level heat

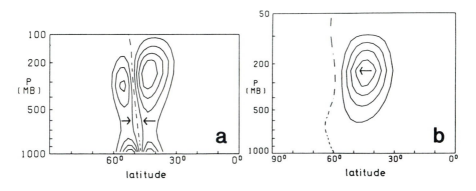

FIG. 7.20   Similar to Figure 7.19 except for horizontal eddy momentum fluxes: ($[u'v'] \cos^2 \phi$). (a) Linear result without $\cos^2$ factor. (b) Nonlinear result. From *J. Atmos. Sci.*, Simmons and Hoskins (1977, 1978) by permission of American Meteorological Society.

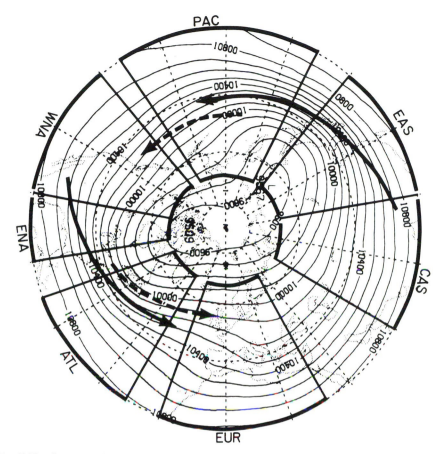

FIG. 7.21 Sectors used to make averages shown in Figures 7.22 and 7.23. Each sector is outlined with a dark border; lighter contours are of 250 hPa geopotential height. The arrows with solid shafts are the main jets while the dashed arrows are "storm tracks" as indicated by maxima in the band pass (2.5 to 6 day periods) height variance. From *J. Atmos. Sci.*, Lau (1978) by permission of American Meteorological Society.

flux reaches its maximum value here. In the eastern oceans and western continents, many eddies are at a decaying stage and the low-level heat flux maximum is greatly diminished, while the lower stratospheric heat flux is still maintained (though further poleward). However, in the upper troposphere, the heat flux is equatorward. These equatorward fluxes do not appear on zonal averages (Figure 4.31) nor time averages at 500 hPa (Figure 6.19). The temperature gradient declines downstream in response. Meridional heat fluxes are not equivalent to baroclinic conversions, but some general conclusions are possible. The heat flux sector averages show a clear progression of baroclinic growth in the jet entrance region that extends into higher levels and moves to a higher latitude during the mature stage, followed by very small baroclinic growth (decay?) in the region near the west coasts of the continents.

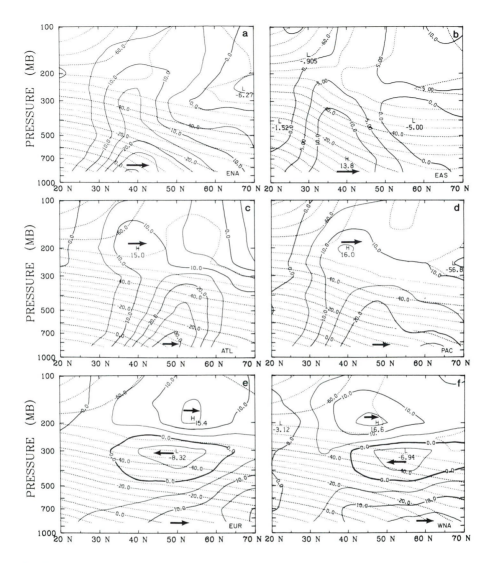

Fig. 7.22   Meridional cross-sections of transient eddy heat flux (solid contours) and time-mean temperature (dotted lines) where zonal averaging is across sectors defined in figure 7.21. The label in the lower right of each panel refers to the indicated sector. The diagrams are displayed to facilitate comparison during parallel periods in the life cycles of frontal systems. From Figure 5.19, most eddies form in the **ENA** and **EAS** sectors. Most eddies decay in the **EUR** and **WNA** sectors. Contour intervals are 5 K m s^{-1} and 5 K. From *J. Atmos. Sci.*, Lau (1978) by permission of American Meteorological Society.

The eddy momentum fluxes found by Lau (1978) are less similar to the theoretical results. A cyclic pattern of eddy momentum fluxes, shown in Figure 7.23, is more difficult to discern. The problem may lie with the choice of longitudinal ranges for the sectors or with the difficulty in establishing a proper climatology for this field (§2.6). The strongest eddy momentum convergence occurs over western North America and

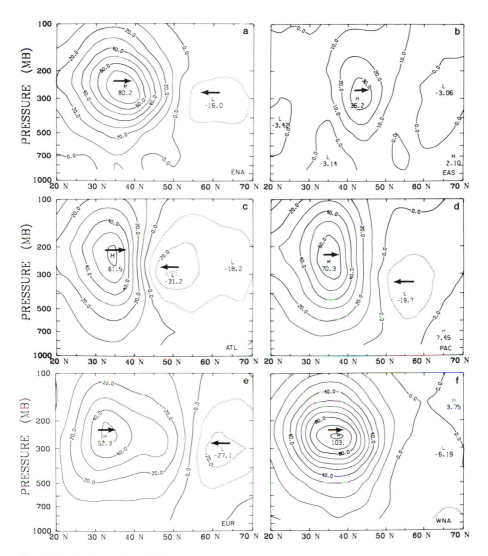

Fig. 7.23 Similar to Figure 7.22, except for transient eddy poleward momentum flux. Heavy arrows show the direction of the flux. The flux convergence is greatest where the contours are most closely spaced in the horizontal. Contour interval 10 m^2 s^2. From *J. Atmos. Sci.*, Lau (1978) by permission of American Meteorological Society.

the Atlantic. Despite these problems, several features are consistent with the theoretical studies. On the southern side of the eddy track the momentum flux is poleward, while a much weaker area of equatorward flux is found on the north side. In between, the momentum convergence increases when moving from the cyclogenesis region to the next sector downstream. The maximum convergence is near the tropopause level.

Lau and Lau (1984) composite observations of frontal cyclones that form near Japan. The data is combined so that a "day 0" is defined on the basis of the start of a cold-air outbreak behind a developing low. They averaged eleven frontal cyclones on this basis. Energy conversions are calculated over volumes whose shape and location shift with the developing, composite low. Figure 7.24a shows estimates of the total eddy energy, the baroclinic energy conversion C_A and the barotropic energy conversion between the high-frequency transient eddy and the long wave present in the composite. The latter conversion is subdivided into averages over eastern C_K^E and western C_K^W parts of the moving domain. The figure compares favorably with Figure 7.17.

Randel and Stanford (1985) report on an observational study intended to be directly comparable to the Simmons and Hoskins (1978) results. They chose a ten-day period of Southern Hemisphere data during which a rather zonally oriented flow develops into a strong, rather regular, wavenumber 6 pattern, followed by a rather zonal flow again. Because the wavenumber 6 pattern is so regular in this case, zonal averages are taken around complete latitude circles. Figure 7.24b shows the total eddy energy and the energy conversions at different times. The date 13 December 1979 is labelled "day 9" to facilitate comparison with Figure 7.17; this is the time of maximum eddy energy. The energy conversions again compare well with theoretical model calculations. As was seen in the theoretical studies, the wave's heat flux is initially large in the lower troposphere but becomes largest at progressively higher levels.

7.4   Eliassen–Palm fluxes

In many places, this book relates various atmospheric properties to the eddy heat and momentum fluxes. The two fluxes can be combined in a consistent fashion (dynamically speaking). Eliassen and Palm (1961) were perhaps the first to do so and to use the resulting variable to diagnose the source of energy for mountain lee waves. In honor of their original work, the quantity in question is known as the "Eliassen-Palm flux," or EP flux. The quantity has been applied to larger scale atmospheric motions by Andrews and McIntyre (1976, 1978) and independently by Boyd (1976). In turn, these works grew from and generalize related studies by Charney and Drazin (1961), Dickinson (1969), Holton (1984), and others. Boyd (1976) is somewhat easier to follow than Andrews and McIntyre (1976) because he does not engage a Lagrangian formalism nor does he invoke subtle concepts like wave action. Edmon et al. (1980) discuss the implications of the work by these authors in a very readable paper that forms the basis of some of the discussion here. Later work has attempted to generalize the EP flux to three-dimensional mean flows (Plumb, 1990).

The EP flux combines the eddy heat and momentum fluxes so that the interaction between the eddies and zonal mean state is more clear than if either flux is considered separately. The use of the EP flux in large scale dynamics has been most successful in applications to the stratosphere (e.g., §7.4.6). Applications to the troposphere have

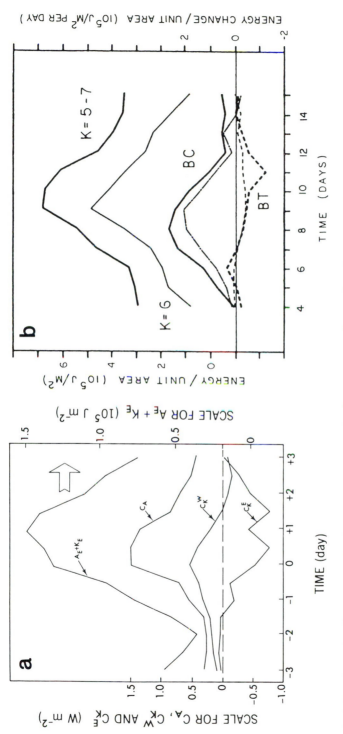

FIG. 7.24 Time series of energy conversions in the atmosphere plotted in a format similar to Figure 7.17. (a) Energy conversions for several storms that form near the east coast of Asia. Data composites are relative to a particular stage in the life cycle of each winter storm. The top curve shows eddy total energy. C_A indicates baroclinic energy conversion. The barotropic energy conversion C_K is split into averages over the eastern and western parts of the domain as indicated by E and W superscripts, respectively. Panel reproduced from Lau and Lau (1984, *Mon. Wea. Rev.*). (b) Energy conversions averaged for all storms during a particular period when a wavenumber 6 pattern developed, was predominate, and decayed. The thick curves show data for wavenumbers 5 through 7; the thin curves show data for wavenumber 6 alone. The top curves are total eddy energy. The two curves labeled BC indicate baroclinic conversion. Barotropic conversion is given by the dashed curves labeled BT. Southern Hemisphere spring conditions; "day 9" corresponds to 13 December 1979. U.S. NMC data. Panel reproduced from Randel and Stanford (1985 *J. Atmos. Sci.*). Both diagrams reproduced by permission of American Meteorological Society.

359

been most useful at elucidating the evolution of the synoptic scale waves (§7.4.5). However, as shown in §7.4.3, the relation between EP flux divergence and the zonal mean flow is inferior to the "traditional" approach (§4.1 and §4.2), which relies primarily upon eddy momentum convergences. Other appealing features of the EP flux formalism are illustrated in §7.4.1, §7.4.2, and §7.4.5.

The *quasi-geostrophic* definition of the EP flux is that $\mathbf{F} = (F^y, F^p)$ where the components are defined as

$$(F^y, F^p) = \left(-[u'v'], \frac{f[v'\theta']}{\frac{\partial[\theta]}{\partial p}} \right) \tag{7.66}$$

On a "β-plane" (where f is a linear function of latitude), F^y is proportional to the momentum flux and F^p is proportional to the heat flux divided by a measure of static stability. Andrews and McIntyre (1976) and Boyd (1976) extend $\mathbf{F}$ to spherical geometry. Eliassen and Palm (1961) express $\mathbf{F}$ for an "ageostrophic" case, as do Andrews and McIntyre (1978). The so-called "EP Theorem" states that the divergence of $\mathbf{F}$ is zero for steady conservative wavelike disturbances upon a zonal wind, or that $\nabla \cdot \mathbf{F} = 0$. The nondivergence can be shown to apply for certain special conditions.

Profiles of $\mathbf{F}$ can relate interaction between eddies and a zonal mean state. Edmon et al. (1980) sketch a couple of illustrations of this interaction. Four examples are now discussed.

7.4.1 *Example 1: Mean meridional circulations*

It is possible to derive an equation that is related to the "Kuo-Eliassen" equation (6.50), whereupon the divergence of $\mathbf{F}$ is expressed as one of the right-hand side forcing terms.

The zonal momentum equation in Cartesian coordinates is used with pressure as the vertical coordinate. After applying a zonal average,

$$\frac{\partial}{\partial t}[u] + \frac{\partial}{\partial y}[uv] + \frac{\partial}{\partial p}[u\omega] - f[v] = R$$

where R is a friction term. The nonlinear terms are partitioned into zonal mean and eddy contributions.

$$\frac{\partial}{\partial t}[u] + \frac{\partial}{\partial y}\left([u][v]\right) + \frac{\partial}{\partial p}\left([u][\omega]\right) =$$
$$f[v] - \frac{\partial}{\partial y}[u'v'] - \frac{\partial}{\partial p}[u'\omega'] + R \tag{7.67}$$

We neglect ω terms now. Meridional advection scales out later. A stream function defined as

$$[v] = \frac{\partial \psi}{\partial p} \qquad \text{and} \qquad [\omega] = -\frac{\partial \psi}{\partial y} \tag{7.68}$$

is substituted. Thus (7.67) becomes

$$\frac{\partial}{\partial t}[u] + \frac{\partial}{\partial y}\left([u]\frac{\partial \psi}{\partial p}\right) = f\frac{\partial \psi}{\partial p} - \frac{\partial}{\partial y}[u'v'] + R \tag{7.69}$$

Applying a zonal average to the thermodynamic equation obtains

$$\frac{\partial}{\partial t}[\theta] + \frac{\partial}{\partial y}\left([v][\theta]\right) + \frac{\partial}{\partial p}\left([\omega][\theta]\right) =$$
$$- \frac{\partial}{\partial y}[v'\theta'] - \frac{\partial}{\partial p}[\omega'\theta'] + Q \tag{7.70}$$

where Q is a diabatic heating rate. For the quasi-geostrophic system the term involving ω' is neglected in favor of the term involving $[\omega]$. That term is further simplified by the nondivergence of stream function velocities

$$\frac{\partial}{\partial y}\left([v][\theta]\right) + \frac{\partial}{\partial p}\left([\omega][\theta]\right) = [\omega]\frac{\partial[\theta]}{\partial p} + [v]\frac{\partial[\theta]}{\partial y} \tag{7.71}$$

Using (7.71) and substituting the stream functions (7.69) obtains

$$\frac{\partial}{\partial t}[\theta] + \frac{\partial[\theta]}{\partial y}\frac{\partial \psi}{\partial p} = -\frac{\partial \psi}{\partial y}\frac{\partial[\theta]}{\partial p} - \frac{\partial}{\partial y}[v'\theta'] + Q \tag{7.72}$$

At this stage a transformed velocity field is introduced

$$v^\natural = [v] - \frac{\partial}{\partial p}\left(\frac{[v'\theta']}{\frac{\partial[\theta]}{\partial p}}\right) \tag{7.73}$$

$$\omega^\natural = [\omega] - \frac{\partial}{\partial y}\left(\frac{[v'\theta']}{\frac{\partial[\theta]}{\partial p}}\right) \tag{7.74}$$

These transformed velocities may seem peculiar at first glance. In actuality, they are the Lagrangian motion of parcels in the meridional plane. The velocities expressed as heat flux terms approximate the "Stokes drift" associated with motions around latitudinally and vertically varying waves. The particular form here is based on Lagrangian motion in the $[\theta]$ equation for small amplitude perturbations.

Stokes drift was introduced in §6.2.4 (Figure 6.11c). Mathematical derivations of Stokes drift can be found in various fluid mechanics texts. Meteorological derivations exist in the literature, too (e.g., Dunkerton, 1978). Stokes drift is a correction to the fluid motion of particles as they traverse some type of *disturbance* superimposed on a mean quantity. In §6.2.4 the disturbance was a long wave; as parcels traverse the long wave they encounter different long wave amplitude (and experience different vertical displacements) between areas of upward and downward motion; consequently, some parcels have a net vertical motion. Most of us have seen a classic example of Stokes drift while watching floating leaves being carried in the direction of surface water waves. In this classic case, the disturbance is the passing gravity waves. The Eulerian mean flow *averaged over the period of the disturbance* is (often) zero. If the Eulerian velocity (measured at a fixed point in space) is zero, the net Lagrangian motion following the parcels is precisely the Stokes drift.

A useful way to derive (7.73) and (7.74) is to first make several simplifying assumptions to the θ equation (7.70). One assumes steady motions ($\partial/\partial t = 0$), conservative motions ($Q = 0$), and quasi-geostrophy (neglect the ω' term and assume, that the zonal mean basic state velocity is $[u]$ only). The last assumption means that since the mean flow has no $[v]$ and $[\omega]$, these motions are only due to the Stokes drift. A subscript s designates the Stokes velocities. Of course, this last assumption also provides a built-in connection between the Stokes drift and the mean meridional circulations derived in §6.3 in addition to the circulation discussed in §6.2.4. Combining the continuity equation for the Stokes motion with the assumptions stated above yields

$$[v]_s \frac{\partial[\theta]}{\partial y} + [\omega]_s \frac{\partial[\theta]}{\partial p} = -\frac{\partial}{\partial y}[v'\theta'] \tag{7.75a}$$

$$[\omega]_s = -\left(\frac{\partial[\theta]}{\partial p}\right)^{-1} \frac{\partial}{\partial y}[v'\theta'] \tag{7.75b}$$

A stream function ψ_s may be defined for the Stokes motion, similar to (7.68). Assuming that the meridional and pressure derivatives of ψ_s are comparable (e.g., as in Figure 6.12a), then the $[v]_s$ term in (7.75a) can be neglected if the surfaces of $[\theta]$ are assumed to be nearly horizontal. The same assumption equates the right-hand side of (7.75b) with a term in (7.74). In order to satisfy a continuity equation for Stokes flow, the heat flux term in (7.73) corresponds to $[v]_s$.

In deriving (7.75), a key assumption is that there be no change in the $[\theta]$ field. This means that any eddy heat flux convergences that attempt to change $[\theta]$ must be cancelled by the mean meridional circulation, ($[v]$, $[\omega]$). By (7.73) and (7.74) the changes to $[\theta]$ caused by the eddy heat fluxes can be thought of as "advection" by the Stokes flow. One could visualize the Stokes flow as attempting to flatten the $[\theta]$ isentropes in the same way that eddy heat fluxes would. To counteract this change, the mean meridional circulation ($[v]$, $[\omega]$) must deform the isentropes in the opposite direction. Two examples illustrate the point.

The Eady (1949) baroclinic instability problem has no meridional variation of $[u]$. As one consequence, plane waves, with troughs oriented north-south are the most unstable solutions. One can easily imagine a solution that has eddy heat flux increasing with p. By examining (7.73) through (7.75), this example is clearly one where $[\omega]_s = 0$ and $[v]_s > 0$. The baroclinic instability mechanism, by having meridional and vertical heat fluxes, can be represented by parcel motions as indicated

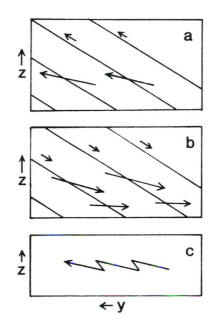

FIG. 7.25   Schematic meridional cross-sections (with higher latitudes on the left). Solid lines are zonal mean isentropes for a quasi-geostrophic wind field $[u]$ that only varies (increases) with height. Arrows cross the isentropes when there are nonzero eddy heat fluxes; baroclinic growth is indicated since the horizontal $[\theta]$ gradient is diminished. The eddy field is independent of meridional direction y. In this case the eddy heat fluxes increase with pressure but do not vary with y. (a) Arrows show eddy motion ahead (east) of the surface trough axis. (b) Arrows show eddy motion ahead of the surface high. (c) Resultant Lagrangian parcel motion from traversing the growing wave.

by the arrows in Figure 7.25. Eddy heat fluxes cause the arrows to cross the isentropes, with vertical and meridional fluxes as indicated. (For baroclinic growth, the arrow shaft must lie between the isentrope and the horizontal.) Clearly, these arrows transport heat horizontally in a way that reduces the meridional temperature gradient at low levels, while increasing the static stability. (Events at the bottom boundary are ignored in this example. In practice, the bottom boundary heat fluxes are quite important to the problem.) At higher levels, where the eddy heat flux is weak, the motions remain closer to isentropic surfaces. Since the angle between the isentropes and the parcel path increases with p, a parcel moving downwards does not move meridionally as far as when it moves upwards. Consequently, all parcels have a net poleward drift as indicated in the figure and consistent with (7.73).

A case where $[\omega]_s \neq 0$ and $[v]_s = 0$ is illustrated in Figure 7.26. The eddy pressure has maximum amplitude on the central latitude (y_0). The amplitudes of the high and low do not vary with height and boundary effects are again neglected. The heat fluxes are greatest at y_0 and vanish along the north and south edges of the domain. Parcel paths are again indicated by short arrows relative to the isentropes. The meridional component of the motion is greater in the middle of the domain, but

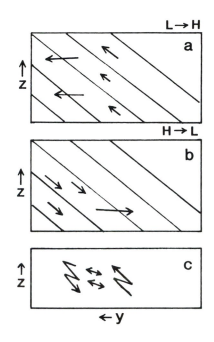

FIG. 7.26   Similar to Figure 7.25 except for a baroclinically growing eddy that has maximum amplitude along $y = 0$ and zero amplitude along the left and right edges of the domain pictured. The eddy amplitude and heat fluxes are assumed to be height invariant. $[u]$ only varies (increases) with height. (a) Arrows show eddy motions ahead of the surface low; cross-isentropic motions principally occur where $y > 0$. (b) Arrows show eddy motions ahead of the surface high; cross-isentropic motions occur mainly where $y < 0$. (c) Resultant Lagrangian trajectories for middle tropospheric parcels.

the parcel path is flatter. As parcels move away from y_0, they do not have as great a vertical component as when they move toward the center. The figure illustrates the sense of the Stokes motion. The reader will notice the similarity to the Stokes drift discussed earlier (§6.2.4; Figure 6.11c). Before leaving these examples, it should be apparent from the qualitative description in Figure 7.26 that (7.73), (7.74), and (7.75) are just an approximation to the actual drift.

Returning to the derivation at hand, one notes that the continuity equation remains

$$\frac{\partial v^\natural}{\partial y} + \frac{\partial \omega^\natural}{\partial p} = 0$$

Writing (7.69) as

$$\frac{\partial [u]}{\partial t} + \frac{\partial}{\partial y}\left([u]\frac{\partial \psi}{\partial p} \right) = f v^\natural + \nabla \cdot \mathbf{F} + R \tag{7.76}$$

using (7.74) to express (7.72) as

$$\frac{\partial[\theta]}{\partial t} + \frac{\partial[\theta]}{\partial y}\frac{\partial\psi}{\partial p} = \omega^{\natural}\frac{\partial[\theta]}{\partial p} + Q \tag{7.77}$$

One next combines (7.76) and (7.77) by first taking the pressure derivative of (7.76)

$$\underbrace{\frac{\partial^2[u]}{\partial p\partial t}}_{(A)} + \underbrace{\frac{\partial^2}{\partial p\partial y}\left([u]\frac{\partial\psi}{\partial p}\right)}_{(B)} = f\frac{\partial v^{\natural}}{\partial p} + \frac{\partial}{\partial p}(\nabla\cdot\mathbf{F}) + \frac{\partial R}{\partial p} \tag{7.78}$$

The meridional derivative of (7.77) multiplied by $1/(\rho f[\theta])$ obtains

$$\frac{1}{\rho f[\theta]}\left\{\underbrace{\frac{\partial^2[\theta]}{\partial y\partial t}}_{(C)} + \underbrace{\frac{\partial}{\partial y}\left(\frac{\partial[\theta]}{\partial y}\frac{\partial\psi}{\partial p}\right)}_{(D)} = \frac{\partial}{\partial y}\left(\omega^{\natural}\frac{\partial[\theta]}{\partial p}\right) + \frac{\partial Q}{\partial y}\right\} \tag{7.79}$$

Subtracting (7.79) from (7.78) leads to cancellation of terms A and C from the thermal wind relation (6.49). Using the chain rule, terms B and D become

$$\mathbf{B} = \frac{\partial}{\partial y}\left([u]\frac{\partial^2\psi}{\partial p^2} + \frac{\partial\psi}{\partial p}\frac{\partial[u]}{\partial p}\right) \tag{7.80}$$

$$\mathbf{D} = \left(\rho f[\theta]\right)^{-2}\frac{\partial\psi}{\partial p}\frac{\partial[\theta]}{\partial y}\frac{\partial}{\partial y}\left(\rho f[\theta]\right) + \frac{\partial}{\partial y}\left(\frac{-1}{\rho f[\theta]}\frac{\partial\psi}{\partial p}\frac{\partial[\theta]}{\partial y}\right) \tag{7.81}$$

The second terms in (7.80) and (7.81) cancel from (6.49). The remaining two terms are neglected by Edmon et al. (1980) from scaling arguments based on the smallness of ageostrophic motions. The result of subtracting (7.79) from (7.78) is therefore

$$f\frac{\partial v^{\natural}}{\partial p} - \frac{1}{\rho f[\theta]}\frac{\partial}{\partial y}\left(\omega^{\natural}\frac{\partial[\theta]}{\partial p}\right) = -\frac{\partial}{\partial p}\nabla\cdot\mathbf{F} - \frac{\partial R}{\partial p} + \frac{1}{\rho f[\theta]}\frac{\partial Q}{\partial y} \tag{7.82}$$

A new stream function $\psi^{\natural}$ can be defined that is related to $\omega^{\natural}$ and $v^{\natural}$ as in (7.68). Making that substitution obtains a second-order partial differential equation for $\psi^{\natural}$ that is analogous to the Kuo-Eliassen equation (6.50). On the right-hand side of (7.82) is forcing by the pressure derivative of friction and the differential diabatic heating just as in (6.50). The forcing by eddy heat and momentum fluxes in (6.50) has been replaced by the pressure derivative of the divergence of the EP flux. A crucial difference from before is that we are solving for $\psi^{\natural}$, not ψ. The former *removes* any eddy flux contribution that is simply cancelled by a mean meridional cell. The intention is to draw a better picture of how the eddies alter the mean flow. This issue is covered further in §7.4.3. Another difference from §6.3 occurs at the bottom boundary of the atmosphere; $\psi^{\natural} \neq 0$ but is proportional to the eddy heat flux there (Pfeffer, 1987; Hayashi, 1985).

7.4.2   Example 2: Potential vorticity

Another connection worth mentioning is how the EP flux relates to the concept of potential vorticity. For adiabatic motion the Ertel potential vorticity is a conserved quantity. That is,

$$\frac{d}{dt}\left(\frac{\zeta + f}{\frac{\partial P}{\partial \theta}}\right) = 0$$

where the Ertel potential vorticity is the quantity inside the parentheses, and ζ is the relative vorticity normal to a θ surface. (Definitions in other coordinates may be found in Pedlosky, 1987, pp.38–42.) For quasi-geostrophic motion a similar quantity can be defined from the vertical component vorticity equation (to first order in the Rossby number).

$$\frac{d}{dt}(\zeta + f) = f\frac{\partial \omega}{\partial p} \tag{7.83}$$

A nondimensional form of the vorticity equation is used here. The adiabatic equation is

$$\omega = \frac{-1}{S}\frac{d\theta}{dt} \tag{7.84}$$

where $S = [\partial \theta / \partial p]$ is the average pressure derivative of potential temperature. Consistent with the quasi-geostrophic approximation, one assumes that S is a constant. Thus (7.84) is analogous to (7.8), used earlier. If one differentiates (7.84) with respect to pressure and substitutes the result into (7.83), then to first order,

$$\frac{dq}{dt} \equiv \frac{d}{dt}\left(\zeta + f\left(1 + \frac{\partial}{\partial p}\left\{\frac{\theta}{S}\right\}\right)\right) = 0 \tag{7.85}$$

This form of the potential vorticity equation is analogous to (7.1), and quasi-geostrophic potential vorticity q is again conserved.

One can separate q into zonal mean and eddy parts

$$q = [q] + q'$$

The zonal mean part includes the planetary vorticity, leaving

$$q' = \frac{\partial v'}{\partial x} - \frac{\partial u'}{\partial y} + f\frac{\partial}{\partial p}\left(\frac{\theta'}{S}\right) \tag{7.86}$$

From this equation it is straightforward to show that the divergence of **F** equals the northward flux of q'.

Multiplying (7.86) by v' and using the chain rule,

$$
\begin{aligned}
v'q' =&\frac{1}{2}\frac{\partial}{\partial x}((v')^2) - \frac{\partial}{\partial y}(u'v') + u'\frac{\partial v'}{\partial y} \\
&+ \frac{f}{S}\left(\frac{\partial}{\partial p}(v'\theta') - \theta'\frac{\partial v'}{\partial p}\right)
\end{aligned}
\tag{7.87}
$$

The perturbation velocities are geostrophic and thus are nondivergent; they also satisfy the nondimensional thermal wind relation.

$$
\frac{\partial u'}{\partial x} + \frac{\partial v'}{\partial y} = 0 \qquad \text{and} \qquad \frac{\partial v'}{\partial p} = \frac{\partial \theta'}{\partial x}
\tag{7.88}
$$

Therefore (7.87) becomes

$$
\begin{aligned}
v'q' =&\frac{1}{2}\frac{\partial}{\partial x}((v')^2) - \frac{\partial}{\partial y}(u'v') - \frac{1}{2}\frac{\partial}{\partial x}((u')^2) \\
&+ \frac{f}{S}\left(\frac{\partial}{\partial p}(v'\theta') - \frac{1}{2}\frac{\partial}{\partial x}((\theta')^2)\right)
\end{aligned}
$$

Taking a zonal average around a latitude circle causes the x derivative terms to vanish leaving

$$
[v'q'] = -\frac{\partial}{\partial y}[u'v'] + \frac{f}{S}\frac{\partial}{\partial p}[v'\theta'] \equiv \nabla \cdot \mathbf{F}
\tag{7.89}
$$

7.4.3 Example 3: Zonal mean zonal wind maintenance

Dickinson (1969) shows that the zonal mean flow can be related to the eddies by the following argument. One writes (7.85) in flux form and applies the zonal average. The x derivatives vanish because the integration is around a closed latitude circle. Since (7.85) is for the quasi-geostrophic system, the advecting velocities are geostrophic, hence $[v] = 0$.

$$
\frac{\partial [q]}{\partial t} + \frac{\partial}{\partial y}[v'q'] = 0
$$

Differentiating with respect to y yields

$$
-\frac{\partial^2 [q]}{\partial t \partial y} = \frac{\partial^2}{\partial y^2}[v'q']
\tag{7.90}
$$

From (7.86) and (7.85) one may write

$$
-\frac{\partial [q]}{\partial y} = \mathbf{L}^\star[u]
\tag{7.91}
$$

where $\mathbf{L}^\star = C\partial^2/\partial p^2 + \partial^2/\partial y^2$ and C is a constant. Therefore (7.90), (7.91) and (7.89) are related as follows

$$\frac{\partial \mathbf{L}^\star[u]}{\partial t} = \frac{-\partial^2[q]}{\partial t \partial y} = \frac{\partial^2}{\partial y^2}[v'q'] = \frac{\partial^2}{\partial y^2}(\nabla \cdot \mathbf{F}) \qquad (7.92)$$

One might be tempted to assume that the pressure derivatives in $\mathbf{L}^\star$ can be neglected, in which case the divergence of the EP flux will increase $[u]$ up to integrative constants. However, that assumption is misleading (Pfeffer, 1987). One can get a sense for the problems by examining the quasi-geostrophic form of (7.76), where advecting velocities are geostrophic, leading to $[v] = 0$ once again. Then

$$\frac{\partial[u]}{\partial t} = f[v]_s + \nabla \cdot \mathbf{F} + R$$

where (7.73) has been used. It is clear from this equation that part of the EP flux divergence will be cancelled by the Stokes drift. The cancellation should be obvious immediately since heat fluxes do not appear in the zonal momentum equation! Pfeffer (1987) shows that the first two terms on the right-hand side are large, but nearly cancel. In order to eliminate the explicit dependence on $[v]_s$, one obtains the more complicated operator in (7.92). The sensitivity of the operator $\mathbf{L}^\star$ is discussed in Pfeffer (1987). For the general case (spherical geometry and variable S) $\mathbf{L}^\star$ has variable coefficients. Pfeffer shows that the local derivative of $[u]$ depends strongly on the vertical pattern of the right-hand side of (7.92) and not on the local value of $\nabla \cdot \mathbf{F}$, especially in the higher latitudes of the troposphere. The sensitivity arises from C, which is inversely dependent on static stability S. S is small in the troposphere. That sensitivity probably explains why $\nabla \cdot \mathbf{F}$ and $\partial[u]/\partial t$ are not highly correlated in the troposphere. Baldwin et al. (1985) find the average correlations to be 0.34 in the upper troposphere of the Northern Hemisphere middle latitudes during a winter period. In Hartmann et al. (1984), the best correlations are in the middle stratosphere and near the tropopause; only a few grid points exceed 0.6. In the stratosphere, S is much larger, and one is further away from the surface boundary, so the EP flux divergence method works better; more details are given in §7.4.6. In the troposphere, Pfeffer shows that the traditional reliance upon eddy momentum convergence is a better way to diagnose eddy changes to the zonal mean flow. What one can conclude about (7.92) is that it is necessary for the eddies to transport potential vorticity in order for them to alter the zonal mean zonal flow.

7.4.4   Example 4: Eddy energy propagation

One may draw a connection between the EP flux and such subtle concepts as wave action and group velocity. To illustrate, one can begin by linearizing (7.85) about a known, zonal mean state.

$$\frac{\partial q'}{\partial t} + [u]\frac{\partial q'}{\partial x} + v'\frac{\partial [q]}{\partial y} = 0 \tag{7.93}$$

Multiplying (7.93) by q' and taking the zonal average,

$$\frac{1}{2}\frac{\partial}{\partial t}[(q')^2] + [v'q']\frac{\partial [q]}{\partial y} = 0 \tag{7.94}$$

where the x derivatives again vanish. Rearranging and employing (7.89) provides

$$\frac{\partial A}{\partial t} \equiv \frac{\partial}{\partial t}\left(\frac{[q'q']}{2\frac{\partial [q]}{\partial y}}\right) = -[v'q'] = -\nabla \cdot \mathbf{F} \tag{7.95}$$

where A is defined by (7.95). Edmon et al. (1980) refer to A as the "EP wave activity." A is related to (but differs from) the wave action. (Wave action is usually defined as the density of total energy divided by the frequency of the wave.) The expression (7.95) also assumes that $\partial [q]/\partial y$ does not vanish. In problems where it does vanish, such as Eady (1949), then Edmon et al. cast the definition of A in terms of air parcels whose meridional displacements are produced by the eddies (as outlined by Bretherton, 1966). It is worth noting that (7.95) is a special case of the so-called "generalized Eliassen-Palm relation"

$$\frac{\partial A^\star}{\partial t} + \nabla \cdot \mathbf{F} = D \tag{7.96}$$

derived by Andrews and McIntyre (1976). In (7.96) D represents frictional and other diabatic effects and $A^\star$ has a slightly different definition from A because the quasi-geostrophic assumption is lifted. One could view $\nabla \cdot \mathbf{F}$ in (7.96) as being the flux form of the transport of $A^\star$ in the meridional plane. In other words, $\mathbf{F}$ is proportional to the Rossby wave group velocity. That interpretation will be used in §7.4.5. A similar but related proportionality is now demonstrated for a special case.

To simplify the mathematics, the quasi-geostrophic system of equations is assumed to be incompressible. Matters are further simplified by letting $[u]$ be a constant. This leaves the eddy potential vorticity equation

$$\frac{\partial q'}{\partial t} + [u]\frac{\partial q'}{\partial x} + \beta\frac{\partial \psi}{\partial x} = 0 \tag{7.97}$$

From (7.85),

$$q' = \nabla^2\psi + \frac{f}{S}\frac{\partial \theta}{\partial p} = \nabla^2\psi + \epsilon\frac{\partial^2\psi}{\partial p^2} \tag{7.98}$$

where $\epsilon = f^2/(gS^2)$ and is inversely proportional to the Brunt-Väisälä frequency. The second equality in (7.97) comes from a quasi-geostrophic formulation of the eddy potential temperature in terms of the stream function.

$$\theta = \frac{f}{gS} \frac{\partial \psi}{\partial p} \tag{7.99}$$

This formulation required two further assumptions (Grotjahn, 1987). First, the simplest linear balance relation is assumed: $\psi = f\Phi$, where Φ is eddy geopotential and f is assumed constant. Second, the global horizontal average geopotential is assumed to vary as $p^{\kappa-1}$ where $\kappa = R/C_p$.

One assumes the eddies to be three-dimensional plane waves capable of propagating in any direction. The eddy stream function is defined as

$$\psi = \psi_0 \exp(i(kx + ly + mp - \sigma t)) \tag{7.100}$$

Substituting (7.100) and (7.98) into (7.97) gives

$$-\{-\sigma + k[u]\}(k^2 + l^2 + \epsilon m^2) + k\beta = 0$$

Solving for phase frequency obtains

$$\sigma = k[u] - \frac{k\beta}{k^2 + l^2 + \epsilon m^2} \tag{7.101}$$

The solution is a generalization of the well-known Rossby wave phase speed formula to motion in three dimensions. The group velocity is obtainable from (7.101).

$$\mathbf{C}_g \equiv (C_{gx}, C_{gy}, C_{gp}) = \left(\frac{\partial \sigma}{\partial k}, \frac{\partial \sigma}{\partial l}, \frac{\partial \sigma}{\partial m}\right)$$

Therefore,

$$C_{gx} = [u] - \frac{\beta}{k^2 + l^2 + \epsilon m^2} + \frac{2k^2\beta}{(k^2 + l^2 + \epsilon m^2)^2} \tag{7.102}$$

$$C_{gy} = \frac{2kl\beta}{(k^2 + l^2 + \epsilon m^2)^2} \tag{7.103}$$

$$C_{gp} = \frac{2\epsilon km\beta}{(k^2 + l^2 + \epsilon m^2)^2} \tag{7.104}$$

The EP flux is expressed using geostrophic velocities and potential temperature (7.99).

$$F^y = -[u'v'] = \left[\frac{\partial \psi}{\partial x} \frac{\partial \psi}{\partial y}\right] \tag{7.105}$$

$$F^p = \frac{f}{S}[v'\theta'] = \frac{f^2}{gS^2}\left[\frac{\partial \psi}{\partial x} \frac{\partial \psi}{\partial p}\right] \tag{7.106}$$

Substituting (7.100) into (7.105) and (7.106) yields

$$F^y = kl[\psi^2] \tag{7.107}$$

and

$$F^p = \epsilon km[\psi^2] \tag{7.108}$$

Inspection of (7.95) implies a wave activity advecting velocity defined as

$$\mathbf{C}^\star = (0, C_y, C_p) = \frac{\mathbf{F}}{A} = (0, \frac{F^y}{A}, \frac{F^p}{A}) \tag{7.109}$$

Substituting (7.99) into the definition of q' in (7.98) gives

$$[q'q'] = (k^2 + l^2 + \epsilon m^2)^2 [\psi^2] \tag{7.110}$$

Substituting (7.110) into the definition of A in (7.95) yields

$$C_y = \frac{2kl}{(k^2 + l^2 + \epsilon m^2)^2} \frac{\partial[q]}{\partial y} \tag{7.111}$$

$$C_p = \frac{2km\epsilon}{(k^2 + l^2 + \epsilon m^2)^2} \frac{\partial[q]}{\partial y} \tag{7.112}$$

Comparing (7.112) and (7.111) with (7.104) and (7.103), it follows that

$$(C_{gy}, C_{gp}) = \beta \left(\frac{\partial[q]}{\partial y}\right)^{-1} (C_y, C_p) \tag{7.113}$$

If the dominant term in $\partial[q]/\partial y$ is the Coriolis term, then $C_y = C_{gy}$ and $C_p = C_{gp}$. The importance of (7.113) is that the components of the EP flux have the same proportionality as the vertical and meridional components of Rossby wave group velocity. Thus, $\mathbf{F}$ gives the direction and relative magnitude of the group velocity components projected onto the (y, p) plane. Palmer (1982) shows that a relationship like (7.113) still holds for less restricted equations than the example developed here.

7.4.5   EP flux diagrams for frontal cyclones

The background provided by the previous subsections allows interpretation of frontal cyclone evolution in terms of the distribution of EP fluxes. Figures 7.27 and 7.28 show plots of the EP flux vector (plotted as an arrow) and contours of $\nabla \cdot \mathbf{F}$. The vertical component of the arrows arises from meridional eddy heat fluxes; the horizontal component is from eddy momentum fluxes. In these figures a vertically pointing $\mathbf{F}$ vector means F^p is *negative*.

The arrows are scaled by the cosine of latitude so that the convergence of $\mathbf{F}$ is reflected in the visual convergence of the arrows. Baldwin et al. (1985) state that this scheme emphasizes the fluxes in the lower troposphere, due to an $\exp(-z/H)$ scaling. They chose to plot the fluxes using $\ln p$ as the vertical coordinate and with the exponential height factor scaled out. These changes sacrifice the appearance of nondivergent arrows along the zero contour of $\nabla \cdot \mathbf{F}$, but heat and momentum flux components of $\mathbf{F}$ are scaled equally and can be directly compared.

Figure 7.27 shows the EP flux and its divergence for the theoretical models discussed in §7.3.1. The flux changes with the eddy life-cycle as follows. (a) Initially the flux is concentrated near the surface and is vertically pointing, so the baroclinic energy conversion dominates at the early stage. The pattern looks a lot like the linear solution in Figure 7.27a. (b) Later, however, the heat fluxes extend through a greater depth, shifting the maximum convergence (negative contour values) to higher altitudes. It is tempting to assume that the convergence of the vertically pointing arrows (due to eddy heat flux variation with height) will weaken $\partial[u]/\partial z$ more in the lower troposphere than in the upper troposphere, initially. The weakened shear might be further assumed to diminish $[u]$ in the upper troposphere, where $\mathbf{F}$ arrows converge. However, the connection between heat fluxes and changes in $[u]$ requires care in the troposphere (§7.4.3), and this assumption may not be entirely true. For example, Grotjahn and Lai (1991; Figure 7) find that the low-level shear decreases, but the jet is *not* changed in the early stages of their simulation. (c) As the eddy matures and begins to decay, the arrows begin to tilt equatorward, mainly in the upper troposphere, indicating momentum divergence and convergence. The positive contour values extend from the surface to the upper troposphere, but close inspection of the direction of the arrows implies that only the upper positive maximum is due to eddy momentum convergence; the lower positive values are due to the stronger heat flux at mid-levels compared with the surface. The stronger negative area moves equatorward and upward during its life cycle; it comes from convergence of eddy momentum flux out of the subtropics, implying reduced $[u]$ there. Unlike the convergence of vertically pointing arrows (due to heat fluxes), the convergence of these horizontally pointing arrows (due to momentum fluxes) is probably an indicator of $[u]$ changes.

The movement of the center of maximum negative values reflects the motion ("group velocity") of the Rossby wave. The movement of this center reflects what was shown in the amplitude profiles: the eddy amplitude "moves" vertically from having maximum value at the surface initially to having maximum amplitude in the upper troposphere for the mature storm. Recalling Figure 7.18, the eddy initially has large amplitude at the tropopause and at the surface. As the eddy matures, the amplitude near the tropopause is most enhanced.

The net effect, averaged over the life cycle, is shown in Figure 7.27d. Over the life cycle the jet stream is reduced above and equatorward of the eddy. Judging from Figure 7.27c, the arrows imply reduction of the zonal wind speed at the jet axis and enhancement of velocity on the poleward side. So, the zonal mean jet is either smeared

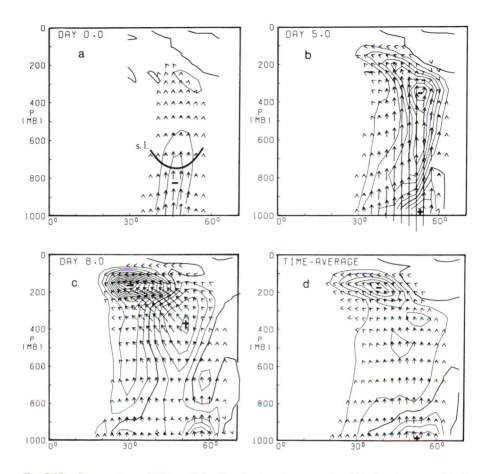

FIG. 7.27 Cross-sections of Eliassen-Palm flux (**F**; plotted as vectors) and EP flux divergence ($\nabla \cdot \mathbf{F}$; contours). Vertical arrows are eddy heat fluxes, horizontal arrows are eddy momentum fluxes as in equation (7.66). This figure shows results for the nonlinear model of Simmons and Hoskins (1980) where the initial condition has a $[u]$ jet at 45 N. (a) Solution for equations linearized about $[u]$. Nonlinear solutions at (b) day 5 and (c) day 8 using contour interval 4×10^{15} m^3. (c) Time-mean cross-sections for an unspecified period over the life cycle, with contour interval 1.5×10^{15} m^3. The units of EP flux divergence differ from (7.66) because the integration around a latitude circle with respect to mass requires division by the acceleration of gravity. From *J. Atmos. Sci.*, Edmon et al. (1980) by permission of American Meteorological Society.

out or displaced poleward. These inferred changes to the jet stream are consistent with the model results reported in §7.3.1.

How does the theoretical distribution of EP fluxes compare with observed EP fluxes? To facilitate comparison one must first isolate the appropriate phenomena. Randel and Stanford (1985) designed a study to directly compare observations with the nonlinear simulations of Simmons and Hoskins (1978). Selected EP flux diagrams are presented in Figure 7.28 using the same convention as in the previous figure. (There are differences. One difference is they derive wind and temperature fields from geopotential height data. Another is they use a 1–2–1 filter in latitude to smooth the geopotential heights.) They chose a wintertime period during which the Southern Hemisphere proceeded from a strongly zonal flow to a prominent wavenumber 6 and back to a strong zonal flow. The Southern Hemisphere was chosen because the stationary long wave pattern is weak; those long waves make the comparison with Figure 7.27 in the Northern Hemisphere (e.g., Edmon et al. , 1980) problematic. One caution about the Southern Hemisphere data is that the first guess field can dominate the data (§2.4); since the first guess comes from the model, one should expect some built-in similarity between two sets of EP flux diagrams. However, other observational studies (Edmon et al., 1980; Hartmann et al. , 1984) show a similar time-average pattern. In short, the evolution of the EP flux shown in Figure 7.28 is generally quite similar to the theoretical model. Wave activity propagates upward, then equatorward. Vertical shear decreases at low levels, but the jet maximum is little changed. By the end of the period the $[u]$ jet has moved about five degrees poleward.

7.4.6   *Stratospheric sudden warming*

As stated in §7.4.3, the divergence of the EP flux has proven to be a more valuable indicator of mean flow changes in the stratosphere than in the troposphere. Nonetheless, EP fluxes were used (§7.4.5) to illuminate some aspects of frontal cyclone dynamics in the troposphere, most notably the vertical propagation of wave activity. A related discussion of stratospheric wave life cycles can be found in Randel (1989). EP fluxes have been usefully applied to other phenomena that are peculiar to the stratosphere. One example, stratospheric sudden warming, is the subject of this subsection.

During the winter and spring, occasional periods of rapid warming occur in the polar middle stratosphere. Because the polar region warms more rapidly than the surrounding midlatitudes, the latitudinal temperature gradient is greatly diminished or even reversed over the course of a few days. From thermal wind balance, the polar night jet is greatly diminished in strength. If the warming is extreme, an easterly jet may be established. Such a warming event often marks the change from winter to summer. From Figure 3.11, one recalls that the warmest temperatures in the middle stratosphere are at the summer pole, the coldest at the winter pole. Several reviews of sudden warming observations have been published (e.g., Labitzke, 1982). Mechoso (1989) focuses on the final transition from winter to summer conditions and compares the Northern and Southern Hemispheres.

Our understanding of the dynamics behind the sudden warming has improved markedly over the past two decades. An early theory was developed by Matsuno (1971). He used a spherical coordinates model consisting of an eddy potential vorticity equation, similar to (7.93), and a zonal mean flow potential vorticity tendency equation. These equations are forced by a time dependent eddy amplitude at the bottom of the domain (10 km elevation), which has no momentum flux. Because the

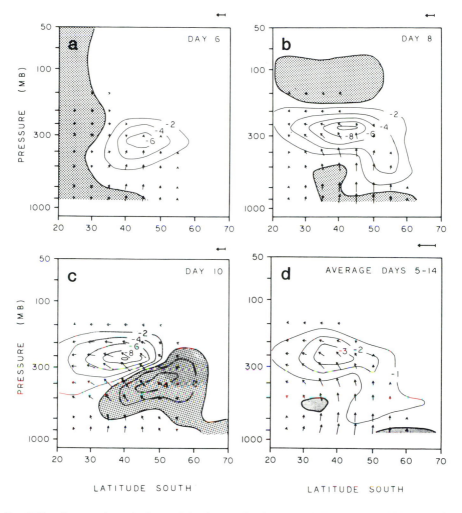

FIG. 7.28 Cross-sections of Eliassen-Palm flux and its divergence similar to Figure 7.27, except for observations of transient eddies during a special period (as in Figure 7.24b). Southern Hemisphere data from a period when an approximately zonal flow develops into a strong wavenumber 6 pattern that later decays. Figures 7.28b and 7.28d may be directly compared with Figures 7.27c and 7.27d, respectively. U.S. NMC data. From *J. Atmos. Sci.*, Randel and Stanford (1985) by permission of American Meteorological Society.

waves propagate upwards, and eastward, the trough and ridge axes have vertical tilt upstream relative to $[u]$. From our analysis of baroclinic instability, such an upstream tilt has eddy heat fluxes down the gradient of the zonal mean temperature. Thus, the forced waves are warming the middle stratosphere. To preserve thermal wind balance, a mean meridional circulation is set up (rising at the pole, sinking in midlatitudes), which decelerates $[u]$. This model reveals much about the sudden warming. Additional properties have been diagnosed with the aid of EP fluxes.

O'Neill and Taylor (1979) carefully examined not only heat fluxes, but also momentum fluxes in the middle stratosphere during three warming events. They found eddy momentum divergence occurred during part of each event, consistent with the slowing down of $[u]$. Coupled with the poleward heat flux, no mean meridional circulation could maintain the $[u]$ field, and the zonal flow rapidly gained easterly momentum. During two periods of marked $[u]$ acceleration, the eddy momentum flux at 20 hPa was opposite to the flux at 300 hPa. However, the eddy momentum field was not always consistent with changes in $[u]$. Palmer (1981a,b) reported similar lack of consistency. During two periods of marked $[u]$ acceleration, O'Neill and Taylor found the eddy momentum flux convergence at 20 hPa and divergence at 300 hPa. These features have been largely explained by the three-dimensional propagation of the long waves.

Vertical propagation was briefly discussed in §7.2.5. Charney and Drazin (1961) derived a quantity similar to a refractive index to describe the vertical attenuation of long waves. As stated before, the index allows only the longest waves to penetrate into the middle stratosphere. Figure 7.29 illustrates how the vertical propagation differs between wavenumbers 1 and 3. The amount of time to propagate from the middle troposphere to middle stratosphere is 1 to 4 days for these waves (Randel, 1987). The discussion in §7.2.5 assumed a very simple mean flow, but the atmosphere contains jets. One can gain a sense for the influence of shears by generalizing the refractive index (7.65). The β term in (7.65) is the meridional gradient of zonal mean potential vorticity, $[q]_y$. O'Neill and Youngblut (1982) and others have noted that the $-\partial^2[u]/\partial y^2$ term in $[q]_y$ plays a key role in the vertical propagation. During winter, there may be a minimum $[u]$ lying between the strong winds of the subtropical and polar night jets (e.g., 45 S, 200 hPa in Figure 3.15b). Such a minimum in $[u]$ can create a minimum in $[q]_y$ and thus in the refractive index. Ray paths will be refracted away from such a minimum with the result that the ray paths may be split, some heading upwards and equatorward, some upwards and poleward. As pointed out in §7.4.4, the EP fluxes should reveal such a split, which they do.

The EP fluxes show the vertical propagation and also the stratospheric mean flow changes very nicely. As discussed in §7.4.3, divergence of the EP flux should accelerate the flow; convergence should decelerate the flow. Figure 7.30a illustrates possible EP flux patterns during typical winter conditions and Figure 7.30b shows a possible pattern during a sudden warming. Figure 7.30 is schematic; paths based on observed EP fluxes can be seen in Palmer (1981a), O'Neill and Youngblut (1982), and elsewhere. The typical situation is similar to the pattern in the previous figure: EP vectors point upward then equatorward. Figure 7.30b shows the wave activity being split, and gives some sense for how the group velocity (§7.4.4) varies with height. The variation of **F** can mean that the wave energy moves more slowly at a higher latitude (point A) than at a lower latitude (point B). The variation in group speed may explain the reversal of eddy momentum flux between troposphere and stratosphere. A simple conceptual model illustrates the point. One might assume that the long wave

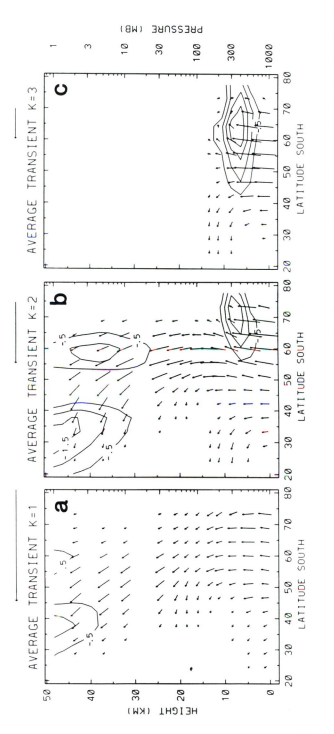

Fig. 7.29 EP flux diagrams for individual long waves in the Southern Hemisphere based on 5-year averages of winter periods. U.S. NMC data from June through September during 1982 through 1986 are used. (a) Wavenumber 1; (b) wavenumber 2; (c) wavenumber 3. Only the lowest wavenumbers penetrate far into the stratosphere. Higher wavenumbers (e.g., Figure 7.28d) are similar to (c). Contour interval is 0.5 ms^{-1} per day. Figure courtesy of W. Randel, personal communication.

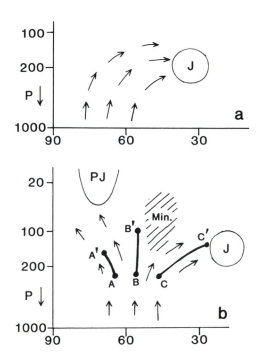

Fig. 7.30 Schematic meridional cross-sections for two Northern Hemisphere winter conditions. Arrows show EP vectors as in the previous diagrams. (a) Typical winter pattern, similar to Figure 7.27d. **J** marks location of subtropical jet. (b) Situation during stratospheric sudden warming. Polar night (**PJ**) and subtropical jets shown; in between, hatched area shows region of minimum in $[u]$, $[q]_y$ and refractive index. Letters **A**, **B**, and **C** illustrate possible ray paths inferred from EP flux vectors, including distance traveled over a fixed time period.

has no horizontal tilt at 200 hPa; the trough axis is oriented north-south. Since the long wave has a poleward heat flux (**F** points upward), then the long-wave trough must tilt towards the west with height. As the wave moves vertically, it will appear to move towards the east when viewed on, say, the 20 hPa surface. Since the upward motion is less at higher latitudes, the wave will not move as far at the higher latitudes. The result is that a northwest-to-southeast tilt appears at 20 hPa, even though the wave doing the forcing has no tilt. The effect is hard to draw, but easy to show with a tilted sheet of paper representing the trough axis. Raising the "midlatitude" edge of the paper faster than the "polar" edge creates the horizontal tilt. This northwest-to-southeast tilt implies EP vectors that point north, an obvious feature in Figure 7.30b. In this possible sudden warming scenario, both the heat and momentum fluxes contribute to the convergence of **F** and the deceleration of $[u]$.

Figure 7.31 illustrates the sudden warming event in synoptic terms. The horizontal tilts in the long waves shown match the horizontal components of the EP vectors given in Figure 7.30b. The figure shows the splitting of the wave and the direction of the eddy momentum fluxes (double-shafted arrows). Since the wave takes several

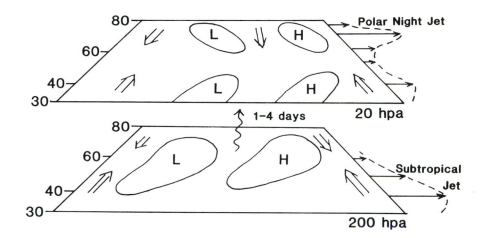

FIG. 7.31   Schematic example of the horizontal wave structure at two pressure levels during a stratospheric sudden warming. Diagram is consistent with Figure 7.30b. Lower level (200 hPa) long wave has eddy momentum (double-shafted arrows) convergence. Length of arrows on right edge is proportional to $[u]$. Wavy arrows indicate the upward propagation of the long wave over a few days. At the upper level (20 hPa), the long wave has split and the northern pair of lows has eddy momentum flux down the gradient of $[u]$. Horizontal tilts are intended to match the information in Figure 7.30b; other long wave properties, including phase information between the two levels, is arbitrary.

days to reach 20 hPa, no phase relationship is intended between the patterns at 20 and 200 hPa.

As mentioned above, EP flux concepts have been applied to other atmospheric phenomena. The principal value of the EP flux diagnostics lies in the ways that these fluxes unify various pieces of synoptic and dynamic information. It therefore is appropriate to conclude our observational and theoretical presentations with this unifying concept.

8

APPLICATIONS

Had the sun revolved round the earth, and not the earth on its axis, the air over the torrid zone, and particularly about the equator, would have been in effect stagnant; and in other zones the winds would have had little variation either in strength or direction.

Daulton, 1793

We now begin to conceive what a powerful machine the atmosphere must be; and, though it is apparently so capricious and wayward in its movements, here is evidence of order and arrangement. [The atmosphere] is therefore as obedient to law as the steam-engine to the will of its builder.

Maury, 1855

Mythologies of ancient cultures are often rooted in some factual material. For example, several pre-Columbian cultures believed that the god Quetzalcoatl created four lesser gods, one for each direction of the wind, each the bringer of a different type of weather. The kernel of truth here is the reasonable, though approximate, relationship between type of weather and surface wind direction. The expression of the evidence in terms of gods is the language of those ancient cultures, somewhat as mathematics is the language of modern science.

My undergraduate physical anthropology instructor used to stress that a field was not a "science" unless it could *predict* a result. He was obsessed with the issue, perhaps because the study of the past offers fewer opportunities to predict the future. Meteorologists do not share his anxiety since weather prediction is both routine and accurate. His test of the "scientificness" of a discipline is too narrow, but it does provide a premise for this chapter. The narrowness is illustrated by another example: the weather adage "Red sky at night, sailor's delight; red sky at morning, sailor take warning" is a prediction of sorts. However, most people would not consider that adage scientific, but at best empirical.

This chapter will be largely empirical. I shall use first and second person forms of address to distinguish the speculative nature of my remarks in this chapter from the third person forms used elsewhere in this book. Speculations can be fun, occasionally are useful, and benefit from some creativity. Before there can be creativity, there must be knowledge. Just as a good artist must know his subject before commencing to draw, so too must a scientist gather comprehensive knowledge first. The preceding chapters have provided us with a basic knowledge of the structures and mechanisms comprising the general circulation. A creative use for this knowledge is to attempt to predict the circulations that could arise in various scenarios. This chapter illustrates two applications that are relevant to meteorologists and that provide an opportunity to review several conclusions reached earlier in this book.

8.1   Speculations on a qualifying exam problem

When I took my Ph.D. qualifying exam more than 15 years ago, I was asked what the weather would be like if the Earth rotated in the opposite direction, from east to west. Some variant of this question has been used in many exams before and since. The question is simple to state, but hard to verify. It provides the student with plenty of latitude to make an impression, favorable or not.

Given the commonplace usage of general circulation models, it is tempting to just reverse the sign of the Coriolis parameter and spin up a GCM from a rest state. The result of such an experiment would be of uncertain relevance because the oceanic circulation would be much different if the Earth's rotation were reversed. The ocean not only handles some of the necessary poleward heat transport, it also seems to play a key role in the locations of cyclogenesis (§4.5.1.3; §5.8) regions and the ICZ (§6.7). I suppose one could make a rough guess about the ocean surface: replace the western boundary currents with eastern boundary currents, etc. If the model included the diurnal cycle then that would have to be reversed as well. This revised GCM experiment might give a reasonable estimate of the anticipated circulation. Even so, there are two problems. First, as far as I know, the revised experiment has never been tried, so we cannot consult it before proceeding here. Second, I suspect that one would learn less by dealing with computer code and numeric output, than by trying to reason inductively, even if the intuition turned out to be wrong. Even if one were to perform the revised GCM experiment, it still would be crucial to perform the thought experiment first. The thought experiment provides you with far more powerful hypotheses to test than "I wonder what would happen if . . . ." From this context, it is clear that the latter hypothesis is ambiguous and does not give you a basis to check the model results. I believe that this point is important and not always communicated to students who work with GCMs.

In our thought experiment, we begin by speculating about the zonal mean circulation that results from reversing the rotation. Aside from the fact that this book is organized around that partitioning, we can anticipate that the zonal average circulation for the exam question would look pretty similar to the real Earth's atmosphere. The zonal variations, however, could be very different. One would find Hadley and Ferrel cells in similar locations (§3.3), and there would be subtropical and stratospheric jets similar to those shown in this book (§3.3). The most obvious difference would be that the [u] velocities would be reversed: the subtropical jet would be easterly and westerlies would occur at low levels of the tropics. Geopotential heights would still be higher along the equator than at the poles due to the solar heating (§1.2; §3.4). The Coriolis force would be directed toward the left in the Northern Hemisphere, instead of the right, necessitating the reversed direction of the geostrophic winds. The mean meridional cells would not be affected in that way since [v] and [ω] are ageostrophic motions. The trade winds would be northwesterly in the Northern Hemisphere and southwesterly in the Southern Hemisphere. Some subtler differences might arise. Perhaps the zonal mean subtropical jet would be weaker in the Northern Hemisphere, but unchanged in the Southern Hemisphere due to changes in the zonal variations.

The more interesting parts of this question concern changes to the zonally varying parts of the general circulation. A simple assumption is that the circulations would be mirror images: weather on the east and west sides of each ocean basin would be swapped. I think the differences would be much more interesting than that.

The mountains of eastern Asia, the intense tropical convection in the vicinity of Indonesia, and the cyclogenesis near the Kuro Shio coordinate to make the Asian subtropical jet much stronger than the others (§6.5; §7.2.4). For a reversed rotation, the Kuro Shio is gone and the "eastern boundary current" is many thousands of kilometers away. The mountain ranges remain, but the flow is down their western sides, which have a much different shape than the eastern sides. Strong convection is likely to persist over Indonesia, though the northern summer monsoon would be directed oppositely to the trade winds. These changes suggest that the coordination would be lost, and presumably, the impetus behind the dominant Asian subtropical jet would also be lost. It is not clear where the decrease of the Asian jet could be made up. North America has mountains in a favorable location, and presumably cyclogenesis would prevail off the California coast, but the tropical convection may not be as greatly concentrated in the eastern Pacific as it now is in the west. The western Pacific has large islands that locally enhance the convection, whereas the eastern Pacific does not. Also, the Rockies help block cold continental air from reaching the west coast, a fact that could reduce the contrast between frontal-cyclone air masses and possibly the cyclone intensity. In the northeastern Atlantic, several features could conceivably coordinate. Monsoonal type flows could make the western Sahara a wetter place than at present. However, the Alps are not comparable in areal extent to the Tibetan plateau.

A variety of other changes might occur in the Northern Hemisphere. The time-mean long waves would still be prominent only in the Northern Hemisphere. To the extent that topography dominates the forcing (§7.2.1), then long wave troughs would prefer the west coast of North America and somewhere over eastern Europe and western Asia. The former would be supported by transient eddies (§7.2.3); the latter might not. Cyclogenesis in the northeastern Atlantic could broaden the European trough further west. Tropical heating (§7.2.2) in the eastern Pacific would probably enhance the North America trough; it is unclear whether the increased convection in western Africa would play a significant role in the long wave.

Since the Southern Hemisphere extratropics are quite zonally symmetric (§5.1 through §5.5), I assume that there would be little change other than the swapping mentioned above. An intriguing prospect is the Amazon basin. With high mountain barriers *upstream* of the low-level flow, the Amazon basin could be a far drier place if the rotation reversed.

8.2   Applications to "global change"

A quite different question to examine is how the general circulation would change under one of the "global change" scenarios (doubling CO_2; increasing soot and dust; deforesting the humid tropics; etc.). Some scenarios may interfere: the effects of doubling CO_2 are likely to be contrary to some changes caused when human activities also increase particulate concentrations in the atmosphere. Unfortunately, the prevailing view is that global change is a holistic science that links several subject areas: meteorology, oceanography, glaciology, biology, and geology. Only two pieces of the puzzle have been considered to any degree in this book, so our conclusions must be limited. However, we can make progress by speculating about what is likely *not* to change.

For the case of doubling CO_2, the scenario is often labelled "greenhouse warming" or something similar. The most basic tenet of doubling CO_2 seems to be that the climate may grow warmer. While local deviations exist, the *global* average temperature change since the turn of the century is consistent with recent double-CO_2 simulations by GCMs. There is proxy evidence for a connection between elevated temperature and elevated CO_2 from ice core data. However, the ice core data cannot tell which causes which. Active biological processes, usually not included in GCM simulations of greenhouse warming, may moderate the temperature rise. Nonetheless, a common belief is that the global average temperatures may rise. Another common belief is that arctic temperatures may rise more than equatorial temperatures. We consider these possibilities in turn.

One trivial notion is that weather patterns might simply shift poleward in response to the warming. This prospect seems doubtful. For one thing, the midlatitude frontal cyclones tend to form where the subtropical jets predominate. The location of the jets is complicated by topographic and diabatic effects, in addition to nonlinear interaction with the waves themselves. Nonetheless, a simple application of angular momentum conservation (§6.2.3.2) would lead to baroclinically unstable shears at a latitude determined much more by the Earth's rotation rate than by the mean temperature. Perhaps increased surface temperature may lead to even more vigorous tropical convection (say, by permitting greater moisture content from the nonlinearity of the Clausius-Clapeyron equation). Two competing effects may result: a stronger Asian subtropical jet (§7.2.4) may reach baroclinically unstable shears at a lower latitude, but convective heating may be more widespread across tropical latitudes.

One possible consequence of the greater surface temperature increases in polar regions is a reduction of the meridional temperature gradient. However, the data is missing a key element since it is based upon time average fields. The models produce so much data that it is generally unfeasible to store daily data from decades of simulated time. The key element is the source of the smaller gradient. The weakened meridional gradient may be due to changes in radiative properties; this *leads to weaker* baroclinic development, and thus weaker frontal cyclones. An alternative is that the weaker gradient *follows from stronger* frontal cyclones that transport greater amounts of heat than current systems. Obviously, the weather experienced, and the impacts upon agriculture, are very different for these two possibilities. The "baroclinic adjustment" arguments (§6.4) imply that weaker temperature gradients follow the passage of stronger frontal systems. However, on a seasonal time average, stronger meridional temperature gradients occur when the systems are more vigorous. The baroclinic adjustment results would seem to imply less vigorous frontal systems on average.

Changes to the time-means are less important than changes to the variability. My point is easily illustrated by the winter of 1984–1985 in the southeastern United States. In December the temperature had been above normal; in January a couple of frontal systems brought unusually severe cold for a week or so. The average temperatures in Florida were about normal, but the citrus farmers reportedly lost about 80% of their trees! Such examples are plentiful and may explain why some locales have had their crop climate zones recently revised to *colder* values. A longer historical data record is likely to capture more extreme events.

Warmer temperatures allow the atmosphere to hold greater amounts of moisture, especially in the tropics. If the result of this change is more cloud cover, then the heating may indeed be offset by other radiative changes (e.g. reflectivity changes). Recent satellite observations confirm that clouds cool the climate on a global average (§5.1). However, the global mean is largely dictated by the middle latitude clouds associated with frontal cyclones (§6.1). In the tropics, clouds have an approximate balance between reflecting sunlight (a cooling effect) and trapping terrestial radiation (a warming effect) because the thunderstorm anvil clouds become very broad.

As stated above, my discussion here has been limited to properties that may resist change under greenhouse warming. Many interesting climate *change* scenarios have been deduced from climate model simulations. Even these very sophisticated models can miss some possibilities. For example, the tropical changes may have unforeseen effects upon the subtropics. It has long been thought that the intensity and the distribution of tropical cyclones are linked to sea surface temperatures. As our example, assume that ocean temperatures off Baja California warm by just a few degrees; this change could have a major impact on southern California. Presently, the California summer is tediously dry and clear, except on rare occasions when moisture from a dying hurricane in the eastern Pacific reaches the midlatitude westerlies. These occasions can bring heavy rainfall during brief periods of our summer and early fall. If global warming increases the domain of these hurricanes while leaving the westerlies unchanged, then summer rain could become a regular feature in California. This example is unforeseen by GCMs because no existing climate model forms hurricanes. The discussion brings us full circle to the first issue raised in Chapter 1. Namely, the point at which one stops including phenomena in the definition of the general circulation is somewhat arbitrary. I believe that all the most important items have been addressed in this book.

Appendix

AVERAGING

Reality is not an exhibit for man's inspection, labelled "Do not touch." Science, like art, is not a copy of nature but a recreation of her. We remake by the act of discovery, in the poem or in the theorem.

Bronowski, 1956

Bronowski could have been talking about the act of investigation that goes *before* the discovery. To some extent the type of answer one gets is limited by the way in which the question is asked. To discuss the "General Circulation" implies the use of a model: that global scale motions are emphasized while global patterns of radiation, moisture, and chemistry are included only as they relate to the dynamics. Partitioning the atmosphere this way defines and limits the questions and answers allowed in this book.

In the subject of the general circulation this is most apparent when the question is how to divide up the atmosphere. There are many ways to divide a pie, some are better than others.

Evidence can be found in this book that dividing along zonal average versus nonzonal average has limited applicability to the real atmosphere. Nonetheless, that same distinction is a fundamental part of the structure of this book. The division is made for several reasons, but two are most important. First, this book must be based upon the General Circulation literature available; the exisiting materials are overwhelmingly organized in the same way. So, one hardly has any choice. Second, it is easier to draw and grasp a two-dimensional picture on a flat page than would be a four-dimensional image. The essential elements of the atmosphere are more easily seen. Zonal averaging accomplishes the purposes of this book, as long as the reader is reminded at key moments of pitfalls to avoid.

The act of dividing the atmosphere into groups of zonal averages and deviations is to employ a model. In doing so, one places an important distinction between groups.

A particular model is a specialized tool in the way that a screwdriver is. A screwdriver can turn screws but has limited use as a hammer. More to the point, with a screwdriver one tends to connect objects by attaching screws rather than hammering nails. The connections drawn throughout this book can viewed in a similar light.

A.1 Basic definitions

Various types of averages are applied throughout the book. The purpose is often to isolate various types of phenomena. Researchers typically average in horizontal space directions, in time and occasionally in the vertical. This appendix defines the convention of symbols used everywhere in this text.

Square brackets denote average over longitude.

$$[q] = \frac{1}{L} \int q dx$$

where q is any dependent variable and L is the circumference of a latitude circle for a complete zonal average. A deviation from this average is denoted by a single prime superscript: $q' \equiv q - [q]$.

An overbar indicates time averaging.

$$\bar{q} = \frac{1}{\tau} \int_{t_1}^{t_2} q dt$$

where $\tau = \int_{t_1}^{t_2} dt$. Departures from a time average are denoted by a double prime superscript: $q'' \equiv q - \bar{q}$.

A curly overbar specifies vertical averaging.

$$\widehat{q} = \frac{1}{P_0} \int_0^{P_0} q dP$$

where P_0 is the surface pressure, and pressure is the vertical coordinate. Departures from a vertical average are indicated with an asterisk: $q^* \equiv q - \widehat{q}$.

A.2 Some general rules

An average of a single quantity's deviation from that average is zero. In other words, $[q']$ equals zero, but $[q'']$ is not necessarily zero. The average of the product of two deviations is not necessarily zero either: $[q'q'] \neq 0$ unless $q' = 0$ everywhere. The average of a product of an average times its deviation separate: $[[q]q'] = [q][q'] = 0$. Different averages are not always independent though that assumption is occasionally made in this text for convenience, i.e., $\widehat{[q]} \neq [(\widehat{q})]$; here the vertical average includes P_0, which may be a function of longitude. Finally, successive averages are redundant.

Averages can be combined in several ways to partition a field. To illustrate the phenomena selected by each partition, a longitudinal average is made first, then time, then pressure.

$$u = [u] + u'$$

$$u = \overline{[u]} + [u]'' + \overline{u'} + (u')''$$

$$u = \underbrace{\overline{[u]}}_{(1)} + \underbrace{\overline{[u]}^*}_{(2)} + \underbrace{\overline{[u]''}}_{(3)} + \underbrace{\left([u]''\right)^*}_{(4)} + \underbrace{\overline{(u')}}_{(5)} + \underbrace{\overline{(u')}^*}_{(6)} + \underbrace{\overline{(u')''}}_{(7)} + \underbrace{\left((u')''\right)^*}_{(8)}$$

Each numbered term relates to a specific process:

1. Net mean mass shift or flux; a net average velocity.
2. The time-mean meridional cell flux (the only variations allowed are in the vertical and meridional).
3. Time-varying, zonal average mass flux.
4. Time-varying meridional cell flux.
5. Flux due to standing barotropic eddies (only zonal variation allowed; no time change).
6. Flux due to standing baroclinic eddies (must have vertical and zonal variations; no time change).
7. Flux due to transient barotropic eddies.
8. Flux due to transient baroclinic eddies.

Mean meridional cells are the circulations (like the supposed Hadley and Ferrel cells) that do not vary with longitude but do vary with height. The eddies include all phenomena that vary with longitude. If these eddies vary with height, they are "baroclinic"; if they do not, they are "barotropic." To be strictly proper, baroclinic and barotropic labels should refer to the type of energy conversion that predominates for a given eddy. However, the structure adopted by an eddy is different for the two modes of energy conversion. In a crude manner, a vertical average will often select mainly storms that have the structure associated with the barotropic conversion. Similarly, calculating vertical deviations tends to select the baroclinically developing eddies. "Transient" phenomena are those that vary with time, while "standing" phenomena are time- invariant.

Phenomena can be gathered into groups by selectively combining and applying the averages and deviations. For example, all eddy phenomena can be isolated by merely selecting deviations from a zonal average. To see more sophisticated examples of this grouping, all three averages are applied to the velocity covariance uv, where

u is zonal and v is meridional velocity. It is useful to break up this covariance into different contributions. The resultant phenomena being isolated depend a lot upon the order in which the averages are applied.

We want to partition $\widehat{[uv]}$. A zonal, then a time, and finally a pressure average are taken. The steps may be easier to follow if one takes an indirect approach. Partitioning u into zonal mean and zonal deviation first:

$$u = [u] + u'$$

Next, a time partitioning, but only to the *first* term on the right hand side.

$$[u] = \overline{[u]} + [u]''$$

The first term on the right side of this equation is split according to the pressure average:

$$\overline{[u]} = \widehat{\overline{[u]}} + \overline{[u]}^*$$

Now substituting for the first term in each of these three equations obtains:

$$u = \widehat{\overline{[u]}} + \overline{[u]}^* + [u]'' + u' \qquad (A.1)$$

The same set of operations for v yield:

$$v = \widehat{\overline{[v]}} + \overline{[v]}^* + [v]'' + v' \qquad (A.2)$$

Now that u and v are expanded, (A.1) and (A.2) are multiplied and then the averages are again applied to all the terms. Many of the terms cancel. All terms involving the product of a mean and its deviation vanish. For example, when three averages are applied:

$$\widehat{\overline{\left[\widehat{\overline{[u]}}\,[v]''\right]}} = \widehat{\overline{[u]}}\,\widehat{\overline{\left[[v]''\right]}} = 0$$

Similarly, other mixed terms vanish. One is left with products of "like" terms. That is:

$$\widehat{\overline{[uv]}} = \underbrace{\widehat{\overline{[u]}}\,\widehat{\overline{[v]}}}_{(1)} + \underbrace{\widehat{\overline{[u]}^*\,\overline{[v]}^*}}_{(2)} + \underbrace{\widehat{\overline{[u]''[v]''}}}_{(3)} + \underbrace{\widehat{\overline{[u'v']}}}_{(5)} \qquad (A.3)$$

Multiplying by the proper factor $(RL\tau P_0/g)$, the five terms in (A.3) become fluxes. They are considered as such.

One can interpret (A.3) in light of the physical phenomena isolated by each term. In order to better understand how the interpretation is deduced, *one always works from the innermost operation (be it an average or deviation) toward the outermost operation prior to the multiplication.* Using term (4) as an example, the innermost operation is a zonal average; so the phenomena must not vary with longitude. The next operation is a time deviation, hence the phenomena must be zonally symmetric and vary with time. The quantities are next multiplied before any further averages are taken so there is no further restriction upon the phenomena included in term (4).

- *Term (1)* is the flux of relative angular momentum through a vertical "wall" around a latitude circle and during the time τ.

- *Term (2)* is that part of the angular momentum flux due to a *net mass shift*.

- *Term (3)* isolates phenomena that have departures from the vertical of the time and zonal averages. This term isolates the *time or climatological mean meridional circulation*.

- *Term (4)* is either or both of *(a)* the *time varying meridional cells* and *(b)* the time variation of the *net mass flux*.

- *Term (5)* selects all transverse waves. It includes both *(a) standing and transient barotropic eddies* and *(b) standing and transient baroclinic eddies*.

These brackets, bars, and primes are sometimes confusing when seen for the first time. Much of the mystery about how to use these symbols can be removed if one remembers that these symbols merely represent integrals and local subtractions. The symbols for averages are integrals. The symbols for deviations denote subtracting the integrated value from the actual value at a local point.

REFERENCES

Anderson, D.L.T., 1983: The oceanic general circulation and its interaction with the atmosphere. In: *Large-Scale Dynamical Processes in the Atmosphere*, B. Hoskins and R. Pearce, eds. Academic Press, New York, 305–336.

Andrews, D.G., 1987: Transport mechanisms in the middle atmosphere: an introductory survey. In: *Transport Processes in the Middle Atmosphere*, G. Visonti and R. Garcia, eds., D. Reidel, Dordrecht, 169–181.

Andrews, D.G., 1990: On the forcing of time-mean flows by transient, small-amplitude eddies. *J. Atmos. Sci.*, **47**, 1837–1844.

Andrews, D.G., and M.E. McIntyre., 1976: Planetary waves in horizontal and vertical shear: The generalized Eliassen-Palm relation and the mean zonal acceleration. *J. Atmos. Sci.*, **33**, 2031– 2048.

Andrews, D.G. and M.E. McIntyre., 1978: Generalized Eliassen-Palm and Charney-Drazin theorems for waves on axisymmetric flows in compressible atmospheres. *J. Atmos. Sci.*, **35**, 175–185.

Arakawa, A., and S. Moorthi, 1982: "Baroclinic (and Barotropic) Instability with Cumulus Heating." *Proc. WMO Programme on Research in Tropical Meteorology*, Tsukuba, WMO, V43–V62.

Arkin, P.A., V.E. Kousky, J.E. Janowiak, and E.A. O'Lenic, 1986: "Atlas of the Tropical and Subtropical Circulation Derived from National Meteorological Center Operational Analyses." NOAA Atlas No. 7, NOAA/NWS, Silver Spring, 60 pp.

Arking, A., 1991: The radiative effects of clouds and their impacts on climate. *Bull. Amer. Meteor. Soc.*, **72**, 795–813.

Baer, F., 1972: An alternate scale representation of atmospheric energy spectra. *J. Atmos. Sci.*, **29**, 649–664.

Baer, F., 1974: Hemispheric spectral statistics of available potential energy. *J. Atmos. Sci.*, **31**, 932–941.

Baer, F., and J. J. Tribbia, 1977: On complete filtering of gravity modes through nonlinear initialization. *Mon. Wea. Rev.*, **105**, 1536–1539.

Baines, P.G., 1976: The stability of planetary waves on a sphere. *J. Fluid Mech.*, **73**, 193–213.

Baker, W.E. and R.J. Curran, (eds.) 1985: "Report of the NASA Workshop on Global Wind Measurements." A. Deepak Publ., Hampton, 60 pp.

Baldwin, M.P., H.J. Edmon and J.R. Holton, 1985: A diagnostic study of eddy-mean flow interactions during FGGE SOP–1. *J. Atmos. Sci.*, **42**, 1838–1845.

Bannon, P.R., 1991: Aspects of a rotating shear flow over a mountain ridge. *J. Atmos. Sci.*, **48**, 211–216.

Barkstrom, B.R., E.F. Harrison, and R.B. Lee, III, 1990: Earth radiation budget experiment, preliminary seasonal results. *EOS Trans.*, **71**, No. 9 (February 27) 279, 299, 304–305.

Barry, R.G., 1980: Meteorology and climatology of the seasonal sea ice zone. *Cold Regions Sci. Technol.*, **2**, 134.

Bates, J.R., 1977: Dynamics of stationary ultra-long waves in middle latitudes. *Quart. J. Roy. Meteor. Soc.*, **103**, 397–430.

Bauer, K.G., 1976: A comparison of cloud motion winds with coinciding radiosonde winds. *Mon. Wea. Rev.*, **104**, 922–931.

Baumhefner, D.P. and P.R. Julian, 1975: The initial structure and resulting error growth in the NCAR GCM produced by simulated, remotely sensed temperature profiles. *Mon. Wea. Rev.*, **103**, 273–284.

Becker, R.J., 1979: The global distribution of radiative heating and its relation to the large-scale features of the general circulation. Ph.D. thesis, University of Maryland, 201 pp.

Bengtsson, L., M. Kanamitsu, P. Kållberg, and S. Uppala, 1982: FGGE research at ECMWF. *Bull. Amer. Meteor. Soc.*, **63**, 277–303.

Bennet, A.F., 1978: Poleward heat fluxes in the Southern Hemisphere oceans. *J. Phys. Oceanog.*, **8**, 785–798.

Benton, G.S. and M.A. Estoque, 1954: Water-vapor transfer over the North American continent. *J. Meteor.*, **11**, 462–477.

Benzi, R., B. Saltzman and A.C. Wiin-Nielsen (eds.), 1986: "Anomalous Atmospheric Flows and Blocking." *Adv. Geophys.*, **29**, Academic Press, Orlando, 459 pp.

Berlyand, T.G., and L.A. Strokina, 1980: Zonal cloud distribution on the Earth. *Meteor. Gidrol.*, **3**, 15–23.

Beyer, W.H., 1984: *CRC Standard Mathematical Tables*, 27th Ed., CRC Press, Boca Raton, 615 pp.

Bigelow, F.H., 1900: "Report on the international cloud observations." Report of the Chief of the Weather Bureau, 1898–99, Vol. II. Washington, 787 pp.

Bigelow, F.H., 1902: Studies of the statics and kinematics of the atmosphere in the United States. *Mon. Wea. Rev.*, **30**, 13–19, 80–87, 117–125, 163–171, 250–258, 304–311, 347–354.

Bjerknes, J., 1969: Atmospheric teleconnections from the equatorial Pacific. *Mon. Wea. Rev.*, **97**, 163–172.

Bjerknes, V., 1937: Application of line integral theorems to the hydrodynamics of terrestrial and cosmic vortices. *Astrophys. Norv.*, **2**, No. 6, 263–339.

Bjørheim, K., P. Julian, M. Kanamitsu, P. Kållberg, P. Price, S. Tracton, S. Uppala, 1981: "FGGE III–B Daily Global Analyses, Part I, December 1978–February 1979." ECMWF Tech. Rept., 368 pp.

Blackburn, M., 1985: Interpretation of ageostrophic winds and implications for jet stream maintenance. *J. Atmos. Sci.*, , **42**, 2604– 2620.

Blackmon, M.L., J.M. Wallace, N.-C. Lau and S.L. Mullen, 1977: An observational study of the Northern Hemisphere wintertime circulation. *J. Atmos. Sci.*, **34**, 1040–1053.

Blake, D.W., T.N. Krishnamurti, S.V. Low-Nam, and J.S. Fein, 1983: Heat low over the Saudi Arabian desert during May 1979 (summer MONEX). *Mon. Wea. Rev.*, **111**, 1759–1775.

Blinova, E.N., 1976: On the annual mean temperature distribution in the Earth's atmosphere taking into account continents and oceans. *Izv. Akad. Nauk. SSR, Ser. Geogr., Geofiz.*, **XI**, 3–13. (In Russian.)

Boer, G.J. and T.G. Shepherd, 1983: Large-scale two- dimensional turbulence in the atmosphere. *J. Atmos. Sci.*, **40**, 164–184.

Bolin, B., 1950: On the influence of the earth's orography on the general character of the westerlies. *Tellus*, **2**, 184–195.

Boyd, J., 1976: The noninteraction of waves with the zonally averaged flow on a spherical earth and the interrelationships of eddy fluxes of energy, heat and momentum. *J. Atmos. Sci.*, **33**, 2285–2291.

Branstator, G., 1983: Horizontal energy propagation in a barotropic atmosphere with meridional and zonal structure. *J. Atmos. Sci.*, **40**, 1689–1708.

Bretherton, F.P., 1966: Critical layer instability in baroclinic flows. *Quart. J. Roy. Meteor. Soc.*, **92**, 325–334.

Brewer, A.M., 1949: Evidence for a world circulation provided by the measurements of helium and water vapor distribution in the stratosphere. *Quart. J. Roy. Meteor. Soc.*, **75**, 351–363.

Bronowski, J., 1956: *Science and Human Values*, Harper & Row Publ., New York, 119 pp. (second ed. 1965)

Brooks, C.E.P., 1927: The mean cloudiness over the Earth. *Mem. Roy. Meteorol. Soc.*, **1**, 127–138.

Brown, J.A., 1969: A numerical investigation of hydrodynamic instability and energy conversions in the quasi-geostrophic atmosphere: part II. *J. Atmos. Sci.*, **26**, 366–375.

Bruce, R.E., L.D. Duncan, and J.H. Pierluissi, 1977: Experimental study of the relationship between radiosonde temperatures and satellite-derived temperatures. *Mon. Wea. Rev.*, **105**, 493–496.

Bryan, K., 1962: Measurements of meridional heat transport by ocean currents. *J. Geophys. Res.*, **67**, 3403–3413.

Bryan, K. and I.J. Lewis, 1979: A watermass model of the World ocean. *J. Geophys. Res.*, **84**, 2503–2517.

Bryden, H.L. and M.M. Hall, 1980: Heat transport by ocean currents across 25°N latitude in the Atlantic Ocean. *Science*, **207**, 884–886.

Buchan, A., 1889: Report on atmospheric circulation." *Report on the Scientific Results of the Exploring Voyage of the H.M.S. Challenger, 1873–76*. Physics and Chemistry, **2**, H.M. Stationery Off. 1868 reference cited by Shaw, 1926.

Budyko, M.I., 1963: *Atlas Teplovogo Balansa Zemnogo Shara (Atlas of Heat Balance of the Earth's Surface)*. Moscow, Glavnaia Geofiz. Observ., 69 pp. plus charts.

Burridge, D., 1982: In: The "CAGE" Experiment. A Feasibility Study, UNESCO JSC/CCCO Liaison Panel.

Buzyna, G., R.L. Pfeffer, and R. Kung, 1984: Transition to geostrophic turbulence in a rotating differentially heated annulus of fluid. *J. Fluid Mech.*, **145**, 377–403.

Cai, M., and M. Mak, 1990a: On the basic dynamics of regional cyclogenesis. *J. Atmos. Sci.*, **47**, 1417–1442.

Cai, M., and M. Mak, 1990b: Symbiotic relation between planetary and synoptic-scale waves. *J. Atmos. Sci.*, **47**, 2953–2968.

Campbell, G.G. and T.H. Vonder Haar, 1980: "Climatology of Radiation Budget Measurements from Satellites." Atmos. Sci. Paper No. 323. Dept. of Atmos. Sci., Colorado State Univ., Fort Collins.

Carleton, A.M., 1981: Monthly variability of satellite-derived cyclonic activity for the Southern Hemisphere winter. *J. Climatol.*, **1**, 21–38.

Cehelsky, P., and K.K. Tung, 1991: Nonlinear baroclinic adjustment. *J. Atmos. Sci.*, **48**, 1930–1947.

CES, 1989: "Our Changing Planet: The FY 1990 Research Plan." U. S. Global Change Program. Report by Committee on Earth Science for the Off. of Science and Technology Policy. U. S. Govt. Print. Off., 183 pp.

Chahine, M.T., 1970: Inverse problems in radiative transfer: Determination of atmospheric parameters. *J. Atmos. Sci.*, **27**, 960–967.

Chang, C.-P., and T.N. Krishnamurti, eds., 1987: *Monsoon Meteorology*, Oxford Univ. Press, New York, 544 pp.

Chang, C.-P., and R.T. Williams, 1974: On the short-wave cutoff of CISK. *J. Atmos. Sci.*, **31**, 830-833.

Charney, J.G., 1947: The dynamics of long waves in a baroclinic westerly current. *J. Meteor.*, **4**, 135–162.

Charney, J.G., 1962: "Integration of the primitive and balance equations." In: Proc. Int'l Symp. Num. Wea. Pred., J. Meteor. Soc. Japan, Tokyo, 131–152.

Charney, J.G., 1971a: Tropical cyclogenesis and the formation of the intertropical convergence zone. In: *Mathematical Problems in Geophysical Fluid Dynamics*, W.H. Ried, ed. Lectures in Applied Math **13**, Amer. Math. Soc., 355–368. (Analysis repeated in Charney, 1973.)

Charney, J.G., 1971b: Geostrophic turbulence. *J. Atmos. Sci.*, **28**, 1087–1095.

Charney, J.G., 1973: Planetary fluid dynamics. In: *Dynamic Meteorology*, P. Morel, ed. D. Reidel, Boston, 97–351.

Charney, J.G. and A. Eliassen, 1949: A numerical method for predicting the perturbations of the middle latitude westerlies. *Tellus*, **1**, 38–54.

Charney, J.G. and P.G. Drazin, 1961: Propagation of planetary-scale disturbances from the lower into the upper atmosphere. *J. Geophys. Res.*, **66**, 83–109.

Charney, J.G. and M.E. Stern, 1962: On the stability of internal baroclinic jets in a rotating atmosphere. *J. Atmos. Sci.*, **19**, 159–172.

Charney, J.G. and A. Eliassen, 1964: On the growth of the hurricane depression. *J. Atmos. Sci.*, **21**, 68–75.

Charney, J.G., M. Shukla and K.L. Mo, 1981: Comparison of a barotropic blocking theory with observations. *J. Atmos. Sci.*, **38**, 762–779.

Chen, S.-C., and K.E. Trenberth, 1988: Forced planetary waves in the northern hemisphere winter: Wave-coupled orographic and thermal forcings. *J. Atmos. Sci.*, **45**, 682–704.

Chen, T.-C., and A. Wiin-Nielsen, 1978: On nonlinear cascades of atmospheric energy and enstrophy in a two-dimensional spectral index. *Tellus*, **30**, 313–322.

Chen, T.-C., and A.C. Wiin-Nielsen, 1993: *Fundamentals of Atmospheric Energetics.* Oxford Univ. Press, New York, *in press.*

Clapp, P.F., 1964. Global cloud cover for seasons using TIROS nephanalyses. *Mon. Wea. Rev.*, **92**, 495–507.

Daley, R., 1991: *Atmospheric Data Analysis.* Cambridge Univ. Press, Cambridge, 457 pp.

Daley, R., and T. Mayer, 1986: Estimates of global analysis errors from the Global Weather Experiment observational network. *Mon. Wea. Rev.*, **114**, 1642–1653.

Dalton, J., 1793: *Meteorological Observations and Essays.* Harrison and Crosfield for Baldwin and Cradock, Manchester. 195 pp. (Second edition publ. 1834; 244 pp.)

Danielsen, E.F., 1982: A dehydration mechanism for the stratosphere. *Geophys. Res. Lett.*, **9**, 605–608.

Defant, A., 1921: Die zirkulation der atmosphare in den gemassigten breiten der erde. *Geogr. Ann.*, **3**, 209–266.

Del Genio, A.D. and R.J. Suozzo, 1987: A comparative study of rapidly and slowly rotating dynamical regimes in a terrestrial general circulation model. *J. Atmos. Sci.*, **44**, 973–986.

Descartes, R., 1637: *Discourse on Method, Optics, Geometry, and Meteorology.* [Translated by P.J. Olscamp, Bobbs- Merrill Co. Inc., Indianapolis, 1965; See Meteorology, Fourth Discourse: of Winds.]

Dickinson, R.E., 1969: Theory of planetary wave-zonal flow interaction. *J. Atmos. Sci.*, **26**, 73–81.

Dickinson, R.E., 1980: Planetary Waves; Theory and Observations. In: "Orographic Effects in Planetary Flows." GARP Publ. No. 23, WMO, 53–84.

Dobson, G.M.B., 1930: Observations of the amount of ozone in the Earth's atmosphere and its relation to other geographical conditions—Part IV. *Proc. Roy. Soc. London*, **A129**, 411–433.

Dobson, G.M.B., 1956: Origin and distribution of polyatomic molecules in the atmosphere. *Proc. Roy. Soc. London*, **A236**, 187–193.

Donner, L.J. and H.-L. Kuo, 1984: Radiative forcing of stationary planetary waves. *J. Atmos. Sci.*, **41**, 2849–2868.

Dorman, C.E. and R.H. Bourke, 1979: Precipitation over the Pacific ocean, 30°S to 60°N. *Mon. Wea. Rev.*, **107**, 896–910.

Döös, B.R., 1962: The influence of exchange of sensible heat with the earth's surface on the planetary flow. *Tellus*, **14**, 133–147.

Dove, H.W., 1837: *Meteorologische Untersuchungen*. Berlin, Sandersche Buchhandlung, 344 pp.

Dove, H.W., 1862: *Law of Storms.*, Second edition, Longman, Roberts and Green, London, 273 pp. First edition publ. 1852 as *Verbreitung der Warme*.

Dunkerton, T., 1978: On the mean meridional mass motions of the stratosphere and mesosphere. *J. Atmos. Sci.*, **35**, 2325–2333.

Dutton, J.A. and D.R. Johnson, 1967: The theory of available potential energy and a variational approach to atmospheric energetics. In: *Adv. Geophys.*, **12**, Academic Press, New York, 333–436.

Eady, E.T., 1949: Long waves and cyclone waves. *Tellus*, **1**, 33–52.

Edmon, H.J., 1978: A reexamination of limited-area available potential energy budget equations. *J. Atmos. Sci.*, **35**, 1655–1659.

Edmon, H.J., B.J. Hoskins and M.E. McIntyre, 1980: Eliassen-Palm cross sections for the troposphere. *J. Atmos. Sci.*, **37**, 2600–2616. Erratum: **38**, 1115.

Egger, J., 1976: Comments on "Zonal and eddy forms of available potential energy equations in pressure coordinates" by G.J. Boer. *Tellus*, **28**, 377–378.

Eliassen, A., 1952: Slow thermally or frictionally controlled meridional circulation in a circular vortex. *Astrophys. Norv.*, **5**, No. 2, 19–60.

Eliassen, A., and E. Palm, 1961: On the transfer of energy in stationary mountain waves. *Geofys. Publ.*, **22**, No. 3, 1–23.

Emden, R., 1913: Uber strahlungsgleichgewicht und atmospharische strahlung. *Sitz. Bayerische Akad. Wiss.*, Math-Phys. K1., **1**, 55.

Errico, R.M., 1989: "Theory and Application of Nonlinear Normal Mode Initialization." NCAR Technical Note, NCAR/TN–344+IA, 137 pp.

ETAC, 1971: "Northern Hemisphere Cloud Cover." ETAC, Project 6168, USAF, Washington, DC, 20 pp.

Exner, F.M., 1925: *Dynamische Meteorologie*, Vienna, J. Springer, 421 pp.

Fein, J.S., and P.L. Stephens, eds., 1987: *Monsoons*, J. Wiley and Sons, New York, 632 pp.

Ferrel, W., 1856: An essay on the winds and the currents of the ocean. *Nashville J. Medicine and Surgery*, **11**, 287–301. Reprinted (1882) in *Popular essays on* *the movements of the atmosphere.* Prof. Pap. Signal Serv. No. 12, Washington, 7–19.

Ferrel, W., 1859: The motions of fluids and solids relative to the Earth's surface. *Math. Monthly,* **1**, 140–147, 210–216, 300–307, 366–372, 397–406.

Ferrel, W., 1893: *A Popular Treatise on the Winds: Comprising the General Motions of the Atmosphere, Monsoons, Cyclones, Tornadoes, Waterspouts, Hail-storms, Etc. Etc.,* second ed., J. Wiley & Sons, New York, 505 pp.

Fjørtoft, R., 1953: On the changes in the spectral distribution of kinetic energy for two-dimensional, nondivergent flow. *Tellus,* **5**, 225–230.

Fjørtoft, R., 1960: On the control of kinetic energy of the atmosphere by external heat sources and surface friction. *Quart. J. Roy. Meteor. Soc.,* **86**, 437–453.

Fowlis, W.W., and R. Hide, 1965: Thermal convection in a rotating fluid annulus: Effect of viscosity on the transition between axisymmetric and non-axisymmetric flow regimes. *J. Atmos. Sci.,* **22**, 541–558.

Frederiksen, J.S., 1978: Instability of planetary waves and zonal flows in two-layer models on a sphere. *Quart. J. Roy. Meteor. Soc.,* , **104**, 841– 872.

Frederiksen, J.S., 1981: Growth and vacillation cycles of disturbances in Southern Hemisphere flows. *J. Atmos. Sci.,* **38**, 1360–1375.

Frederiksen, J.S., 1983: Disturbances and eddy fluxes in Northern Hemisphere flows: Instability of three-dimensional January and July flows. *J. Atmos. Sci.,* **40**, 836–855.

Frederiksen, J.S., and K. Puri, 1985: Nonlinear instability and error growth in Northern Hemisphere three-dimensional flows: Cyclogenesis, onset-of-blocking and mature anomalies. *J. Atmos. Sci.,* **42**, 1374–1397.

Fu, L.L., 1981: General circulation and meridional heat transport of the subtropical South Atlantic determined by inverse methods. *J. Phys. Oceanog.,* **11**, 1171–1193.

Fujita, T.T., 1981: Tornadoes and downbursts in the context of generalized planetary scales. *J. Atmos. Sci.,* **38**, 1511–1534.

Fultz, D., R.R. Long, G.V. Owens, W. Bohan, R. Kaylor, and J. Weil, 1959: Studies of Thermal Convection in a Rotating Cylinder with Some Implications for Large-Scale Atmospheric Motions. *Meteor. Mono.,* **4**, Amer. Meteor. Soc., 104 pp.

Gall, R., 1976: Structural changes of growing baroclinic waves. *J. Atmos. Sci.,* **33**, 374–390.

Gall, R., R. Blakeslee, and R.C.J. Somerville, 1979: Cyclone-scale forcing of ultralong waves. *J. Atmos. Sci.,* **36**, 1692–1698.

Gaster, M., 1962: A note on the relation between temporally increasing and spatially increasing disturbances in hydrodynamic stability. *J. Fluid Mech.,* **14**, 222–224.

Geller, M.A., and M.F. Wu, 1987: Troposphere-stratosphere general circulation statistics. In: *Transport Processes in the Middle Atmosphere,* G. Visconti and R. Garcia, eds., D. Reidel, Dordrecht, 3–17.

Genthon, C., H. Le Trent, J. Jouzel, and R. Sadourny, 1990: Parameterization of eddy sensible heat transports in a zonally averaged dynamic model of the atmosphere. *J. Atmos. Sci.*, **47**, 2475–2487.

Gill, A.E., 1982: *Atmosphere-Ocean Dynamics*. Academic Press, New York, 662 pp.

Goody, R.M., and Y.L. Yung, 1989: *Atmospheric Radiation: Theoretical Basis*. Oxford Univ. Press, New York, 519 pp.

Gordon, A.L., 1986: Interocean exchange of thermocline water. *J. Geophys. Res.*, **91**, 5037–5046.

Gordon, A.L., E. Molinelli, and T. Baker, 1978: Large-scale topography of the southern ocean. *J. Geophys. Res.*, **83**, 3023.

Green, J.S.A., 1970: Transfer properties of the large-scale eddies and the general circulation of the atmosphere. *Quart. J. Roy. Meteor. Soc.*, **96**, 157–185.

Grose, W.L., and B.J. Hoskins, 1979: On the influence of orography on large-scale atmospheric flow. *J. Atmos. Sci.*, **36**, 223–234.

Grotjahn, R., 1979. Cyclone development along weak thermal fronts. *J. Atmos. Sci.*, **36**, 2049–2074.

Grotjahn, R. 1980. Linearized tropopause dynamics and cyclone development. *J. Atmos. Sci.*, **37**, 2396–2406.

Grotjahn, R., 1987: "Frontal Cyclone Structure in Three Dimensions," UCD Atmos. Sci. Group Tech. Rept. 100013, 79 pp.

Grotjahn, R., 1991a: "ECMWF Model Error Growth and Verification in Southern Hemisphere Middle Latitude Regions During Winter." UCD Atmos. Sci. Group Tech. Rept. 100017, 32 pp.

Grotjahn, R., 1991b: Some general results for normal mode instability on a sphere. **58**, 113–121.

Grotjahn, R., M. Chen, and J.J. Tribbia, 1992: Linear instability with Ekman and Interior Friction: Quasi-geostrophic Eigen analysis. *Submitted*.

Grotjahn, R., and P. Kennedy, 1986: "A study of the error due to sampling at 12 and 24 hour intervals." In: Programme on Short- and Medium-Range Weather Prediction Research (PSMP). World Meteorological Organization. PSMP Report Series No. 19, 37–40.

Grotjahn, R., and P.J. Kennedy, 1990: "On Zonal Averages Using Pibal and Radiosonde Station Data." UCD Atmos. Sci. Group Tech Rept. 100016, 22 pp.

Grotjahn, R., and S.-S. Lai, 1989: "Storm tracks over the north Pacific grouped by climate anomalies during 22 winters." UCD Atmos. Sci. Group Tech. Rept. 100015, 22 pp.

Grotjahn, R., and S.-S. Lai, 1991: Nonlinear impacts upon frontal cyclones from dipole surface temperature anomalies. *Meteor. Atmos. Phys.*, **45**, 159–180.

Grotjahn, R., and C.-H. Wang, 1989: On the source of air modified by ocean surface fluxes to enhance frontal cyclone development. *Ocean-Air Inter.*, **1**, 257–288.

Grotjahn, R., and C.-H. Wang, 1990: Topographic linear instability on a sphere for various ridge orientations and shapes. *J. Atmos. Sci.*, **47**, 2249–2261.

Gutowski, W.J., 1985: Baroclinic adjustment and midlatitude temperature profiles. *J. Atmos. Sci.*, **42**, 1733–1745.

Gutowski, W.J., L.E. Branscome, and D.A. Stewart, 1989: Mean flow adjustment during life cycles of baroclinic waves. *J. Atmos. Sci.*, **46**, 1724–1737.

Hack, J.J., W.H. Schubert, D.E. Stevens, and H.-C. Kuo, 1989: Response of the Hadley circulation to convective forcing in the ITCZ. *J. Atmos. Sci.*, **46**, 2957–2973.

Hadley, G., 1735: Concerning the cause of the general trade-winds. *Phil. Trans.*, **29**, 58–62.

Hahn, C.J., S.W. Warren, J. London, R.M. Chervin, and R. Jenne, 1982: "Atlas of Simultaneous Occurrence of Different Cloud Types over the Ocean." NCAR Tech. Note TN-201 + STR., 212 pp.

Halem, M., E. Kalnay-Rivas, W. Baker, and R. Atlas, 1981: "The state of the atmosphere as inferred from the FGGE satellite observing systems during SOP-1." In: *International Conference on Early Results of FGGE and Large-Scale Aspects of Its Monsoon Experiments*, ICSU/WMO, Geneva, Pages 1-27 through 1-41.

Halem, M., E. Kalnay, W. Baker, and R. Atlas, 1982: An assessment of the FGGE satellite observing system during SOP- 1. *Bull. Amer. Meteor. Soc.*, **63**, 407–426.

Halley, E., 1686: An historical account of the trade-winds and monsoons observable in the seas between and near the tropics with an attempt to assign the physical cause of said winds. *Phil. Trans.*, **26**, 61–80.

Harrison, E.F., P. Minnis, B.R. Barkstrom, V. Ramanathan, R.D. Cess, and G.G. Gibson, 1990: Seasonal variation of cloud radiative forcing derived from the Earth Radiation Budget Experiment. *J. Geophys. Res.*, **95D**, 18,687–18,703.

Hart, J.E., 1981: Wavenumber selection in nonlinear baroclinic instability. *J. Atmos. Sci.*, **38**, 400–408.

Hartmann, D.L., 1985: Some aspects of stratospheric dynamics. *Adv. Geophys.*, **28A**, 219–247.

Hartmann, D.L., C.R. Mechoso, and K. Yamazaki, 1984: Observations of wave-mean flow interaction in the Southern Hemisphere. *J. Atmos. Sci.*, **41**, 351–362.

Hasler, A.F., 1981: Stereographic Observations from Geosynchronous Satellites: An Important New Tool for the Atmospheric Sciences, *Bull. Amer. Meteor. Soc.*, **62**, 194–212.

Hasler, A.F., W.C. Skillman, W.E. Shenk, and J. Steranka, 1979: *In situ* aircraft verification of the quality of satellite cloud winds over oceanic regions. *J. Appl. Meteorol.*, **18**, 1481– 1489.

Hastenrath, S., 1980: Heat budget of tropical ocean and atmosphere. *J. Phys. Oceanog.*, **10**, 159–170.

Hastenrath, S., 1985: *Climate and Circulations of the Tropics*, D. Reidel, Dordrecht, 455 pp.

Hayashi, Y., 1985: Theoretical interpretations of the Eliassen-Palm diagnostics of wave-mean flow interaction. Part I. Effects of the lower boundary. *J. Meteor. Soc. Japan*, **63**, 497–512.

Hayashi, Y., 1987: A modification of the atmospheric energy cycle. *J. Atmos. Sci.*, **44**, 2006–2017.

Hayes, J.L., R.T. Williams, and M.A. Rennick, 1987: Lee cyclongenesis: Part I: Analytical solutions. *J. Atmos. Sci.*, **44**, 432–442.

Held, I.M., 1982: On the height of the tropopause and the static stability of the troposphere. *J. Atmos. Sci.*, **39**, 412–417.

Held, I.M., 1983: Stationary and quasi-stationary eddies in the extratropical troposphere: theory. In: *Large-Scale Dynamical Processes in the Atmosphere*, B. Hoskins and R. Pearce, eds. Academic Press, New York, 127–168.

Held, I.M., and A.Y. Hou, 1980: Nonlinear axially symmetric circulations in a nearly inviscid atmosphere. *J. Atmos. Sci.*, **37**, 515–533.

Held, I.M. and B.J. Hoskins, 1985: Large scale eddies and the general circulation of the troposphere. In: *Issues in Atmospheric and Oceanic Modelling, Part A: Climate Dynamics*, S. Manabe, ed., *Adv. Geophys.*, **28**, Academic Press, Orlando, 3–31.

Held, I.M., and M. Ting, 1990: Orographic waves versus thermal forcing of stationary waves: the importance of mean low level wind. *J. Atmos. Sci.*, **47**, 495–500.

Helmhotz, H. v., 1888: Uber atmospharische Bewegungen. *Sitz.-Ber. Akad. Wiss. Berlin*, 647–663. English trans.: Abbe., C., 1893: *The mechanics of the Earth's atmosphere*. Washington, Smithsonian Inst., 78–93.

Helmholtz, H. v., 1889: Uber atmospharische Bewegungen, II. *Sitz.-Ber. Akad. Wiss. Berlin*, 761–780. English trans.: Abbe, C., 1893: *The mechanics of the Earth's atmosphere*. Washington, Smithsonian Inst., 94–111.

Hide, R., and P.J. Mason, 1975: Sloping convection in a rotating fluid. *Adv. in Phys.*, **24**, Academic Press, New York, 47–100.

Hildebrandsson, and Teisserenc de Bort, 1907: *Les bases de la météorologie dynamique,* tome I. Gauthier-Villars et Fils, Paris.

Holopainen, E.O., 1984: Statistical local effect of synoptic-scale transient eddies on the time-mean flow in northern extratropics in winter. *J. Atmos. Sci.*, **41**, 2505–2515.

Holopainen, E.O., and A.H. Oort, 1981: On the role of large- scale transient eddies in the maintenance of the vorticity and enstrophy of the time-mean flow. *J. Atmos. Sci.*, **38**, 270–280.

Holopainen, E.O., L. Rontu, and N.-C. Lau, 1982: The effect of large-scale transient eddies on the time-mean flow in the atmosphere. *J. Atmos. Sci.*, **39**, 1972–1984.

Holton, J.R., 1979: *An Introduction to Dynamic Meteorology* (Second Edition). Academic Press, New York, 391 pp.

Holton, J.R., 1984: Troposphere-stratosphere exchange of trace constituents: The water vapor puzzle. In: *Dynamics of the Middle Atmosphere*, J. R. Holton and T. Matsuno, eds., Terra Sci., Dordrecht, 369–385.

Hoskins, B.J., H.H. Hsu, I.N. James, M. Masutani, P.D. Sardeshmukh, and G. H. White, 1989: "Diagnostics of the Global Atmospheric Circulation Based on ECMWF Analyses 1979–1989." WCRP–27, WMO/TD No. 326, 217 pp.

Hoskins, B.J., I.N. James, and G.H. White, 1983: The shape, propagation and mean-flow interaction of large-scale weather systems. *J. Atmos. Sci.*, **40**, 1595–1612.

Hoskins, B.J., and D.J. Karoly, 1981: The steady linear response of a spherical atmosphere to thermal and orographic forcing. *J. Atmos. Sci.*, **38**, 1179–1196.

Hoskins, B.J., M.E. McIntyre, and A.W. Robertson, 1985: On the use and significance of isentropic potential vorticity maps. *Quart. J. Roy. Meteor. Soc.*, **111**, 877–946.

Houghton, H.G., 1954: On the annual heat balance of the Northern Hemisphere. *J. Meteor.*, **11**, 1–9.

Houghton, J.T., 1977: *The Physics of Atmospheres*, Cambridge Univ. Press, Cambridge, 203 pp.

Houze, R.A., 1982: Cloud clusters and large-scale vertical motions in the tropics. *J. Meteor. Soc. Japan*, **60**, 396–409.

Hsuing, J., R.E. Newell, and T. Houghtby, 1989: The annual cycle of oceanic heat storage and oceanic meridional heat transport. *Quart. J. Roy. Meteor. Soc.*, **115**, 1–28.

Hughes, N.A., 1984: Global cloud climatologies: A historical review. *J. Climate Appl. Meteor.*, **23**, 724–751.

Hunt, B.G., 1979: The influence of the Earth's rotation rate on the general circulation of the atmosphere. *J. Atmos. Sci.*, **36**, 1392–1408.

Hunt, B.G., and S. Manabe, 1968: Experiments with a stratospheric general circulation model, II. Large-scale diffusion of tracers in the stratosphere. *Mon. Wea. Rev.*, **96**, 503–539.

Iribarne, J.V., and H.-R. Cho, 1980: *Atmospheric Physics*, D. Reidel, Dordrecht, 212 pp.

Jaeger, L., 1976: "Monatskarten des Niederschlags fur die ganze Erde." *Berichte des Deutschen Wetterdienstes* Nr. 139, **18**, Im Selbstverlag des Deutschen Wetterdienstes, Offenbach, W. Germany.

James, I., 1983: Some aspects of the global circulation of the atmosphere in January and July 1980. In: *Large-scale Dynamical Processes in the Atmosphere*, B. Hoskins and R. Pearce, eds., Academic Press, London, 5–25.

Jeffreys, H., 1926: On the dynamics of geostrophic winds. *Quart. J. Roy. Meteor. Soc.*, **52**, 85–104.

Johnson, D.R., 1970: The available potential energy of storms. *J. Atmos. Sci.*, **27**, 727–741.

Jonas, P.R., 1981: Laboratory observations of the effects of topography on baroclinic instability. *Quart. J. Roy. Meteor. Soc.*, **107**, 775–792.

Julian, P.R., 1983: "On the use of observation error data in FGGE main level III-b analysis." ECMWF Tech. Mem. No. 76, 24 pp.

Karoly, D.J., and A.H. Oort, 1987: A comparison of southern hemisphere circulation statistics based on GFDL and Australian analyses. *Mon. Wea. Rev.*, **115**, 2033–2059.

Kasahara, A., and W.M. Washington, 1971: General circulation experiments with a six-layer NCAR model, including orography, cloudiness and surface temperature calculations. *J. Atmos. Sci.*, **28**, 657–701.

Kessler, A., 1968: "Globalbilanzen von Klimaelementen." Ber. Inst. Meteor. Klimat. Techn. Univ. Hannover, No. 3, 141 pp.

Kida, H., 1983a: General circulation of air parcels and transport characteristics derived from a hemispheric GCM. Part 1. A determination of advective mass flow in the lower stratosphere. *J. Meteor. Soc. Japan*, **61**, 171–187.

Kida, H., 1983b: General circulation of air parcels and transport characteristics derived from a hemispheric GCM. Part 2. Very long-term motions of air parcels in the troposphere and stratosphere. *J. Meteor. Soc. Japan*, **61**, 510–523.

Kistler, R.E., and D.F. Parrish, 1982: Evolution of the NMC data assimilation system: September 1978–January 1982. *Mon. Wea. Rev.*, **110**, 1335–1346.

Klein, W.H., 1958: The frequency of cyclones and anticyclones in relation to the mean circulation. *J. Meteor.*, **15**, 98–102.

Kochin, N.E., ed., 1936: *Dynamic Meteorology*, Centr. Admin United Hydrometeor Service, USSR, 602 pp.

Köppen, W., 1882: Die monatlichen Barometerschwankungen, deren geographische Verbreitung und Beziehungen zu anderen Phenomenen. *Ann. Hydrogr.*, **10**, 275–289.

Kornegay, F.C., and D.G. Vincent, 1976: Kinetic energy budget analysis during interaction of tropical storm Candy (1968) with an extratropical frontal system. *Mon. Wea. Rev.*, , **104**, 849–859.

Krishnamurti, R., T.N. Krishnamurti, and H.S. Bedi, 1989: Reduction of systematic errors in a global model from parameterization of momentum transport by cumulus convection. *J. Meteor. Soc. Japan*, **67**, 1035–1045.

Krishnamurti, R., and Y. Zhu, 1991: Momentum flux measurements in turbulent convection with an imposed shear. In preparation.

Krishnamurti, T.N., 1971: Tropical east-west circulations during the northern summer. *J. Atmos. Sci.*, **28**, 1342–1347.

Krishnamurti, T.N., 1985: Summer monsoon experiment—A review. *Mon. Wea. Rev.*, **113**, 1590–1626.

Kung, E.C., 1966: Large-scale balance of kinetic energy in the atmosphere. *Mon. Wea. Rev.*, **94**, 627–640.

Kung, E.C., and H. Tanaka, 1983: Energetics analysis of the global circulation during the special observing periods of FGGE. *J. Atmos. Sci.*, **40**, 2575–2592.

Kupffer, 1829: Uber die mittlere temperatur der luft und des bodens auf einige punkten des ostlichen Russlands, *Pogg. Ann.* (Incomplete citation in Shaw, 1926; p 291.)

Kuo, H.-L., 1949: Dynamic instability of two-dimensional nondivergent flow in a barotropic atmosphere. *J. Meteor.*, **6**, 105–122.

Kuo, H.-L., 1951: Vorticity transfer as related to the development of the general circulation. *J. Meteor.*, **8**, 307–315.

Kuo, H.-L., 1956: Forced and free meridional circulations in the atmosphere. *J. Meteor.*, **13**, 561–568.

Labitzke, K., 1982: On the interannual variability of the middle stratosphere during northern winters. *J. Meteor. Soc. Japan*, **60**, 124–139.

Lamb, H.H., 1972: *Climate Present; Past and Future. Vol. 1: Fundamentals and Climate Now.* London, Chapman and Hall/Methnen.

Lau, K.-M., and P.H. Chan, 1986: Aspects of the 40–50 day oscillation during the northern summer as inferred from outgoing longwave radiation. *Mon. Wea. Rev.*, **114**, 1354–1367.

Lau, K.-M., G.J. Yang, and S.H. Shen, 1988: Seasonal and intraseasonal climatology of summer monsoon rainfall over East Asia. *Mon. Wea. Rev.*, **116**, 18–37.

Lau, N.-C., 1978: On the three-dimensional structure of the observed transient eddy statistics of the Northern Hemisphere wintertime circulation. *J. Atmos. Sci.*, **35**, 1900–1923.

Lau, N.-C., 1984: "Circulation statistics based upon FGGE level III-B analyses produced by GFDL." NOAA data report ERL GFDL-5. 427 pp.

Lau, N.-C., 1986: "The influences of orography on large- scale atmospheric flow simulated by a general circulation model." *Proc. Int'l Symp. on the Qinghai-Xizang Plateau and Mountain Meteor.* Amer. Meteor. Soc., Beijing, 241–269.

Lau, N.-C., and E.O. Holopainen, 1984: Transient eddy forcing of the time-mean flow as identified by geopotential tendencies. *J. Atmos. Sci.*, **41**, 313–328.

Lau, N.-C., and K.-M. Lau, 1984: The structure and energetics of midlatitude disturbances accompanying cold-air outbreaks over east Asia. *Mon. Wea. Rev.*, **112**, 1310–1327.

Lau, N.-C., and A.H. Oort, 1981: A comparative study of observed northern hemisphere circulation statistics based on GFDL and NMC analyses. Part I: The time-mean fields. *Mon. Wea. Rev.*, **109**, 1380–1403.

Lau, N.-C., and A.H. Oort, 1982: A comparative study of observed northern hemisphere circulation statistics based on GFDL and NMC analyses. Part II: Transient eddy statistics and the energy cycle. *Mon. Wea. Rev.*, **110**, 889–906.

Leach, A., 1984: Some correlations between the large-scale meridional eddy momentum transport and zonal mean quantities. *J. Atmos. Sci.*, **41**, 236–245.

Leach, H., 1981: Thermal convection in a rotating fluid: Effects due to bottom topography. *J. Fluid Mech.*, **109**, 75–87.

Le Marshall, J.F., and G.A.M. Kelly, 1981: A January and July climatology of the southern hemisphere based on daily numerical analyses 1973–1977. *Aust. Met. Mag.*, **29**, 115–123.

Le Marshall, J.F., G.A.M. Kelly, and D.J. Karoly, 1985: An atmospheric climatology of the southern hemisphere based on ten years of daily numerical analyses (1972–82): I Overview. *Aust. Met. Mag.*, **33**, 65–85.

Le Mone, M.A., G.M. Barnes, and E.J. Zipser, 1984: Momentum flux by lines of cumulonimbus over tropical oceans. *J. Atmos. Sci.*, **41**, 1914–1932.

Li, G.-Q., R. Kung, and R.L. Pfeffer, 1986: An experimental study of baroclinic flows with and without two-wave bottom topography. *J. Atmos. Sci.*, **43**, 2585–2599.

Lighthill, J., and R.P. Pearce, eds., 1981: *Monsoon Dynamics*, Cambridge Univ. Press, Cambridge, 735 pp.

Lilly, D.K., and P.J. Kennedy, 1973: Observation of a stationary mountain wave and its associated momentum flux and energy dissipation. *J. Atmos. Sci.*, **30**, 1135–1152.

Lindzen, R.S., A.Y. Hou, and B.F. Farrell, 1982: The role of convective model choice in calculating the climate impact of doubling CO_2. *J. Atmos. Sci.*, **39**, 1189–1205.

Lindzen, R.S., and A.Y. Hou, 1988: Hadley circulations for zonally averaged heating centered off the equator. *J. Atmos. Sci.*, **45**, 2416–2427.

Liou, K.-N., 1980: *An Introduction to Atmospheric Radiation.* Academic Press, New York, 392 pp.

Liou, K.N., and S.S. Ou, 1983: Theory of equilibrium temperature in radiation-turbulent atmospheres. *J. Atmos. Sci.*, **40**, 214–229.

Lockyer, W.J.S., 1910: *Southern Hemisphere Surface-Air Circulation.* Solar Physics Committee, London, H.M. Stationery Office, 109 pp.

London, J., 1957: "A study of the atmospheric heat balance." Final Report, Contract AF19(122)-165 (AFCRC-TR- 57-287, New York University, [ASTIN 117227]), 99 pp.

Lorenz, E.N., 1955: Available potential energy and the maintenance of the general circulation. *Tellus*, **7**, 157–167.

Lorenz, E.N., 1962: Simplified dynamic equations applied to the rotating basin experiments. *J. Atmos. Sci.*, **19**, 39–51.

Lorenz, E.N., 1963: The mechanics of vacillation. *J. Atmos. Sci.*, , **20**, 448–464.

Lorenz, E.N., 1967: *The Nature and Theory of the General Circulation of the Atmosphere.* World Meteorological Organization, 161 pp.

Lorenz, E.N., 1969: Atmospheric predictability as revealed by naturally occurring analogues. *J. Atmos. Sci.*, **26**, 636–646.

Lorenz, E.N., 1979: Forced and free variations of weather and climate. *J. Atmos. Sci.*, **36**, 1367–1376.

Machenhauer, B., 1977: On the dynamics of gravity oscillations in a shallow water model, with application to normal mode initialization. *Contrib. Atmos. Phys.*, **50**, 253–271.

Mahlman, J.D., D.G. Andrews, D.L. Hartmann, T. Matsuno, and R. G. Murgatroyd, 1984: Transport of trace constituents in the stratosphere. In: *Dynamics of the Middle Atmosphere*, J. R. Holton and T. Matsuno, eds., Terra Sci., Dordrecht, 387–416.

Mak, M., 1982: On moist quasi-geostrophic baroclinic instability. *J. Atmos. Sci.*, **39**, 2028–2037.

Mak, M., 1986: Reply. *J. Atmos. Sci.*, **43**, 317–318.

Manabe, S., and R. Strickler, 1964: Thermal equilibrium of the atmosphere with a convective adjustment. *J. Atmos. Sci.*, **21**, 361–385.

Manabe, S., and T.B. Terpstra, 1974: The effects of mountains on the general circulation of the atmosphere as identified by numerical experiments. *J. Atmos. Sci.*, **31**, 3–42.

Manabe, S. and R.T. Wetherald, 1967: Thermal equilibrium of the atmosphere with a given distribution of relative humidity. *J. Atmos. Sci.*, **24**, 241–259.

Margules, M. 1903. Uber die Energie der Sturme. *Jahrb. Zentralanst. Meteor., Vienna*, 1–26. English trans.: Abbe, C., 1910: *The mechanics of the Earth's atmosphere*, 3rd Coll. Washington, Smithsonian Inst., 533–595.

Matsuno, T., 1971: A dynamical model of the stratospheric sudden warming. *J. Atmos. Sci.*, **28**, 1479–1494.

Maury, M.F., 1855: *The Physical Geography of the Sea*. New York, Harper, 287 pp.

McIntyre, M.E., 1987: Dynamics and tracer transport in the middle atmosphere: An overview of some recent developments. In: *Transport Processes in the Middle Atmosphere*, G. Visconti and R. Garcia, eds., D. Reidel, Dordrecht, 267–296.

McWilliams, J.C., 1980: An application of equivalent modons to atmospheric blocking. *Dyn. Atmos. Oceans*, **5**, 43–66.

Mechoso, C.R., 1989: The final warming of the stratosphere. In: *Dynamics, Transport and Photochemistry in the Middle Atmosphere of the Southern Hemisphere*, A. O'Neill, ed., Kluwer, Dordrecht, 55–69.

Meinardus, W., 1934: Die Niederschlagsverteilung auf der Erde. *Meteor. Z.*, **51**, 345–350.

Merkine, L.-O., 1977: Convective and absolute instability of baroclinic eddies. *Geophys. Astrophys. Fluid Dyn.*, **9**, 129–157.

Merrill, R.T., W.P. Menzel, W. Baker, J. Lynch, and E. Legg, 1991: A report on the recent demonstration of NOAA's upgraded capability to derive cloud motion satellite winds. *Bull. Amer. Meteor. Soc.*, **72**, 372–376.

Miller, D.B., A.L. Booth, and R.E. Miller, 1970: "Automated method of estimating total cloud amount from mesoscale satellite data." Preprints, *Proc. Symp. Tropical Meteorology*, University of Hawaii, 291–306.

Miller, D.B., and R.G. Feddes, 1971: "Global Atlas of Relative Cloud Cover." U.S. Nat'l. Envr. Sat. Serv. and USAF Envr. Tech. Appl. Cntr., AD 739434-Rep. No. 1, Washington D.C., 14 pp. plus charts.

Mills, J., 1770: *An Essay on the Weather; with Remarks on the Shepherd of Banbury's Rules for Judging of it's Changes and Directions for Preserving Lives and Buildings from the Fatal Effects of Lightning*. S. Hooper, London, 108 pp.

Minzner, R.A., W.E. Shenk, R.D. Teagle, and J. Steranka, 1978: Stereographic Cloud Heights from Imagery of SMS/GOES Satellites, *Geophys. Res. Lett.*, **5**, 21–24.

Mitchell, H.L., and J. Derome, 1983: Blocking-like solutions of the potential vorticity equation: Their stability at equilibrium and growth at resonance. *J. Atmos. Sci.*, **40**, 2522–2536.

Miyakoda, K., G.D. Hembree, R.F. Strickler, and I. Shulman, 1972: Cumulative results of extended forecast experiments I. Model performance for winter cases. *Mon. Wea. Rev.*, **100**, 836–855.

Mohr, T., 1984: "Towards an improved global observing system." In: Seminar/Workshop 1984 Data Assimilation Systems and Observing System Experiments with Particular Emphasis on FGGE. Vol. I. ECMWF, 129–144.

Möller, F., 1951: Vierteljahrskarten des Niederschlags für die ganze Erde. *Petermanns Geograph. Mitt.*, **95**, 1–7.

Monin, A.S., 1986: *An Introduction to the Theory of Climate*, D. Reidel, Dordrecht, 261 pp.

Morel, P., M. Desbois, and G. Szejwach, 1978: A new insight into the troposphere with the water vapor channel of Meteosat. *Bull. Amer. Meteor. Soc.*, **59**, 711–714.

Mudrick, S.E., 1974: A numerical study of frontogenesis. *J. Atmos. Sci.*, **31**, 869–892.

Murgatroyd, R.J., and F. Singleton, 1961: Possible meridional circulations in the stratosphere and mesosphere. *Quart. J. Roy. Meteor. Soc.*, , **87**, 125–135.

Namias, J., and P.F. Clapp, 1949: Confluence theory of the high tropospheric jet stream. *J. Meteor.*, **6**, 330–336.

Nappo, C.J., J.Y. Caneill, R.W. Furman, F.A. Gifford, J.C. Kaimal, M.L. Kramer, T.J. Lockhart, M.M. Pendergast, R.A. Pielke, D. Randerson, J.H. Shreffler, and J.C. Wyngaard, 1982: The workshop on the representativeness of meteorological observations, June 1981, Boulder, Colo. *Bull. Amer. Meteor. Soc.*, **63**, 761–764.

Newell, R.E., and S. Gould-Stewart, 1981: A stratospheric fountain? *J. Atmos. Sci.*, **38**, 2789–2796.

Newell, R.E., J.W. Kidson, D.G. Vincent, and G.J. Boer, 1972: *The General Circulation of the Tropical Atmosphere and Interaction with Extratropical Latitudes*. Vol. 1, Massachusetts Institute of Technology, Boston, 258 pp.

Newell, R.E., J.W. Kidson, D.G. Vincent, and G.J. Boer, 1974: *The General Circulation of the Tropical Atmosphere and Interaction with Extratropical Latitudes*. Vol. 2, Massachusetts Institute of Technology, Boston, 371 pp.

Newell, R.E., D.G. Vincent, T.G. Dopplick, D. Ferruzza, and J. W. Kidson, 1970: The energy balance of the global atmosphere. In: *The Global Circulation of the Atmosphere*. G.A. Corby, ed., Roy. Meteor. Soc., London, 42–90.

Newton, C.W., 1972: Southern hemisphere general circulation in relation to global energy and momentum balance requirements (Chapter 9). In: *Meteorology of the Southern Hemisphere*. C.W. Newton, ed., Amer. Meteor. Soc., Boston, 215–246.

Nigam, S., I.M. Held, and S.W. Lyons, 1988: Linear simulation of the stationary eddies in a GCM. Part II: The "mountain" model. *J. Atmos. Sci.*, **45**, 1433–1452.

North, G.R., 1975: Analytical solution to a simple climate model with diffusive heat transport. *J. Atmos. Sci.*, **32**, 1301–1307.

NRC (National Research Council), 1975: *Understanding Climatic Change: A Program for Action*, U.S. Committee for the GARP, National Academy of Sciences, Washington, D.C., 239 pp.

Oberbeck, A., 1888: Uber die Bewegungserscheinungen der Atmosphare. *Sitz.-Ber. Akad. Wiss. Berlin*, 383–395, 1129–1138. English trans.: Abbe, C., 1893: *The mechanics of the Earth's atmosphere*. Washington, Smithsonian Inst., 177–197.

O'Brien, J.J., 1970: Alternative solutions to the classical vertical velocity problem. *J. Appl. Meteor.*, **9**, 197–203.

Ohring, G., 1990: The 1989 IAMAP symposium on the Earth's radiation budget. *Bull. Amer. Meteor. Soc.*, **71**, 1455–1457.

O'Neill, A., and B.F. Taylor, 1979: A study of the major stratospheric warming of 1976/77. *Quart. J. Roy. Meteor. Soc.*, **105**, 71–92.

O'Neill, A., and C.E. Youngblut, 1982: Stratospheric warmings diagnosed using the transformed Eulerian-mean equations and the effect of the mean state on wave propagation. *J. Atmos. Sci.*, **39**, 1370–1386.

Oort, A.H., 1964: On estimates of the atmospheric energy cycle. *Mon. Wea. Rev.*, **92**, 483–493.

Oort, A.H., 1978: Adequacy of the rawinsonde network for global circulation studies tested through numerical model output. *Mon. Wea. Rev.*, **106**, 174–195.

Oort, A.H., 1983: "Global Atmospheric Circulation Statistics, 1958–1973." U.S. Dept. of Commerce, NOAA prof. paper No. 14, Rockville, 180 pp.

Oort, A.H., 1985: Balance conditions in the Earth's climate system. *Adv. Geophys.*, **28A**, 75–98.

Oort, A. H., and J. P. Peixóto, 1974: The annual cycle of the energetics of the atmosphere on a planetary scale. *J. Geophys. Res.*, **79**, 2705–2719.

Oort, A.H., and J.P. Peixóto, 1983: Global angular momentum and energy balance requirements from observations. In: *Theory of Climate*, B. Saltzman, ed., *Adv. Geophys.*, **25**, Academic Press, New York, 355–490.

Oort, A.R., and E.M. Rasmussen, 1971: "Atmospheric Circulation Statistics," U.S. Department of Commerce, NOAA prof. paper No. 5, Rockville, 323 pp.

Opsteegh, J.D., and A.D. Vernekar, 1982: A simulation of the January standing wave pattern including the effects of transient eddies. *J. Atmos. Sci.*, **39**, 734–744.

Palmén, E., and C.W. Newton, 1969: *Atmospheric Circulation Systems, Their Structure and Physical Interpretation*. Academic Press, New York, 603 pp.

Palmer, T.N., 1981a: Diagnostic study of a wavenumber–2 stratospheric sudden warming in a transformed Eulerian-mean formalism. *J. Atmos. Sci.*, **38**, 844–855.

Palmer, T.N., 1981b: Aspects of stratospheric sudden warmings studied from a transformed Eulerian-mean viewpoint. *J. Geophys. Res.*, **86**, 9679–9687.

Palmer, T.N., 1982: Properties of the Eliassen-Palm flux for planetary scale motions. *J. Atmos. Sci.*, **39**, 992–997.

Paltridge, G.W., and C.M.R. Platt, 1976: *Radiative Processes in Meteorology and Climatology*, Elsevier, Amsterdam, 318 pp.

Pedlosky, J., 1964: An initial value problem in the theory of baroclinic instability. *Tellus*, **16**, 12–17.

Pedlosky, J., 1972: Limit cycles and unstable baroclinic waves. *J. Atmos. Sci.*, **29**, 53–63.

Pedlosky, J., 1987: *Geophysical Fluid Dynamics*, Second edition, Springer-Verlag, New York, 710 pp.

Peng, M.S., and R.T. Williams, 1986: Spatial instability of the barotropic jet with slow streamwise variation. *J. Atmos. Sci.*, **43**, 2430–2442.

Peng, M.S., and R.T. Williams, 1987: Spatial instability of a baroclinic current with slow streamwise variation. *J. Atmos. Sci.*, **44**, 1681–1695.

Petterssen, S., 1956: *Weather Analysis and Forecasting* (Second edition). McGraw-Hill, New York, 428 pp.

Pfeffer, R.L., 1981: Wave-mean flow interactions in the atmosphere. *J. Atmos. Sci.*, **38**, 1340–1359.

Pfeffer, R.L., 1987: Comparison of conventional and transformed Eulerian diagnostics in the troposphere. *Quart. J. Roy. Meteor. Soc.*, **113**, 237–254.

Pfeffer, R.L., G. Buzyna, and R. Kung, 1980: Relationships among eddy fluxes of heat, eddy temperature variances and basic state temperature parameters in thermally driven rotating fluids. *J. Atmos. Sci.*, **37**, 2577–2599.

Pfeffer, R.L., R. Kung, and G. Li, 1989: Topographically forced waves in a thermally driven rotating annulus of fluid — Experiment and linear theory. *J. Atmos. Sci.*, **46**, 2331–2343.

Phillips, N.A., 1956: The general circulation of the atmosphere: a numerical experiment. *Quart. J. Roy. Meteor. Soc.*, **82**, 123–164.

Pierrehumbert, R.T., 1984: Local and global baroclinic instability of zonally varying flow. *J. Atmos. Sci.*, **41**, 2141–2162.

Pierrehumbert, R.T., 1986: Spatially amplifying modes of the Charney baroclinic-instability problem. *J. Fluid Mech.*, **170**, 293–317.

Plumb, R.A., 1983: A new look at the energy cycle. *J. Atmos. Sci.*, **40**, 1669–1668.

Plumb, R.A., 1990: A nonacceleration theorem for transient quasi-geostrophic eddies on a three-dimensional time-mean flow. *J. Atmos. Sci.*, **47**, 1825–1836.

Polavarapu, S.M., and W.R. Peltier, 1990: The structure and nonlinear evolution of synoptic scale cyclones: Life cycle simulations with a cloud-scale model. *J. Atmos. Sci.*, **47**, 2645–2672.

Price, P.G., 1975: A comparison between available potential and kinetic energy estimates for the Southern and Northern Hemispheres. *Tellus*, **27**, 443–452.

Ramage, C.S., 1971: *Monsoon Meteorology*, Academic Press, New York, 296 pp.

Ramage, C.S., S.J.S. Khalsa, and B.N. Meisner, 1981: The central Pacific and near-equatorial convergence zone. *J. Geophys. Res.*, **86**, 6580–6598.

Ramanathan, V., L. Callis, R. Cess, J. Hansen, I. Isaksen, W. Kuhn, A. Lacis, F. Luther, J. Mahlman, R. Reck, and M. Schlesinger, 1987: Climate-chemical interactions and effects of changing atmospheric trace gases. *Rev. Geophys.*, **25**, 1441–1482.

Randel, W.J., 1987: A study of planetary waves in the southern winter troposphere and stratosphere. Part I: Wave structure and vertical propagation. *J. Atmos. Sci.*, **44**, 917–935.

Randel, W.J., 1989: A comparison of the dynamic life cycles of tropospheric medium-scale waves and stratospheric planetary waves. In: *Dynamics, Transport and Photochemistry in the Middle Atmosphere of the Southern Hemisphere*, A. O'Neill, ed., Kluwer, Dordrecht, 91–109.

Randel, W.J., and J.L. Stanford, 1985: The observed life cycle of baroclinic instability. *J. Atmos. Sci.*, **42**, 1364–1373.

Rasool, S.I., 1964: Cloud heights and night-time cloud cover from TIROS radiation data. *J. Atmos. Sci.*, **21**, 152–156.

Remer, L.A., 1991: "Cloud-Radiative Feedbacks in Tropical Convection." Ph D. Thesis, Atmos. Sci. Section, Dept. of Land, Air and Water Resources, Univ. Calif. Davis, 129 pp.

Rhines, P.B., 1975: Waves and turbulence on a beta plane. *J. Fluid Mech.*, **69**, 417–443.

Riehl, H., and J.S. Malkus, 1958: On the heat balance of the equatorial trough zone. *Geophysica*, **6**, 503–538.

Rind, D., 1986: The dynamics of warm and cold climates. *J. Atmos. Sci.*, **43**, 3–24.

Roads, J.O., 1982: Forced, stationary waves in a linear, stratified, quasi-geostrophic atmosphere. *J. Atmos. Sci.*, **39**, 2431–2449.

Roemmich, D., 1980: Estimation of meridional heat flux in the North Atlantic by inverse methods. *J. Phys. Oceanog.*, **10**, 1972–1973.

Rosen, R.D., D.A. Salstein, J.P. Peixóto, A.H. Oort, and N.- C. Lau, 1985: Circulation statistics derived from level III-b and station-based analyses during FGGE. *Mon. Wea. Rev.*, **113**, 65–88.

Rossow, W.B., 1989: "Status of the international satellite cloud climatology project." Rept. of 10th session JSC, WMO/TD – No. 314. Appendix C. 91 pp.

Rossow, W.B., and R.A. Schiffer, 1991: ISSCP cloud data products, *Bull. Amer. Meteor. Soc.*, **72**, 2–20.

Sadler, J.C., 1969: Average cloudiness in the tropics from satellite observations, International Indian Ocean Expedition. *Meteor. Monogr.* No. 2, East West Center Press, Honolulu, 22 pp.

Saltzman, B., 1963: A generalized solution for the large-scale, time-average perturbations in the atmosphere. *J. Atmos. Sci.*, **20**, 226–235.

Saltzman, B., 1968: Surface boundary effects on the general circulation and macro-climate: a review of the theory of the quasi-stationary perturbations in the atmosphere. *Meteor. Mono.*, **8**, 4–19.

Saltzman, B., 1970: Large-scale atmospheric energetics in the wavenumber domain. *Rev. Geophys. Space Phys.*, **8**, 289–302.

Sardeshmukh, P., and B.J. Hoskins, 1988: The generation of global rotational flow by steady idealized tropical divergence. *J. Atmos. Sci.*, **45**, 1228–1251.

Schiffer, R.A., and W.B. Rossow, 1983: The international satellite cloud climatology project (ISSCP): The first project of the World Climate Research Programme. *Bull. Amer. Meteor. Soc.*, **64**, 779–784.

Schiffer, R.A., and W.B. Rossow, 1985: ISCCP global radiance data set: A new resource for climate research. *Bull. Amer. Meteor. Soc.*, **66**, 1498–1505.

Schneider, E.K., 1977: Axially symmetric steady-state models of the basic state for instability and climate studies. Part II: nonlinear calculations. *J. Atmos. Sci.*, **34**, 280–296.

Schneider, E.K., 1990: Linear diagnosis of stationary waves in a general circulation model. *J. Atmos. Sci.*, **47**, 2925–2952.

Schneider, E.K., and R.S. Lindzen, 1976: The influence of stable stratification on the thermally driven tropical boundary layer. *J. Atmos. Sci.*, **33**, 1301–1307.

Schneider, E.K., and R.S. Lindzen, 1977: Axially symmetric steady-state models of the basic state for instability and climate studies. Part I: linearized calculations. *J. Atmos. Sci.*, **34**, 263–279.

Schutz, C., and W.L. Gates, 1971: "Global climatic data for surface, 800 mb and 400 mb: January." Rept. for Adv. Res. Projects Agency, Rand, Santa Monica, R–915–ARPA, 21 pp.

Schutz, C., and W.L. Gates, 1972: "Global climatic data for surface, 800 mb and 400 mb: July." Rept. for Adv. Res. Projects Agency, Rand, Santa Monica, R–1029–ARPA, 22 pp.

Sellers, W.D., 1966: *Physical Climatology*. Chicago, Univ. Chicago Press, 272 pp.

Semtner, A.J., and R.M. Chervin, 1992: Ocean general circulation from a global eddy-resolving model. *J. Geophys. Res.*, **97**, 5493–5550.

Shaw, N., 1926: *Manual of Meteorology, Vol. I, Meteorology in History*. Cambridge Univ. Press, Cambridge, 337 pp. (2nd ed: 1932, note page 291.)

Shea, D.J., 1986: "Climatological Atlas: 1950–1979. Surface Air Temperature, Precipitation, Sea-level Pressure and Sea-Surface Temperature (45 S – 90 N)." NCAR Tech. Note NCAR/TN-269+STR. 35 pp. plus charts.

Shepherd, T.G., 1987: A spectral view of nonlinear fluxes and stationary-transient interaction in the atmosphere. *J. Atmos. Sci.*, **44**, 1166–1178.

Shine, K.P., 1987: The middle atmosphere in the absence of dynamical heat fluxes. *Quart. J. Roy. Meteor. Soc.*, **113**, 603–633.

Shutts, G.J., 1987: Some comments on the concept of thermal forcing. *Quart. J. Roy. Meteor. Soc.*, , **113**, 1387–1394.

Simmons, A.J., and B.J. Hoskins, 1977: Baroclinic instability on the sphere: solutions with a more realistic tropopause. *J. Atmos. Sci.*, **34**, 581–588.

Simmons, A.J., and B.J. Hoskins, 1978: The life cycles of some non-linear baroclinic waves. *J. Atmos. Sci.*, **35**, 414–432.

Simmons, A.J., and B.J. Hoskins, 1980: Barotropic influences on the growth and decay of nonlinear baroclinic waves. *J. Atmos. Sci.*, **37**, 1679–1684.

Smagorinsky, J., 1953: The dynamical influence of large-scale heat sources and sinks on the quasi-stationary mean motions of the atmosphere. *Quart. J. Roy. Meteor. Soc.*, **79**, 342–366.

Smagorinsky, J., 1967: The role of numerical modeling. *Bull. Amer. Meteor. Soc.*, **48**, 89–93.

Smagorinsky, J., S. Manabe, and J.L. Holloway, Jr., 1965: Numerical results from a nine-level general circulation model of the atmosphere. *Mon. Wea. Rev.*, **93**, 727–768.

Smith, P.J., 1969: On the contribution of a limited region to the global energy budget. *Tellus*, **21**, 202–207.

Smith, P.J., D.G. Vincent, and H.J. Edmon, 1977: The time dependence of reference pressure in limited region available potential energy budget equations. *Tellus*, **29**, 476–480.

Smith, R.B., 1979: The influence of mountains on the atmosphere. *Adv. Geophys.*, **21**, B. Saltzman, ed., Academic Press, New York, 87–230.

Smith, W.L., 1985: "Satellite observed thermodynamics during FGGE." In: *Proceedings of the First Nat'l Workshop on the Global Weather Experiment*, Vol. 2, Part I, NAS, Washington, D.C. 3–18.

Smith, W.L., H.M. Woolf, C.M. Hayden, D.Q. Wark, and L.M. McMillin, 1979: The TIROS-N operational vertical sounder. *Bull. Amer. Meteor. Soc.*, **60**, 1177–1187.

Sparkman, J.K., and G.J. Smidt, 1985: "An operational ASDAR system." In: *Proceedings of the NASA Symposium on Global Wind Measurements*, W.E. Baker and R.J. Curran, eds. A. Deepak Publ., Hampton, 115–118.

Stanton, T.E., 1911: The mechanical viscosity of fluids. *Proc. Roy. Soc. London* **A85**, 366.

Starr, V.P., 1948: An essay on the general circulation of the Earth's atmosphere. *J. Meteor.*, **5**, 39–43.

Starr, V.P., and R.M. White, 1951: A hemispherical study of the atmospheric angular-momentum balance. *Quart. J. Roy. Meteor. Soc.*, **77**, 215–225.

Starr, V.P., J.P. Peixóto, and R.G. McKean, 1969: Pole-to-pole moisture conditions for the IGY. *Pure Appl. Geophys.*, **75**, 300–331.

Steranka, J., L.J. Allison, and V V. Salomonson, 1973: Application of Nimbus-4 THIR 6.7 μm observations to regional and global moisture and wind field analyses. *J. Appl. Meteor.*, **12**, 386–395.

Stevens, D.E., 1983: On symmetric stability and instability of zonal mean flows near the equator. *J. Atmos. Sci.*, **40**, 882–893.

Stewart, T.R., and C.M. Hayden, 1985: "A FGGE water vapor wind data set." In: *Proceedings of the NASA Symposium on Global Wind Measurements*, W.E. Baker and R.J. Curran, eds. A. Deepak Publ., Hampton, 119–122.

Stokes, G.G., 1847: On the theory of oscillating waves. *Trans. Cambridge Phil. Soc.*, **8**, 441–455.

Stone, P.H., 1972: A simplified radiative-dynamical model for the static stability of rotating atmospheres. *J. Atmos. Sci.*, **29**, 405–418.

Stone, P.H., 1973: The effect of large-scale eddies on climatic change. *J. Atmos. Sci.*, **30**, 521–529.

Stone, P.H., 1978: Baroclinic adjustment. *J. Atmos. Sci.*, **35**, 561–571.

Stone, P.H., S.J. Ghan, D. Speigel, and S. Rambaldi, 1982: Short-term fluctuations in the eddy heat flux and baroclinic stability of the atmosphere. *J. Atmos. Sci.*, **39**, 1734–1746.

Stone, P.H., and M.-S. Yao, 1987: Development of a two-dimensional zonally averaged statistical-dynamical model. Part II: The role of eddy momentum fluxes in the general circulation and their parameterization. *J. Atmos. Sci.*, **44**, 3769–3786.

Sutcliffe, R.C., 1951: Mean upper contour patterns of the northern hemisphere — the thermal-synoptic viewpoint. *Quart. J. Roy. Meteor. Soc.*, **77**, 435–440.

Swinbank, R., 1985: The global atmospheric angular momentum balance inferred from analyses made during the FGGE. *Quart. J. Roy. Meteor. Soc.*, **111**, 977–992.

Taljaard, J.J., 1972: Physical features of the Southern Hemisphere (Chapter 1) and Synoptic Meteorology of the Southern Hemisphere (Chapter 8). In: *Meteorology of the Southern Hemisphere*, C. W. Newton, ed., *Meteor. Mono.*, **13**, Amer. Meteor. Soc., Boston, 1–8 and 139–214.

Taljaard, J.J., H. van Loon, H.L. Crutcher, and R.L. Jenne, 1969: "Climate of the Upper Air: Southern Hemisphere, Vol. 1, Temperatures, Dew Points, and Heights at Selected Pressure Levels." NAVAIR 50–1C–55, Naval Weather Service Command, Washington, D.C., 135 pp.

Tanaka, H.L., and E.C. Kung, 1988: Normal mode energetics of the general circulation during the FGGE year. *J. Atmos. Sci.*, **45**, 3723–3736.

Tanner, R.W. (ed.), 1990: *Monthly Climatic Data for the World*, NOAA/NCDC, Asheville NC., U. S. Dept. of Commerce.

Taylor, G.I., 1916: Conditions at the surface of a hot body exposed to the wind. *Brit. Adv. Com. Aero. Rep. and Memor.*, 272.

Taylor, R.C., 1973: "An Atlas of Pacific Islands Rainfall." Rept. HIG-73-9, Hawaii Inst. of Geophys., 7 pp. plus charts.

Teisserenc de Bort, L.P., 1902: Variations de la température de l'air libre dans la zone comprise entre 8 km et 13 km d'altitude. *C. R. Hebd. Seances Acad. Sci.*, **134**, 987–989.

Thekaekara, M.P., 1976: Solar irradiance: Total and spectral and its possible variations. *Appl. Opt.*, **15**, 915–920.

Thomas, A.R., 1975: "Quality control of upper air data from a mesoscale network." In: Third Symposium on Meteorological Observations and Instrumentation, Washington, D.C., 10–13 February. Amer. Meteor. Soc., Boston, MA, 115– 122.

Thomson, J., 1857: "Grand currents of atmospheric circulation." British Assoc. Meeting, Dublin.

Townsend, R.D., and D.R. Johnson, 1985: A diagnostic study of the isentropic zonally averaged mass circulation during the first GARP global experiment. *J. Atmos. Sci.*, **42**, 1565–1579.

Trenberth, K.E., 1991: Storm tracks in the Southern Hemisphere. *J. Atmos. Sci.*, , **48**, 2159–2178.

Trenberth, K.E., and J.G. Olson, 1988: An evaluation and intercomparison of global analyses from the National Meteorological Center and the European Centre for Medium Range [sic] Weather Forecasts. *Bull. Amer. Meteor. Soc.*, **69**, 1047– 1057.

Valdes, P.J., and B.J. Hoskins, 1989: Linear stationary wave simulations of the time-mean climatological flow. *J. Atmos. Sci.*, **46**, 2509–2527.

Vallis, G.K., 1988: Numerical studies of eddy transport properties in eddy-resolving and parameterized models. *Quart. J. Roy. Meteor. Soc.*, **114**, 183–204.

Van Loon, H., 1966: On the annual temperature range over the southern oceans. *Geogr. Rev.*, **56**, 495–515.

Van Loon, H., 1972: Temperature in the Southern Hemisphere, (Chapter 3), Pressure in the Southern Hemisphere (Chapter 4), Wind in the Southern Hemisphere (Chapter 5), and Cloudiness and precipitation in the Southern Hemisphere (Chapter 6). In: *Meteorology of the Southern Hemisphere*, C. W. Newton, ed., *Meteor. Mono.*, **13**, Amer. Meteor. Soc., Boston, 25–111.

Van Loon, H., and D.J. Shea, 1988: A survey of the atmospheric elements at the ocean's surface south of 40°S. In: *Antarctic Ocean and Resources Variability*, D. Sahrhage, ed., Springer-Verlag, Berlin, 2–20.

Van Mieghem, J., 1956: The energy available in the atmosphere for conversion into kinetic energy. *Beitr. Phys. Atmos.*, **29**, 129–142.

Van Mieghem, J., 1957: Energies potentielle it interne convertibles en energies cinetique dans l'atmosphere. *Beitr. Phys. Atmos.*, **30**, 5–17.

Vincent, D.G., 1968: Mean meridional circulations in the Northern Hemisphere lower stratosphere during 1964 and 1965. *Quart. J. Roy. Meteor. Soc.*, **94**, 333–349.

Vincent, D.G., and L.N. Chang, 1973: Some further considerations concerning energy budgets of moving systems. *Tellus*, **25**, 224–232.

Vowinckel, E., 1962: Cloud amount and type over the Arctic. *Publ. Meteor.*, No. 5, Arctic Meteorology Research Group, McGill University, AFRCL 62-663, 26 pp.

Wahr, J.M., and A.H. Oort, 1984: Friction- and mountain-torque estimates from global atmospheric data. *J. Atmos. Sci.*, **41**, 190–204.

Wallace, J.M., 1978a: "Maintenance of the zonally averaged circulation: a eulerian perspective." In: *The General Circulation: Theory, Modelling and Observations.* NCAR Summer Colloquium, Boulder, 15–24.

Wallace, J.M., 1978b: Trajectory slopes, countergradient heat fluxes and mixing by lower stratospheric waves. *J. Atmos. Sci.*, **35**, 554–558.

Wallace, J.M., and D.S. Gutzler, 1981: Teleconnections in the geopotential height field during the Northern Hemisphere winter. *Mon. Wea. Rev.*, **109**, 785–812.

Warren, S.G., C.J. Hahn, J. London, R.M. Chervin, and R.L. Jenne, 1986: "Global Distributions of Total Cloud Cover and Cloud Type Amounts Over Land." NCAR Tech. Note TN–273+STR; DOE/ER/60085–H1.

Warren, S.G., C.J. Hahn, J. London, R.M. Chervin, and R.L. Jenne, 1988: "Global Distributions of Total Cloud Cover and Cloud Type Amounts Over the Ocean." NCAR Tech. Note TN–317+STR; DOE/ER–0406.

Warren, S.G., J. London, and C.J. Hahn, 1991: Cloud hole over the United States? *Bull. Amer. Meteor. Soc.*, **72**, 237–238.

Washington, W.M. and C.L. Parkinson, 1986: *An Introduction to Three-Dimensional Climate Modeling*, Univ. Sci. Books, Mill Valley, 422 pp.

White, R.M., 1951: The meridional flux of sensible heat over the northern hemisphere. *Tellus*, **3**, 82–88.

Whittaker, L.M., and L.H. Horn, 1982: "Atlas of Northern Hemisphere Extratropical Cyclone Activity, 1958–1977," Univ. of Wisconsin. Dept. of Meteorology Tech. Note. Madison, 65 pp.

Whorf, B.L., 1940: Science and linguistics. *Tech. Rev.*, **42**, 229–231, 247–248. Reprinted in *Language, Thought, and Reality*, MIT Press and J. Wiley & Sons, New York, 207–219.

Williams, G.P., 1978: Planetary circulations: 1. Barotropic representation of Jovian and terrestrial turbulence. *J. Atmos. Sci.*, **35**, 1399–1426.

Williams, G.P., 1979: Planetary circulations: 3. The terrestrial quasi-geostrophic regime. *J. Atmos. Sci.*, **36**, 1409–1435.

Winston, J.S., A. Gruber, T.I. Gray, M.S. Varnadore, C.L. Earnest, and L.P. Mannello, 1979: *Earth-Atmosphere Radiation Budget Analyses Derived from NOAA Satellite Data June 1974–February 1978*. Vols. 1 and 2. Dept. of Commerce, Washington, D.C.

WMO (World Meteorological Organization), 1990: "Scientific Plan for the Global Energy and Water Cycle Experiment." WCRP–40, WMO/TD No. 376, Geneva 83 pp.

WMO (World Meteorological Organization), 1991: "The Global Climate Observing System." WCRP–56, WMO/TD No. 412, Geneva, 83, pp.

Young, E.C., 1972: *Partial Differential Equations: An Introduction*, Allyn and Bacon, Boston, 346 pp.

Youngblut, C., and T. Sasamori, 1980: The nonlinear effects of transient and stationary
    eddies on the winter mean circulation. Part I: Diagnostic analysis. *J. Atmos. Sci.*,
    37, 1944–1957.
Zipser, E.J., 1974: A review of the Newell et al. (1972) reference above. *Bull. Amer.
    Meteor. Soc.*, **55**, 324–326.

INDEX